# Mesenchymal Stem Cell in Veterinary Sciences

Mudasir Bashir Gugjoo • Amar Pal
Editors

# Mesenchymal Stem Cell in Veterinary Sciences

Springer

*Editors*
Mudasir Bashir Gugjoo
Division of Veterinary Clinical Complex
FVSc & AH, SKUAST-Kashmir
Jammu and Kashmir, India

Amar Pal
Division of Surgery
Indian Veterinary Research Institute
Izatnagar, India

ISBN 978-981-15-6036-1 ISBN 978-981-15-6037-8 (eBook)
https://doi.org/10.1007/978-981-15-6037-8



This Springer imprint is published by the registered company Springer Nature Singapore Pte Ltd.
The registered company address is: 152 Beach Road, #21-01/04 Gateway East, Singapore 189721, Singapore

# Preface

Tissues and/or organs are constituted by building blocks, the cells. These cells are either undifferentiated stem cells or differentiated tissue-specific cells. The miniscule number of available undifferentiated stem cells act as a reserve to maintain normal cell turn-over and thereby, tissue homeostasis. The differentiated cells perform the tissue-specific function. For a substantial and effective cellular therapy, the cells have to be sufficient in quantity and compatible with the recipient immune system, besides expressing as per the tissue requirements. The differentiated cells taken from the particular tissue source often are limited in number, have little or no proliferation potential and elicit the immune response in the recipient upon allogeneic or xenogeneic transplantation. Contrarily, the stem cells especially mesenchymal stem cells (MSCs) can be harvested from almost all the body tissue sources including the foetal membranes and have good proliferation and trans-differentiation potential. MSCs tend to have little tendency to elicit an immune response in the recipient even upon allogeneic or xenogeneic applications. These cells have minimal teratogenic and ethical issues normally associated with the pluripotent stem cells. However, MSCs unlike that of pluripotent stem cells tend to have limited proliferation potential.

The advancement of stem cell biology is increasing by each passing day and has found its way into the clinical applications under regenerative medicine, a branch of tissue engineering. Due to the limited understanding in the cellular characteristics, *ipso facto* MSCs therapeutic applications currently come under the advanced therapy medicinal products (ATMPs). The role of microenvironment in cellular fate remains to be understood. Such an understanding shall aid in their definitive utilization in regenerative medicine.

There are many books that discuss stem cells and their features and have invariably focused on their applications in humans. However, the book on stem cells especially mesenchymal stem cells and their applications in veterinary species is very much lacking. The pressing need for a book that provides a comprehensive introduction to the field and summarizes its recent progress in veterinary sciences has given rise to this book. This current work discusses the mesenchymal stem cell features and their applications in veterinary species. The book starts with introduction to the stem cells. In the introductory chapters, the stem cell properties and types are discussed followed by the details of the mesenchymal stem cell sources, isolation, culturing and characterization, and preservation. This is followed by a chapter on the effect of different cues on mesenchymal stem cells properties. Thereafter, the species-specific mesenchymal stem cell features including the complexities in their characteristics, besides, their potential applications have been discussed. Finally, the book is concluded with the summary and future research avenues of MSCs.

We thank all the contributors for their hard work and valuable contributions.

Jammu and Kashmir, India — Mudasir Bashir Gugjoo
Izatnagar, India — Amar Pal

# Acknowledgement

With all the humility I thank almighty "*Allah*" the most Beneficent, the most Merciful, who blessed me with limitless strength and favourable circumstances, to face and pass through all odds successfully at this juncture.

I feel a great sense of elation and fulfilment to submit this tiny drop of work in the ocean of human knowledge to benefit the differently abled creatures surrounding and helping us in many ways. The completion of this work would not have been possible without the help and support of many people.

I would like to appreciate the contribution of all those teachers from childhood till now who played an important role in my life to reach this place without which the realization of the present work into practice would have been a dream. Apart from the teaching faculty, schools/institutes that I attended like Miranda Public High School, Srinagar, SP Higher Secondary School, Srinagar, SKUAST-Jammu and Indian Veterinary Research Institute, Izatnagar, during the course have been the ocean that taught me to swim in all the spheres of life and reach to the stage. A special thanks go to all the stem cell scientists who burn themselves and provide the literature that makes the material in the book.

I am highly thankful to Science and Engineering Research Board (SERB), Government of India that provided the financial support and infused zeal in me to conduct the stem cell research and come out with this piece. I really feel elated to convey my gratitude to the SKUAST-Kashmir under which I am currently working. This institute has provided me the platform to express my potential to the best of my capacities. The faculty members of the institute have provided great company through the journey.

Literary skills grope in the dark when it comes to express what I owe to my family. These mean achievements to which I lay claim today have all precipitated from the sincere prayers and silent sacrifices of my Pappa and Mummy. The blessings, love, guidance and affection provided by my sister and brother-in-law deserve a special mention. I admit myself being incapable to pay regards in words to the affection, blessings, encouragement and dedication of my wife that made many difficult tasks easier for me. In recent times, the happiest moment and encouragement of my life has probably been my daughter.

I am highly thankful to Dr Bhavik Sawhney, Editor-Biomedicine, Springer Nature for his guidance and timely help in giving final shape to this book. I admit the affections and encouragement provided to me by my extended family. This list is obviously incomplete but allows me to submit that the omissions are inadvertent and I once again record my deep-felt gratitude to all those associated with me in this endeavour.

Finally, I supplicate to *Allah* (SWT), the One who listens and responds to supplications, "Oh *Allah*, guide me to the right path, the path of those on whom You have bestowed Your favors, not of those who have incurred Your wrath, nor of those who have gone astray and let not me die except in the state of pure submission to You alone". *Aameen*.

# About the Book

During the last few years, a large volume of information has been generated on stem cell types, their biological nature, *in vitro* and *in vivo* behaviour and potential application in research and therapy. This book makes a systematic collection of the scientific information scattered in the literature on different aspects of stem cell research. The focus of the book is on mesenchymal stem cells as they have been investigated most extensively for therapeutic application in animals. The contents of the book have been designed to include all the relevant information about the mesenchymal stem cells of animal origin. The initial chapter covers the historical aspect and chronology of developments in the field of stem cell research. The details about the basic properties and nature of the stem cells are provided. A comprehensive account of the materials and methods used for isolation of mesenchymal stem cells from different sources, culture expansion, characterization and long-term storage has been given in the subsequent chapters. The book also provides a logical appraisal of the mechanisms involved in therapeutic efficacy, immunomodulation and anti-inflammatory properties of mesenchymal stem cells. The book also highlights knowledge about the interaction between the stem cells and their environment and factors responsible for the maintenance of stemness. A chapter in the book has been dedicated to genetic engineering relevant to stem cell research for their better exploitation for research and therapy in animals. The major section of the book is devoted to the recount of the research efforts and their outcomes towards the potential application of mesenchymal stem cells in regenerative research and therapy in important species of domestic and pet animals including sheep, goat, cattle, buffalo, cat, dog and horse. The book also presents an abridgement of challenges and future prospects of stem cell research and application in medicine in general and veterinary sciences in particular.

# Contents

# Editors and Contributors

## About the Editors

**Mudasir Bashir Gugjoo** is an Assistant Professor at the Faculty of Veterinary Sciences & Animal Husbandry, Sher-e-Kashmir University of Agricultural Sciences & Technology, Kashmir, India. He has received his PhD in mesenchymal stem cell translational studies from the Indian Veterinary Research Institute, Izatnagar. He has received various awards including the Best Faculty Teacher Award. His lab focuses on the isolation and characterisation of mesenchymal stem cells and their therapeutic applications, currently being funded by Science and Engineering Research Board (SERB), Government of India. He has published more than 60 papers in various reputed peer-reviewed national and international journals, authored or co-authored a number of book chapters and is a newspaper columnist. His work has been cited in various standard books on Veterinary Medicine, Radiology and Tissue Engineering, and in numerous international research papers. He is a member of various scientific societies and serves on the editorial boards of several journals. He is also a referee for a number of peer-reviewed journals including the Journal of Cellular Physiology, Cellular Reprogramming and Biomedicine and Pharmacotherapy, among others.

**Amar Pal** is currently working as a Principal Scientist & Head of the Division of Surgery at Indian Veterinary Research Institute, Izatnagar, India. He is also in-charge of the Referral Veterinary Polyclinic and the Anatomy Section at the Indian Veterinary Research Institute, Izatnagar, India. He has done pioneering research work in the areas of veterinary anaesthesia, pain management, urolithiasis, orthopaedic surgery, and stem cell biology and therapy in animals. He has developed and registered the designs of six orthopaedic devices for animals and eight veterinary mobile apps. His work has been cited in standard works on veterinary surgery, anaesthesia and pharmacology, and in countless international research papers. He has received several awards, including the Chancellor's Gold Medal, Vice-Chancellor's Gold Medal, and Best Teacher Award. He is a fellow of the National Academy of Veterinary Sciences, Indian Society for Veterinary Surgery and Indian Society for Advancement of Canine Practice. Dr Amarpal has published more than 360 research and clinical papers in respected national and international journals and more than 380 abstracts in seminar and conference proceedings. He has also authored two books and contributed several book chapters. Dr Amarpal is Editor-in-Chief of the Indian Journal of Veterinary Surgery and Associate Editor of the Indian Journal of Canine Practice, and is a referee for numerous international journals.

## Contributors

**V. Chandra** Division of Physiology and Climatology, Indian Veterinary Research Institute, Izatnagar, Uttar Pradesh, India

**M.R. Fazili** Division of Veterinary Clinical Complex, FVSc and AH, SKUAST-K, Srinagar, Jammu and Kashmir, India

**Mudasir Bashir Gugjoo** Division of Veterinary Clinical Complex, FVSc & AH, SKUAST-Kashmir, Jammu and Kashmir, India

**D.M. Makhdoomi** Division of Surgery and Radiology, FVSc and AH, SKUAST-K, Srinagar, Jammu and Kashmir, India

**M.S. Mir** Directorate of Extension Education, SKUAST-K, Srinagar, Jammu and Kashmir, India

**Amar Pal** Division of Surgery, Indian Veterinary Research Institute, Izatnagar, Uttar Pradesh, India

**E. Rasool** Division of Surgery and Radiology, FVSc and AH, SKUAST-K, Srinagar, Jammu and Kashmir, India

**R.A. Shah** Division of Animal Biotechnology, FVSc and AH, SKUAST-K, Srinagar, Jammu and Kashmir, India

**G.T. Sharma** Division of Physiology and Climatology, Indian Veterinary Research Institute, Izatnagar, Uttar Pradesh, India

# Abbreviations

| | |
|---|---|
| ALCAM | Activated leukocyte cell adhesion molecule |
| ATMPs | Advanced therapy medicinal products |
| bFGF | Bovine fibroblast growth factor |
| CD | Cluster of differentiation |
| ConA | Concanavalin A |
| COX2 | Cyclooxygenase 2 |
| DAZL | Deleted in azoospermia like |
| DCs | Dendritic cells |
| DMSO | Dimethyl sulfoxide |
| EGF | Epidermal growth factor |
| ESCs | Embryonic stem cells |
| FASL | Fas ligand |
| FBS | Foetal bovine serum |
| FGF-2 | Fibroblast growth factor-2 |
| Fox | Forkhead box protein |
| HGF | Hepatocyte growth factor |
| HMOX1 | Heme oxygenase-1 |
| ICAM-1 | Intracellular adhesion molecule-1 |
| IDO | Indoleamine 2,3 dioxygenase |
| IFNG | Interferon, gamma |
| IFNγ | Interferon gamma |
| IGF-1 | Insulin growth factor-1 |
| IL | Interleukin |
| IL1F10 | Interleukin 1, family member 10 |
| IL6 | Interleukin 6 |
| IL-6R | Interleukin-6 receptor |
| INFAC | Interferon alpha C (INF-alpha C) |
| INFB1 | Interferon, beta 1 |
| iNOS | Inducible nitric oxide synthase |
| iPSCs | Induced pluripotent stem cells |
| JSP.1 | Major histocompatibility complex class I |
| LFA-3 | Lymphocyte functional antigen-3 |
| LPS | Lipopolysaccharide |
| LTBR | Lymphotoxin beta receptor |

| | |
|---|---|
| MAPK | Mitogen-activated protein kinase |
| MHC-II (DSB) | Major histocompatibility complex class II (DS beta antigen) |
| NK | Natural killer cells |
| PDGF | Platelet-derived growth factor |
| PDL1 | Programmed death ligand-1 |
| PDX-1 | Pancreatic and duodenal homeobox-1 |
| PMA | Phorbol 12-myristate 13-acetate |
| $PrP^{c}$ | PRion protein cellular |
| PTGES | Prostaglandin E synthase |
| PTGES1 | Prostaglandin E synthase 1 |
| RNA | Ribonucleic acid |
| STC-1 | Stanniocalcin 1 |
| TGF-α | Transforming growth factor-α |
| TGFβ1 | Transforming growth factor beta 1 |
| TLRs | Toll-like receptors |
| TNF | Tumour necrosis factor |
| TNFα | Tumour necrosis factor alpha |
| TRKA | Tropomyosin receptor kinase A |
| TSG | Tumour suppressor gene |
| VCAM | Vascular cell adhesion molecule-1 |
| VEGF | Vascular endothelial growth factor |

# Introduction to Stem Cells

1

M. B. Gugjoo

**Abstract**

The stem cell concept is as old as the understanding of the origin of life. The concept of origin of life has developed from initial 'spontaneous generation' to the current concept of development through 'zygote formation'. Understanding the concept of the development of life has paved way for the 'stem cell' concept. There are various types of stem cells that develop right from early phase of life to the adult life though with variable potentialities. An overview of the development of stem cell concept has been attempted in the current chapter.

## 1.1 Introduction

The first and foremost attempt to solve mystery around the origin of life and of human development may be attributed to Aristotle (384–322 BC) (Arey 1974). His observation made him to believe that embryonic development occurs in the uterus and that the mother's menstrual blood is the basis, supporting then previous hypothesis of 'spontaneous generation'. Until 1800, the concept behind development of an individual was of a pre-formed individual who exists in the egg or sperm, known as homunculus. The homunculus (egg), however, requires an activation to form an individual, which occurs through sperm. The sperm, thus, stimulates homunculus in the uterus that ultimately leads to the formation of an individual. Leydig later in 1855 hypothesized that life originates from an already existing life, *omne vivum* ex vivo. This hypothesis was extended by Rudolf Virchow with a view that all new cells develop from existing cells, *omnis cellula e cellula* (Oberling 1944). Later on

M. B. Gugjoo (✉)
Division of Veterinary Clinical Complex, FVSc & AH, SKUAST-Kashmir, Jammu and Kashmir, India



M. B. Gugjoo, A. Pal (eds.), *Mesenchymal Stem Cell in Veterinary Sciences*, https://doi.org/10.1007/978-981-15-6037-8_1

the concept of fertilization was developed, wherein an egg after fusion with the sperm leads to the formation of zygote (Needham 1959). The zygote forms the actual basis of life as it gives rise to all the cells required together to make an individual (totipotency) (Sell 2004a, b). Thus, the existence of cell, now popularly known as 'stem cell', carries the potential to develop into the complete individual.

## 1.2 Overview of the Historical Chart of the Stem Cells

The term 'stamzelle or stem cell' was initially used by a German biologist Ernest Haeckel in 1868 for a fertilized egg that develops into a complete organism (Reimer 1868). The term was also used for the single-celled organism that is considered as an ancestor to all the living organisms. The stem cell research actually started with the process of learning the basis of development and regeneration. In 1930, nuclear transplantation was first carried out in Amoeba (Comandon and deFonbrune 1930). In 1950s Briggs and King applied the same technique to replace nucleus of a frog's egg with other cell (Briggs and King 1952). Gurdon later cloned a frog from an egg and the nucleus of small intestine (Gurdon 1974). The experiments confirmed that the developmental changes that occur are through the environmental factors rather than the permanent changes to the genetic code and in the present case factors were present in oocyte of the frog (Bio-Rad Lab Inc.n.d.). This somatic cell nuclear transfer (SCNT) technique, however, proved to be comparably harder in mammals, and for long time the mammalian cells were considered to be far specialized and, thus, cannot be reverted to totipotent state. But as the research in the field advanced, the findings were proved to be wrong and different mammalian clones were developed. The most noted one was in the form of Dolly, developed by fusion of mammary gland epithelial cell with the enucleated oocyte (Wilmut et al. 1997), besides, the others (Forsberg et al. 2002; Gao et al. 2003; Walker et al. 2002). However, the frequency of successful transplants appears to be very less, <1–2% (Sell 2004a, b), giving rise to the question, why? Thus, not all adult cells are capable to give rise to successful transplants and the cell that give rise could actually be the 'stem cell' or 'tissue progenitor cell' (Van der Kooy and Weiss 2000). In 1945 the initial experimental basis of the stem cell was recognized (Owen 1945). Since then stem cell properties are being discussed (Lajtha et al. 1964; Lajtha 1967) with in vivo stem cell governing factors yet to be fully comprehended. The historical aspect of various stem cell types is briefed below.

### 1.2.1 Haematopoietic Stem Cell

The blood stem cell concept was first introduced in Berlin Haematological Society by Russian academician Alexander Maximow on 1 June 1909. He delivered a theoretical lecture that all the blood cells come from same ancestral cell and, thus, introduced the concept of multipotent nature of blood stem cell (Kalra and Tomar 2014). The stem cell search in blood had begun after the bombings in Hiroshima and

Nagasaki in 1945 when radiation exposure led to the haematopoietic failure in residents. It was found that the shielding of the spleen with lead (Jacobson et al. 1950) and transplantation of bone marrow or spleen cells (Lorenz et al. 1951; Jacobson et al. 1951) prevent such a failure. However, it was confirmed later in 1963 when Ernest McCulloch and James Till (Canadian scientists) carried out experiments on mice bone marrow. It was demonstrated that various blood cells originate from a specific class of cells bearing self-renewal and differentiation properties (Till et al. 1964; Till and McCulloch 1961). Even before actual laboratory confirmation of blood stem cell, the human bone marrow transplant was first attempted by E. Donnall Thomas in 1957 that later fetched him Nobel prize in 1990. The first successful transplant of bone marrow, however, was conducted by Robert A. Good in 1968. He transplanted bone marrow of a sibling to an immune-deficient child who later developed into a healthy adult (Thomson et al. 1998a, b). Since then the blood transfusion has become a common procedure in case of leukaemia patients.

### 1.2.2 Embryonic Stem Cell

The embryonic stem cell understanding has started with the development of spontaneous testicular tumour in a terato-carcinoma study. In 1970s, it was observed that such spontaneous tumours, composed of differentiated cell types and undifferentiated stem cell, may be produced following ectopic grafting of these tumour cells in the adults (Solter et al. 1970; Stevens 1970; Stevens 1983; Papaioannou and Rossant 1983). Those undifferentiated stem cells (embryonal carcinoma cells) were correlated to the pluripotent stem cells of an early embryo. This paved the way to harvest stem cells directly from embryos. It was in 1981 when independent researchers, Marlin Evans and Kaufmann of University of Cambridge and Garl Martin of University of California, could successfully derive embryonic stem cells (ESCs) from mouse blastocysts (Evans and Kaufman 1981). Trophoectoderm, however, was demonstrated to harbour stem cells with limited differentiation potential called trophoblast stem cells (Tanaka et al. 1998). ESCs could grow on a feeder layer formed by mitotically inactivated mouse embryonic fibroblast cells. Later on ESCs were also derived from primates (Thomson et al. 1995, 1996). All these studies along with the research conducted on culturing human in vitro fertilized (IVF) embryos (Gardner et al. 1998) paved the way for successful cultivation and maintenance of ESCs into an undifferentiated state (Thomson et al. 1998a, b). It was demonstrated that feeder cell layer for mouse ESCs could be employed for human counterparts, but variability in biochemical pathways and humoral factors exist. In the same year, cultured human ESCs were differentiated into different tissues, viz. gut epithelium, cartilage, bone, muscle, neurons, etc. (Thomson et al. 1998a, b; Shamblott et al. 1998). Considering these factors ESC line maintenance and ability to trans-differentiate has led to the tremendous research in the field especially in relation to the development and repair of damaged or lost tissue (Donovan and Gearhart 2001). Attempts to create ESCs from veterinary species like cattle, sheep, dog and others too started immediately afterwards (Telugu et al. 2010). Due to the limited availability, associated ethical issues, especially in humans and a tendency to give rise to teratoma, ESCs currently have limited in vivo applications (Gugjoo and Amarpal 2018; Gugjoo et al. 2019).

### 1.2.3 Induced Pluripotent Stem Cell

Recently developed transgenic techniques have opened ways to reprogram an adult tissue cell to pluripotency. An adult tissue cell is reprogrammed by insertion of the pluripotency induction genes like Oct4, Sox2, Klf4 and cMyc. These dedifferentiated cells are known as induced pluripotent stem cells (iPSCs) (Pittenger et al. 1999; Cavaleri and Scholer 2003; Chambers and Smith 2004; Takahashi and Yamanaka 2006). The work was seen with such an extensive potential that the scientists *John B. Gurdon* and *Shinya Yamanaka* were awarded 2012 Nobel Prize in medicine. Initially, the work was conducted in lab animals like mice, but now the cells have also been harvested in various animal species like goat (Song et al. 2013; Sandmaier et al. 2015; Chen et al. 2017), sheep (Li et al. 2011; Liu et al. 2012; Sartori et al. 2012; German et al. 2015), cattle (Zhao et al. 2017), dog (Nishimura et al. 2013; Baird et al. 2015) and equine (Nagy et al. 2011; Khodadadi et al. 2012; Breton et al. 2013; Whitworth et al. 2014). The current success rate in harvesting the dedifferentiated cell population is very limited. These cells like ESCs too have tendency to give rise to teratoma and, thus, fail to make a definitive way into the in vivo applications (Gugjoo et al. 2018a, b, 2019).

### 1.2.4 Mesenchymal Stem Cell

The presence of mesenchymal stem cells within the bone marrow may be documented from the work of Goujon (1869). An osteogenic potential of bone marrow was demonstrated after its heterotopic transplantation. The results were confirmed in the transplantation experiments by Biakow (1870). Later on the osteogenesis was shown to be due to the whole bone marrow and not by the chemo-attractants (Danis 1960). These experiments were also conducted by other researchers who demonstrated the same results and found that bone and cartilage formation does occur (Tavassoli and Crosby 1968; Friedenstein et al. 1966; Bruder and Caplan 1990). Friedenstein and co-workers observed that the heterotopic bone formation post-bone marrow cells transplantation was associated with minor subpopulation of such cells that can be differentiated from haematopoietic cells based on their specific features, viz. plastic adherence and fibroblast-like appearance. The fibroblast-like appearance indicated their origin from the stromal compartment of the bone marrow. The same authors also observed that seeding of bone marrow cell suspension had resulted in the establishment of discrete colonies. These colonies being initiated by single cell are known as colony-forming unit fibroblastic (CFU–Fs) (Friedenstein et al. 1970). The concept of stem cells in bone marrow was formally presented by Owen in 1978 (Owen 1978). It was hypothesized that bone marrow harbours a lineage that is analogous to haematopoietic lineage (Till and McCulloch 1980). The in vivo experimental transplantation of such cells had resulted in the formation of numerous skeletal tissues, viz. bone and cartilage. The cells were named as osteogenic stem cell (Friedenstein et al. 1987) or bone marrow stromal stem cell (Owen and Friedenstein 1988). The non-haematopoetic stem cell hypothesis though well-established previously, its importance, however, was felt in 1999 when similar work was published by a commercial company, Osiris Therapeutics Inc. (Pittenger et al. 1999).

The term 'mesenchymal stem cell' has been coined by Caplan who showed that these cell types carry the potential to differentiate into other phenotypes including adipogenic, tenogenic, ligamentous, dermal or others (Caplan 1991, 1994). The term 'mesenchymal stem cell' although widely used has restricted acceptability (Sipp et al. 2018) and shall henceforth be called by that name in this chapter. The stem cell has 'stemness' characteristics depicted by 'self-renewal and differentiation' properties. Mesenchymal stem cell (MSC) most often undergoes culture-dependent senescence (Wagner et al. 2009), which also diminishes its ability to differentiate (Bonab et al. 2006; Lo Surdo et al. 2013). It is also unclear whether ectopically placed MSCs can actually differentiate in vivo or not. The stem cell label is, thus, questionable with respect to MSCs (Bianco 2014). As such these cells are also termed as stromal stem cells (Owen and Friedenstein 1988), mesenchymal progenitor cells (Dennis et al. 1999), skeletal stem cells (Bianco and Gehron 2000), multipotent adult progenitor cells and mesodermal progenitor cells (Jiang et al. 2002). In 2006, International Society for Cellular Therapy that gave specific guidelines for characterization of these cells recommended the name to be given as 'Multipotent Mesenchymal Stromal Cells' (Dominici et al. 2006). The latest work of Caplan, however, recommends these cells to be named as "Medicinal Signaling Cells" based on their ability to modulate host systems. It may, thus, be pointed out that more emphasis on the mechanism of their therapeutic effects may be laid (Caplan 2017).

A huge literature has been piled on MSCs including their properties, isolation protocols and their applications in diverse ailments both under preclinical experimental studies and in the clinical trials. In veterinary science, MSCs have been harvested from numerous adult tissue sources and foetal membranes (Gugjoo and Amarpal 2018; Gugjoo et al. 2018a, 2019). Mesenchymal stem cells (MSCs) have also been harvested from iPSCs (Chow et al. 2017; Xu et al. 2019). Currently, there are wide variations in their tri-lineage differentiation potential, especially in relation to a marked reduction in their chondrogenic and adipogenic lineages as compared to the bone marrow-derived (BM) MSCs (Xu et al. 2019). MSCs in veterinary sciences like human are characterized as per the criteria laid down by International Society for Cellular Therapy (ISCT) (Dominici et al. 2006). The early mesenchymal stem cell in vivo applications had employed non-specific cell fractions. A heterogeneous cell fraction containing tissue cells, leukocytes, endothelial cells, fibroblasts and a small subset of self-renewing multipotent cells had been utilized (Tiwary et al. 2014). However in recent time culture-expanded MSC preparations are being increasingly studied containing good number of MSCs. Further, advancement like specific cell selection methods need to be studied to make roadways to a more purified cell preparation (Devireddy et al. 2017). The cells are even being cryopreserved for their future applications (Gugjoo and Amarpal 2018; Gugjoo et al. 2018a, 2019). Among various veterinary species, MSC clinical studies have been conducted mainly in dog, cat and horse (Quimby and Borjesson 2018; Gugjoo et al. 2019a, b). In horse, the therapeutic applications of stem cells have mainly been aimed at the issues of musculoskeletal system. An initial attempt was made to repair the superficial digital flexor tendon (Smith et al. 2003). In sheep and goat MSCs, in vivo experimental evaluations have been conducted mostly as human translational studies (Gugjoo and Amarpal 2018; Gugjoo et al. 2020). Cattle and buffalo MSCs have been evaluated in experimental studies mainly to address reproductive problems (Gugjoo et al. 2018b). It may be noted that all MSC studies have so far failed to provide any conclusive results.

### 1.2.5 Cancer Stem Cell

The cancer stem cell (CSC) concept was initially given by Cohnheim and Durante about one-and-a-half century back. The proliferation and differentiation potential of CSC was demonstrated to be comparable to that of the embryonic tissues and cancers. Thus, a concept was proposed that embryonic cells present in adults upon stimulation may lead to cancer. More curiosity in this concept was developed with the work of Furth and co-workers (Furth et al. 1937). It was demonstrated that acute myeloid leukaemia (AML) develops in a mouse after transplantation of single undifferentiated leukaemia cell (Furth et al. 1937; Al-Hajj et al. 2003). After a long gap of three decades, in vitro studies supported AML cells' ability to self-renew, proliferate and generate new tumours (Sell 2004a, b). Initially, such studies were performed on tumours in liquid (blood) phase. The first evidence of these stem cells in solid tumours came in 2003 and was reported from breast cancer (Al-Hajj et al. 2003). Thereafter, CSC concept was validated in a wide variety of solid tumours including lung cancer (Kim et al. 2005), melanoma (Fang et al. 2005), prostate cancer (Collins et al. 2005), brain cancer (Piccirillo et al. 2006), colon cancer (O'Brien et al. 2006) and ovarian cancer (Szotek et al. 2006). The oncogenic therapy on the basis of CSC concept is also currently being researched extensively (Gugjoo et al. 2015; Atashzar et al. 2020).

## 1.3 Conclusion

The stem cell concept dates back to the origin of life. The concept of development of life has changed from initial 'spontaneous generation' to the current concept of development through 'zygote formation'. Understanding the development of life has itself given rise to stem cell concept. Various stem cell types including MSCs have been identified with variable differentiation potential. Overall stem cell concept is still in its infancy, and there are lot of intricacies involved. Among these cell types, mostly MSCs are being evaluated in clinical trials. The stem cell research though is being conducted at a very rapid pace; however, the understanding is yet to reach to the level that can harvest the fruitful results.

## References


Al-Hajj M, Wicha MS, Benito-Hernandez A, Morrison SJ, Clarke MF (2003) Prospective identification of tumorigenic breast cancer cells. Proc Natl Acad Sci 100:3983–3988

Arey LB (1974) Developmental anatomy: a textbook and laboratory manual of embryology, 7th edn. W.B. Saunders, Philadelphia

Atashzar MR, Baharlou R, Karami J, Abdollahi H, Rezaei R, Pourramezan F, Zoljalali Moghaddam SH (2020) Cancer stem cells: a review from origin to therapeutic implications. J Cell Physiol 235(2):790–803

Baird AEG, Barsby T, Guest DJ (2015) Derivation of canine induced pluripotent stem cells. Reprod Domest Anim 50(4):669–676

Biakow A (1870) Uber transplantation Von Knochenmark. Centralbl F DMed Wiss 8:371–373

Bianco P (2014) Mesenchymal stem cells. Annu Rev Cell Dev Biol 30:677

Bianco P, Gehron RP (2000) Marrow stromal stem cells. J Clin Invest 105(12):1663–1668

Bio-Rad Laboratories Inclusive. (n.d.)Stem cell basics for life science researchers,p. 2

Bonab MM, Alimoghaddam K, Talebian F, Ghaffari SH, Ghavamzadeh A, Nikbin B (2006) Aging of mesenchymal stem cells in vitro. BMC Cell Biol 7:14

Breton A, Sharma R, Diaz A et al (2013) Derivation and characterization of induced pluripotent stem cells from equine fibroblasts. Stem Cells Dev 22:611–621

Briggs R, King TJ (1952) Transplantation of living nuclei form blastula cells into enucleated frogs' eggs. Proc Natl Acad Sci U S A 38:455–463

Bruder SP, Caplan AI (1990) Terminal differentiation of osteogenic cells in embryonic chick tibia is revealed by monoclonal antibody against osteocytes. Bone 11:189–198

Caplan AI (1991) Mesenchymal stem cell. J Orthop Res 9:641–650

Caplan AI (1994) The mesengenic process. Clin Plast Surg 21:429–435

Caplan AI (2017) Mesenchymal stem cells: time to change the name! Stem Cells Transl Med 6:1445–1451

Cavaleri F, Scholer HR (2003) Nanog: a new recruit to the embryonic stem cell orchestra. Cell 113:551–552

Chambers I, Smith A (2004) Self-renewal of teratocarcinoma and embryonic stem cells. Oncogene 23:7150–7160

Chen H, Zuo Q, Wang Y, Song J, Yang H, Zhang Y, Li B (2017) Inducing goat pluripotent stem cells with four transcription factor mRNAs that activate endogenous promoters. BMC Biotechnol 17:11

Chow L, Johnson V, Regan D, Wheat W, Webb S, Koch P, Dow S (2017) Safety and immune regulatory properties of canine induced pluripotent stem cell derived mesenchymal stem cells. Stem Cell Res 25:221–232

Collins AT, Berry PA, Hyde C, Stower MJ, Maitland NJ (2005) Prospective identification of tumorigenic prostate cancer stem cells. Cancer Res 65:10946–10951

Comandon J, deFonbrune P (1930) Greffe nucleaire totale, simple ou multiple, chez une Amibe. C R Soc Biol Paris 130:744–748

Danis A (1960) A pres une graffe de tissue squelettique osteogene est a petir des cellulis transplantis que se constitute l'os denouvelle formation. Bull Soc Int Chir 6:647–652

Dennis JE, Merriam A, Awadallah A, Yoo JU, Johnstone B, Caplan AI (1999) A quadripotential mesenchymal progenitor cell isolated from the marrow of an adult mouse. J Bone Miner Res 14:700

Devireddy DR, Boxer L, Myers MJ, Skasko M, Screven R (2017) Questions and challenges in the development of mesenchymal stromal/stem cell-based therapies in veterinary medicine. Tissue Eng 23(4):462–470. https://doi.org/10.1089/ten.teb.2016.0451

Dominici M, Le Blanc K, Mueller I, Slaper-Cortenbach I, Marini F, Krause D, Deans R, Keating A, Prockop DJ, Horwitz E (2006) Minimal criteria for defining multipotent mesenchymal stromal cells. The International Society for Cellular Therapy position statement. Cytotherarpy 8(4):315–317
Donovan PJ, Gearhart J (2001) The end of the beginning for pluripotent stem cells. Nature 414:92–97
Evans MJ, Kaufman MH (1981) Establishment in culture of pluripotential cells from mouse embryos. Nature 292:154–156
Fang D, Nguyen TK, Leishear K, Finko R, Kulp AN, Hotz S, Van Belle PA, Xu X, Elder DE, Herlyn M (2005) A tumorigenic subpopulation with stem cell properties in melanomas. Cancer Res 65:9328–9337
Forsberg EJ, Strelchenko NS, Augenstein ML et al (2002) Production of cloned cattle from in vitro systems. Biol Reprod 67:327–333
Friedenstein AJ, Piatetzkey-Shapiro II, Petrakova KV (1966) Osteogenesis of transplants of bone marrow cells. J Embryol Exp Morphol 16:381–390
Friedenstein AJ, Chailakhjan RK, Lalykina KS (1970) The development of fibroblast colonies in monolayer cultures of guinea-pig bone marrow and spleen cells. Cell Tissue Kinet 3:393–403
Friedenstein AJ, Chailakhyan RK, Gerasimov UV (1987) Bone marrow osteogenic stem cells: in vitro cultivation and transplantation in diffusion chambers. Cell Tissue Kinet 20:263–272
Furth J, Kahn MC, Breedis C (1937) The transmission of leukemia of mice with a single cell. Am J Cancer 32:276–282
Gao S, McGarry M, Priddle H et al (2003) Effects of donor oocytes and culture conditions on development of cloned mice embryos. Mol Reprod Dev 66:126–133
Gardner DK et al (1998) Culture and transfer of human blastocysts increases implantation rates and reduces the need for multiple embryo transfers. Fertil Steril 69:84–88
German SD, Campbell KH, Thornton E, McLachlan G, Sweetman D, Alberio R (2015) Ovine induced pluripotent stem cells are resistant to reprogramming after nuclear transfer. Cell Reprogram 17(1):19–27
Goujon E (1869) Recherché experimentales sur les propreites physiologiques de la moelle des os. J L'Anat Physiol 6:399–412
Gugjoo MB, Amarpal (2018) Mesenchymal stem cell research in sheep: current status and future prospects. Small Rumin Res 169:46–56
Gugjoo MB, Amarpal, Kinjavdekar P, Aithal HP, Pawde AM, Bodh D, Sharma GT (2015) Cancer stem cells: concepts and therapeutic applications. Asian J Anim Vet Adv 10(9):509–517
Gugjoo MB, Amarpal, Chandra V, Wani MY, Dhama K, Sharma GT (2018a) Mesenchymal stem cell research in veterinary medicine. Curr Stem Cell Res Ther 13(8):645–657
Gugjoo MB, Amarpal, Fazili MR, Shah RA, Sharma GT (2018b) Mesenchymal stem cell: basic research and potential applications in cattle and buffalo. J Cell Physiol 234(6):8618–8635
Gugjoo MB, Amarpal, Makhdoomi DM, Sharma GT (2019a) Equine mesenchyaml stem cells: properties, sources, characterization and potential therapeutic applications. J Equine Veter Sci 72:16–27
Gugjoo MB, Amarpal, Sharma GT (2019b) Mesenchymal stem cell basic research and applications in dog medicine. J Cell Physiol 234(10):16779–16811
Gugjoo MB, Amarpal, Fazili MR, Shah RA, Mir MS, Sharma GT (2020) Goat mesenchymal stem cell basic research and potential applications. Small Rum Res 183:106045
Gurdon JB (1974) The control of gene expression in animal development. Harvard University Press, Cambridge, MA, pp 3–36
Jacobson LO, Simmons EL, Marks EK, Robson MJ, Bethard WF, Gaston EO (1950) The role of the spleen in radiation injury and recovery. J Lab Clin Med 35:746–770
Jacobson LO, Simmons EL, Marks E, Eldredge JH (1951) Recovery from radiation injury. Science 113:510–511
Jiang Y, Jahagirdar BN, Reinhardt RL, Schwartz RE, Keene CD, Ortiz-Gonzalez XR, Reyes M, Lenvik T, Lund T, Blackstad M, Du J, Aldrich S, Lisberg A, Low WC, Largaespada DA,

Verfaillie CM (2002) Pluripotency of mesenchymal stem cells derived from adult marrow. Nature 418(6893):41–49
Kalra K, Tomar PC (2014) Stem cell: basics, classification and applications. Am Jf Phytomed Clin Therapeut 2(7):919–930
Khodadadi K, Sumer H, Pashaiasl M, Lim S, Williamson M, Verma PJ (2012) Induction of pluripotency in adult equine fibroblasts without c-MYC. Stem Cells Int 2012:429160
Kim CFB, Jackson EL, Woolfenden AE, Lawrence S, Babar I, Vogel S, Crowley D, Bronson RT, Jacks T (2005) Identification of bronchioalveolar stem cells in normal lung and lung cancer. Cell 121:823–835
Lajtha LG (1967) Stem cells and their properties. Proc Can Cancer Conf 7:31–39
Lajtha LG, Gilbert CW, Porteous DD, Alexanian R (1964) Kinetics of a bone-marrow stem-cell population. Ann N Y Acad Sci 113:742–752
Li Y, Cang M, Lee AS, Zhang K, Liu D (2011) Reprogramming of sheep fibroblasts into Pluripotency under a drug-inducible expression of mouse-derived defined factors. PLoS One 6(1):e15947
Liu J, Balehosur D, Murray B, Kelly JM, Sumer H, Verma PJ (2012) Generation and characterization of reprogrammed sheep induced pluripotent stem cells. Theriogenol 77(2):338–346
Lo Surdo JL, Millis BA, Bauer SR (2013) Automated microscopy as a quantitative method to measure differences in adipogenic differentiation in preparations of human mesenchymal stromal cell. Cytotherapy 15:1527
Lorenz E, Uphoff D, Reid TR, Shelton E (1951) Modification of irradiation injury in mice and guinea pigs by bone marrow injections. J Natl Cancer Inst 12:197–201
Nagy K, Sung HK, Zhang P et al (2011) Induced pluripotent stem cell lines derived from equine fibroblasts. Stem Cell Rev 7:693–702
Needham A (1959) History of embryology. Cambridge University Press, New York, NY
Nishimura T, Hatoya S, Kanegi R, Sugiura K, Wijewardana V, Kuwamura M, Tanaka M, Yamate J, Izawa T, Takahashi M, Kawate N, Tamada H, Imai H, Inaba T (2013) Generation of functional platelets from canine induced pluripotent stem cells. Stem Cells Dev 22:2026–2035
O'Brien CA, Pollett A, Gallinger S, Dick JE (2006) A human colon cancer cell capable of initiating tumour growth in immunodeficient mice. Nature 445:106–110
Oberling C (1944) The riddle of cancer. Yale University Press, New Haven, CT
Owen RD (1945) Immunogenetic consequences of vascular anastomoses between bovine twins. Science 102(2651):400–401
Owen M (1978) Histogenesis of bone cells. Calcif Tissue Cells 25:205–207
Owen M, Friedenstein AJ (1988) Stromal stem cells: marrow-derived osteogenic precursors. Ciba Found Symp 136:42–60
Papaioannou VE, Rossant J (1983) Effects of the embryonic environment on proliferation and differentiation of embryonal carcinoma cells. Cancer Surv 2:165–183
Piccirillo SGM, Reynolds BA, Zanetti N, Lamorte G, Binda E, Broggi G, Brem H, Olivi A, Dimeco F, Vescovi AL (2006) Bone morphogenetic proteins inhibit the tumorigenic potential of human brain tumour-initiating cells. Nature 444:761–765
Pittenger MF, Mackay AM, Beck SC, Jaiswal RK, Douglas R, Mosca JD, Moorman MA, Simonetti DW, Craig S, Marshak DR (1999) Multilineage potential of adult human mesenchymal stem cells. Science 284:143–147
Quimby JM, Borjesson DL (2018) Mesenchymal stem cell therapy in cats current knowledge and future potential. J Feline Med Surg 20:208–216
Reimer G (1868) Naturliche Schopfungsgeschichte. Springer, Berlin
Sandmaier SE, Nandal A, Powell A, Garrett W, Blomberg L, Donovan DM, Talbot N, Telugu BP (2015) Generation of induced pluripotent stem cells from domestic goats. Mol Reprod Dev 82(9):709–721
Sartori C, DiDomenico AI, Thomson AJ, Milne E, Lillico SG, Burdon TG, Whitelaw CB (2012) Ovine-induced pluripotent stem cells can contribute to chimeric lambs. Cell Reprogram 14(1):8–19
Sell S (2004a) Stem cells: a handbook of stem cells. Humana Press, Totowa, NJ, p 1

Sell S (2004b) Stem cell origin of cancer and differentiation therapy. Crit Rev Oncol Hematol 51:1–28

Shamblott MJ, Axelman J, Littlefield JW, Donovan PJ, Blumenthal PD, Huggins GR, Gearhart JD (1998) Derivation of pluripotent stem cells from cultured human primordial germ cells. Proc Natl Acad Sci U S A 95(13):726–13,731

Sipp D, Robey PG, Turner L (2018) Clear up this stem-cell mess. Nature 561:455–457

Smith RKW, Korda M, Blunn GW, Goodship AE (2003) Isolation and implantation of autologous equine mesenchymal stem cells from bone marrow into the superficial digital flexor tendon as a potential novel treatment. Equine Vet J 35:99–102

Solter D, Skreb N, Damjanov I (1970) Extrauterine growth of mouse egg-cylinders results in malignant teratoma. Nat Lond 227(503):504

Song H, Li H, Huang M, Xu D, Gu C, Wang Z, Dong F, Wang F (2013) Induced pluripotent stem cells from goat fibroblasts. Mol Reprod Dev 80(12):1009–1017

Stevens LC (1970) The development of transplantable teratocarcinomas from intratesticular grafts of pre- and postimplantation mouse embryos. Dev Biol 21:364–382

Stevens LC (1983) The origin and development of testicular, ovarian and embryo-derived teratomas. In: Silver LM et al (eds) Teratocarcinoma stem cells, vol 10. Cold Spring Harbor Laboratory Press, Cold Spring Harbor, NY, pp 23–36

Szotek PP, Pieretti-Vanmarcke R, Masiakos PT, Dinulescu DM, Connolly D, Foster R, Dombkowski D, Preffer F, Maclaughlin DT, Donahoe PK (2006) Ovarian cancer side population defines cells with stem cell-like characteristics and Mullerian inhibiting substance responsiveness. Proc Nati Acad Sci USA 103:11154–11159

Takahashi K, Yamanaka S (2006) Induction of pluripotent stem cells from mouse embryonic and adult fibroblast cultures by defined factors. Cell 126(4):663–676

Tanaka S, Kunath T, Hadjantonakis A-K, Nagy A, Rossant J (1998) Promotion of trophoblast stem cell proliferation by FGF4. Science 282:2072–2075

Tavassoli M, Crosby WH (1968) Transplantation of marrow to extramedullary sites. Science 161:54–56

Telugu BP, Ezashi T, Roberts RM (2010) Porcine induced pluripotent stem cells analogous to naive and primed embryonic stem cells of the mouse. Int J Dev Biol 54:1703–1711

Thomson JA, Kalishman J, Golos TG, Durning M, Harris CP, Becker RA, Hearn JP (1995) Isolation of a primate embryonic stem cell line. Proc Natl Acad Sci U S A 92:7844–7848

Thomson JA, Kalishman J, Golos TG, Durning M, Harris CP, Hearn JP (1996) Pluripotent cell lines derived from common marmoset (Callithrix jacchus) blastocysts. Biol Reprod 55:254–259

Thomson JA, Itskovitz-Eldor J, Shapiro SS, Waknitz MA, Swiergiel JJ, Marshall VS, Jones JM (1998a) Embryonic stem cell lines derived from human blastocysts. Science 282:1145–1147

Thomson JA, Eldor EJ, Shapiro SS, Waknitz MA, Swiergiel JJ, Marshall VS, Jones JM (1998b) Embryonic stem cell lines derived from human blastocysts. Science 282(5391):1145–1147

Till JE, McCulloch EA (1961) A direct measurement of the radiation sensitivity of normal mouse bone marrow cells. Radiat Res 14:213–222

Till JE, McCulloch EA (1980) Haematopoietic stem cell differentiation. Biochem Biophysic Acta 605:431–459

Till JE, McCulloch EA, Siminovitch L (1964) A stochastic model of stem cell proliferation, based on the growth of spleen Colony-forming cells. Proc Natl Acad Sci U S A 51:29–36

Tiwary R, Amarpal, Aithal HP, Kinjavdekar P, Pawde AM, Singh R (2014) Effect of IGF-1 and uncultured autologous bone-marrow-derived mononuclear cells on repair of osteochondral defect in rabbits. Cartilage 5:43–54

Van der Kooy D, Weiss S (2000) Why stem cells? Science 287:1439–1441

Wagner W, Bork S, Horn P, Krunic D, Walenda T, Diehlmann A, Benes V, Blake J, Huber FX, Eckstein V, Boukamp P, Ho D (2009) Aging and replicative senescence have related effects on human stem and progenitor cells. PLoS One 4:e5846

Walker SC, Shin T, Zaunbrecher GM, Romano JE, Johnson GA, Bazer FW, Piedrahita JA (2002) A highly efficient method for porcine cloning by nuclear transfer using in vitro-matured oocytes. Cloning Stem Cells 4:105–112

Whitworth DJ, Ovchinnikov DA, Sun J, Fortuna PR, Wolvetang EJ (2014) Generation and characterization of LIF-dependent equine induced pluripotent stem cells from adult dermal fibroblasts. Stem Cells Dev 23:1515–1523

Wilmut I, Schnieke AE, McWhir J, Kind AJ, Campbell KH (1997) Viable offspring derived from fetal and adult mammalian cells. Nature 385:810–813

Xu M, Shaw G, Murphy M, Barry F (2019) Induced pluripotent stem cell-derived Mesenchymal stromal cells are functionally and genetically different from bone marrow-derived Mesenchymal stromal cells. Stem Cells 37(6):754–765

Zhao L, Wang Z, Zhang J, Yang J, Gao XF, Wu B, Zhao G, Bao S, Hu S, Liu P, Li X (2017) Characterization of the single-cell derived bovine induced pluripotent stem cells. Tissue Cell 49(5):521–527

# Mesenchymal Stem Cell and Its Properties

# 2

M. B. Gugjoo, Amar Pal, and G. T. Sharma

**Abstract**

Stem cell (SC) concept is one of the important research areas seen to bring new vistas in the biological science and/or medicine. Various stem cell types have been identified including mesenchymal stem cells (MSCs). MSCs are present in almost all the body parts of an individual and carry the specialized stemness property, in addition to the immuno-modulation, plasticity and homing. These cells are studied for their therapeutic applications due to lack of any ethical issue together with a limited ability to develop tumour. However, there is dearth of understanding in their basic specialized physiological properties and as such definitive clinical applications are restricted. The current chapter details various stem cell types and discusses the characteristic properties of the stem cells with special focus on MSCs.

## 2.1 Introduction

The stem cell (SC) concept is one of the most thrilling scientific research areas in twenty-first century. Stem cell research is seen and hoped to bring new dawn in the field of biology and medicine. An individual in their lifespan undergoes through

M. B. Gugjoo
Division of Veterinary Clinical Complex, FVSc & AH, SKUAST-Kashmir, Jammu and Kashmir, India

A. Pal
Division of Surgery, Indian Veterinary Research Institute, Izatnagar, UP, India

G. T. Sharma (✉)
Division of Physiology & Climatology, Indian Veterinary Research Institute, Izatnagar, UP, India


M. B. Gugjoo, A. Pal (eds.), *Mesenchymal Stem Cell in Veterinary Sciences*,
https://doi.org/10.1007/978-981-15-6037-8_2

various developmental stages each characterized by the availability of particular stem cell type. During embryonic stage, stem cells give rise to the complete individual and in post-natal life the cells serve a sort of an internal repair system to maintain homeostasis (Gugjoo and Amarpal 2018; Gugjoo et al. 2019). The inner cell mass of trophoblast during embryogenesis carries embryonic stem cells (ESCs) that give rise to the whole individual (pluripotency). Adult tissue stem cells, present in almost all the body tissues, have restricted differentiation potential (multipotent) (Fick et al. 2012; Ayala-Cuellar et al. 2018). The adult tissue stem cells like mesenchymal stem cells (MSCs) maintain tissue homeostasis through tissue cell and matrix turn-over (Gugjoo et al. 2018a, b, 2019). It may be inferred that developmental progression of an individual brings in reduced differentiation potential of the stem cells.

## 2.2 What Is Stem Cell?

Stem cell (SC) is an unspecialized cell carried by a multi-cellular organism that is able to produce an identical copy of itself through the process of 'self-renewal' and a more specialized cell through the process of differentiation (Till et al. 1964; Till and McCulloch 1961; Ab Kadir et al. 2012). Thus, the two main properties that contribute stemness feature to the stem cell are the 'self-renewal' and the 'differentiation'.

## 2.3 Stem Cell Types

Stem cells are classified in many ways but most commonly based upon the degree of differentiation and source of isolation.

### 2.3.1 Based on the Degree of Differentiation

***Totipotent Cell*** The cell differentiates into any cell type of the body including the foetal membrane like zygote up to morula stage of the embryogenesis.

***Pluripotent Cell*** The cell differentiates into all the cell types of the body barring foetal membranes like embryonic stem cells (ESCs) and induced pluripotent stem cells (iPSCs).

***Multipotent Cell*** The cell gives rise to limited cell types of the body, e.g. mesenchymal stem cells (MSCs).

### 2.3.2 Based on the Source

***Embryonic Stem Cell*** The cell is derived from an embryo and may be totipotent (cells up to the morula stage of embryogenesis) or pluripotent (inner cell mass of the blastocyst of embryo).

***Induced Pluripotent Stem Cell*** The cell is developed by reprogramming of the adult somatic cell through "forced" expression of pluripotency-specific genes.

***Adult Stem Cell*** It is an unspecialized cell present in all the tissues/organs of the body for internal repair or replacement of damaged cells. The self-renewal and the differentiation capacity are restricted in these cells as compared to pluripotent cells.

## 2.4 Mesenchymal Stem Cell

Among various stem cell types, MSCs currently form an important constituent of the regenerative medicine. This can be attributed to their diverse and numerous accessible sources, and the simplicity of isolation procedures with better success rates. These cells have minimal teratogenicity usually seen with the pluripotent stem cells and bear limited ethical issues commonly associated with human ESCs (Zuk et al. 2001; Cardoso et al. 2017). Additional features like ability to immunomodulate and provide anti-inflammatory response, thereby regeneration of tissue without scar formation, make it more suitable for regenerative medicine (Zuk et al. 2001; Stewart and Stewart 2011; Gugjoo et al. 2019). However, MSCs in contrary to ESCs have limited culture lifespan, and after certain culture passages these cells become senescent (Caplan 2012). Even upon in vivo implantation, MSCs (autologous/allogeneic) survive only for certain period (ranging from few hours to as long as 90 days) but not indefinitely (Guest et al. 2010; McDuffee et al. 2012). This questions their self-renewal property (Lopez and Jarazo 2015). In order to extend their lifespan, vector viruses like simian virus 40 (SV40-T) and human papillomavirus have been used to extend their culture lifespan (Klingelhutz et al. 1994). The SV40-T retrovirus may immortalize dog adipose tissue-derived MSCs (Ayala-Cuellar et al. 2018) by reducing the growth inhibition on retinoblastoma protein (pRb) and tumour protein (p53) (DeCaprio et al. 1988; Pipas 2009). These are preliminary studies that need further extensive research in the area to confirm their clinical safety and efficacy.

### 2.4.1 Characteristic Features of MSCs

#### 2.4.1.1 Stemness

The prime stemness property, represented by self-renewal and differentiation, of MSCs is poorly understood and has even been questioned. Only few genes support

MSCs stemness markers. There have to be deep insights into its molecular basis especially through the main transcription factors. So far, not a single key transcription factor to MSCs has been determined unlike pluripotent genes like Oct4, Nanog and Sox2 being attributed to the embryonic stem cells. Several highly expressed genes including nine transcription factors in undifferentiated MSCs have been detected after comparing undifferentiated and tri-lineage-differentiated MSCs. However, individual knockdown of such genes could only partially decrease the cell proliferation or their differentiation (Kubo et al. 2009). Several reasons that may be underlined to such poor understanding are as follows:

1. The cellular heterogeneity with respect to the species, age, sources, culture conditions and number of culture passages (Mareddy et al. 2010; Menicanin et al. 2010).
2. MSCs have finite culture lifespan characterized by ageing and senescence after certain passages. MSCs exhibit morphological abnormalities, attenuated expression of specific surface markers, reduced differentiation potential and proliferation arrest (Baxter et al. 2004; Wagner et al. 2008; Liu et al. 2013).
3. Poor understanding of microenvironment/niche that controls MSCs features. There is need to study the environmental effects. Among various niche cues studied include hypoxia (Drela et al. 2014; Hu et al. 2014; Fotia et al. 2015), available extra-cellular matrix (ECM) (Xiong et al. 2015; Wong et al. 2017), scaffolds (Cesarz and Tamama 2016; Obara et al. 2016), mechanical factors and/ humoral/ growth factors (Terraciano et al. 2007; Bottagisio et al. 2015; Hu et al. 2015). Such cues, however, have been studied in isolation while in vivo microenvironment/niche presents an intricate system.

### Self-Renewal and Differentiation

The self-renewal property keeps the stem cell pool intact, which makes it possible to repair the damaged cells or to replace the non-functioning cells throughout an individual's life (Fig. 2.1). This feature of stem cell differentiates it from progenitor cell as the latter cell although able to differentiate but is unable to self-renew. Stem cell self-renewal occurs through either symmetric or asymmetric divisions. Through symmetric divisions, the parent cell is able to produce identical copies, that is, daughter cell copies entire genome including the epigenetic changes. In asymmetric divisions, two variable cells are produced either through divisional asymmetry or environmental asymmetry (Wilson and Trumpp 2006; He et al. 2009). In divisional asymmetry, cell fate determinants get reorganized prior to division and after division two cells are produced, one identical and other non-identical to the parent cell. In environmental asymmetry, the cell gets exposed to multiple signals from the local environment (micro-environment), which initially leads to symmetric division producing identical copies. One of the daughter stem cell then undergoes asymmetric cell division generating non-identical cells to perform tissue-specific function (Spradling et al. 2001; Ohlstein et al. 2004). This process is known as differentiation and is governed by available micro-environment. The growth factors and cytokines like IGF-1, TGF-β1, BMP-1, etc. may facilitate stem cell proliferation or inhibition

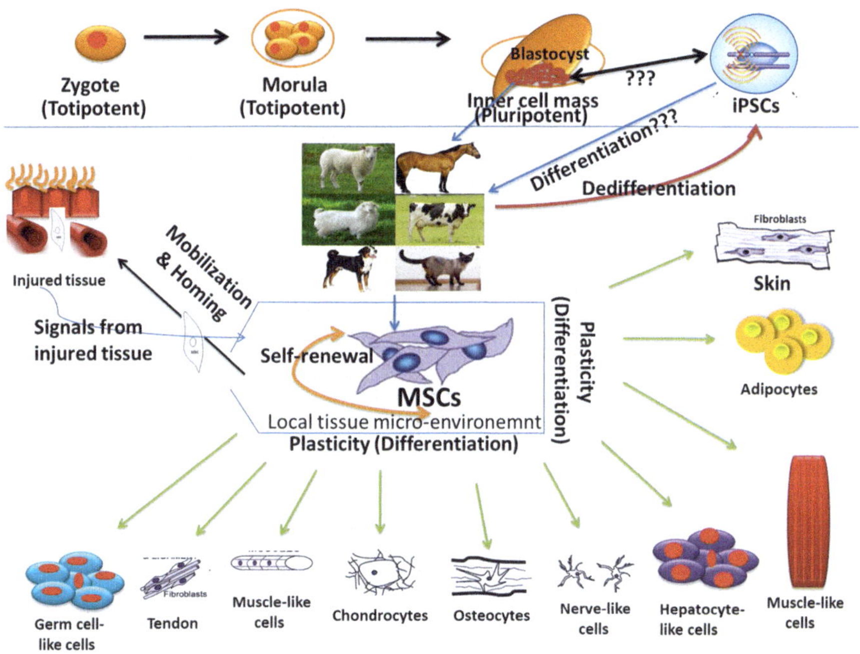


**Fig. 2.1** Diagrammatic representation of development of mesenchymal stem cells (MSCs), their self-renewal, differentiation, and mobilization and homing properties

depending upon the tissue specific requirements (Fuchs et al. 2004; Potten et al. 2009; Spees et al. 2016). The microenvironment/niche tends to play an essential role in maintaining the stem cell self-renewal property. The cells lose their self-renewal potential upon loss of the niche (Xie and Spradling 2000; Mackenzie and Flake 2001; Li and Neaves 2006; Lilly et al. 2011).

The self-renewal potential varies among stem cell types. In culture environment, embryonic stem cells are able to self-renew almost indefinitely while MSCs self-renew only for few generations. The mechanism behind this wonderful property is still ambiguous though some information has been gained in recent times. One of the mechanisms reported is of telomerase, an enzyme required to maintain DNA regions at the chromosomal ends, telomeres, by preventing accumulation of DNA damage due to replication (Greenwood and Lansdorp 2003). It has been seen that telomerase absence decreases self-renewal capacity of haematopoietic stem cells (HSCs) while forced expression is insufficient to transplant such cells indefinitely (Allsopp et al. 2003a, b). The cells that are able to express elements of different signalling pathways like Wnt/ fzd/beta-catenin, sonic, Hedgehog, etc. usually undergo self-renewal divisions (Watt and Hogan 2000; Willert et al. 2003). Besides, over-expression of several proteins like homeobox 4 genes (Hoxb4) and Cdx4 (Kyba et al. 2002; Wang et al. 2005) may also favour self-renewal divisions. All these signalling pathways maintain pleuripotency of the cells and may not be applicable to unipotent or oligopotent cells.

## 2.4.2 Plasticity

Differentiation capability of stem cell type varies. Embryonic stem cell is able to give rise to all the cell types of the three germinal layers while adult stem cells usually have restricted potential. Recent studies have shown an extended differentiation potential of adult stem cells. Mesenchymal stem cells (MSCs) differentiation potential earlier restricted to osteogenic, chondrogenic and adipogenic lineages (Gugjoo et al. 2015) has now been demonstrated to extend to other lineages like myocyte-/cardiomyocyte-like cells (Orlic et al. 2001; Ferrari et al. 1998; Vieira et al. 2010; Wang et al. 2018), neural-like cells (Oh et al. 2011; Ryu et al. 2012; Yan et al. 2015; Kriston-Pál et al. 2017; Li et al. 2017; Zhang et al. 2019) and germ cell-like cells (Ghasemzadeh-Hasankolaei et al. 2014, 2015), among others. This phenomenon is known as trans-differentiation. As such the stem cell fate is not fixed rather changeable or plastic (plasticity) in nature depending upon the available local microenvironment (Pelagalli et al. 2018).

### 2.4.2.1 Anti-Inflammatory and Immuno-Modulation

MSCs are considered as immune privileged or immune evasive cells, with an ability to immuno-modulate and provide anti-inflammatory activities (Caplan and Dennis 2006; Kang et al. 2008; Lee et al. 2011; Kim et al. 2010; Carrade et al. 2012; Prockop and Oh 2012). These cells avoid allogeneic reaction as they are hypo-immunogenic and often lack major histocompatibility complex (MHC)-II and co-stimulatory molecule expression, although variably. These cells prevent T-cell responses through the modulation of dendritic cells and disruption of natural killer (NK) cells (Wang et al. 2009). The cells may additionally modulate the function of $CD8^+$ and $CD4^+$ T-cells (Di Nicola et al. 2002). The cells interact with CD4 and CD8 lymphocytes and upon pro-inflammatory mediator's activation secrete factors that lead to down-regulation of the inflammation (Singer and Caplan 2011). MSCs hampering of T-cell activation and proliferation may also be induced by decreasing the expression of activation markers like CD25, CD38 and CD69 (Le Blanc et al. 2003; Groh et al. 2005). This property of MSCs makes their allogeneic application quite possible (Berner et al. 2013; Gugjoo et al. 2017, 2020). Considering the importance of these properties in regenerative medicine, a separate chapter on the immuno-modulatory and/anti-inflammatory properties of mesenchymal stem cells has been discussed later in the book.

### 2.4.2.2 Stem Cell Migration and Homing

MSCs have characteristic migratory and homing property (Mackenzie and Flake 2001). A significant amount of literature is available that tries to address the mechanism of stem cell migration and homing; properties that explain the basis of stem cell activity at sites beyond their site of implantation or original location. MSCs ability to migrate and home to the target site may be facilitated by humoral factors in addition to the involved unknown mechanisms (Kodeet al. 2009; Guest et al. 2010; McDuffee et al. 2012).

Mobilization of MSCs is activated by down-regulation of their adhesion molecules through the release of cytokines and/or chemokines (Liu et al. 2009). There are various factors relevant to MSCs mobilization including matrix metalloproteinase protein (MMP)-9 and stromal cell-derived factor (SDF)-1α/chemokine (C-X-C motif) ligand (CXCL12) and its receptor CXCR4 (Hristov and Weber 2009). MSCs overexpression of insulin-like growth factor 1 (IGF-1) may induce their massive mobilization via SDF-1α signalling (Haider et al. 2008). In addition to SDF-1α, MSCs may release high mobility group box-1 (HMGB-1) extracellularly after their activation through inflammatory cytokine or by cell necrosis. This protein acts as a chemo-attractant for MSCs (Palumbo and Bianchi 2004; Chavakis et al. 2007). The injury-/ischemia-induced activation of fibroblasts cause an increased migration velocity and invasion of MSCs. Two important components like basic fibroblast growth factor (bFGF) and epithelial neutrophil-activating peptide-78 (ENA-78 or CXCL5) seem to be mainly involved in such activities (Nedeau et al. 2008). Surface receptors like chemokine receptors (CCR1, CCR4, CCR7, CCR9, CXCR5, CXCR6 and CCR10) expressed on human BM-MSCs may be involved in their migration (Wynn et al. 2004; Sordi et al. 2005; Von Luttichau et al. 2005).

MSCs homing to the target tissue sites follow cascade of events. This includes MSCs rolling into the blood vessels, endothelial cell surface adhesion, trans-endothelial migration, extravasation and ultimately their migration through the tissue matrix to the injured target area (Zhuge et al. 2010). Stem cells express numerous surface receptors and cell adhesion molecules, similar to that of leucocytes, that aide in migration and homing of the cells (Luster et al. 2005; Ley et al. 2007). The receptors include integrins, selectins, cell adhesion molecules (CAMs) and chemokine receptors. P-selectin, a cell adhesion molecule, is considered to be involved in MSCs rolling that slows their motion and allows them to adhere to endothelium (Ruster et al. 2006). Various other adhesion molecules that mediate MSC–endothelial cell interaction involve integrins such as α1 down to α5 and β1 up to β4 and CAMs such as VCAM-1 to ICAM-3, ALCAM and CD105 (Minguell et al. 2001; Krampera et al. 2006). These adhesion molecules and their counter-ligands are expressed on MSCs and/or endothelial cells. CD44 may act as a homing molecule for BM-MSCs as it traffics them to the bone through marrow vasculature supported E-selectin interaction (Sackstein et al. 2008). Equine MSCs (eMSCs) homing may be achieved through their expression of the stromal cell-derived factor 1 (SDF-1) and its receptor C-X-C chemokine receptor type 4 (CXCR4) (Angelone et al. 2017). Additionally, adhesion molecules like Galectin-1 and Galectin-3 might play an important role in equine MSCs homing to the inflamed and injured tissues (Reesink et al. 2017).The role of hypoxia in stem cell homing demands thorough investigation as hypoxia activates peripheral blood mononuclear cells to turn to the MSCs (Hopper et al. 2015).

All these studies demonstrate active inflammation to be prime factor involved in directing MSCs migration and homing, but such properties against chronic lesions that are devoid of inflammation currently remain questioned.

## 2.5 Significance of Stem Cell Research

Tissue-specific cells continuously undergo the process of wear and tear and ultimately die. The cells if not replaced may lead to malfunctioning of the tissue system and finally death. Such damaged cells are, thus, being replaced by an internal repair system formed by the stem cells. However, in extensive cases an intrinsic system fails to maintain such homeostasis. In an organ failure case (kidney, heart failure), donors are required who are hardly available that too at right time. Besides, some highly specialized tissues like nervous tissue have very limited internal repair system, which demands the medical support for damage repair. Currently, stem cells are seen to carry the potential to fill this gap in medical field. Further, stem cell differentiation if controlled can be used to produce specific tissue/organs. Such tissue/organs can be used as disease models to study the course of disease and the therapeutic options.

## 2.6 Conclusion

MSCs are important stem cell types that are considered to provide therapeutic effects in regenerative medicine through self-renewal, differentiation, immunomodulation, migration and homing. However, there are various queries that need to be addressed before stem cell application can be made successful.

### 2.6.1 Answers Required

1. How do stem cells proliferate without differentiation while non-stem cells do not?
2. What are the in vivo factors and how do they regulate stem cell self-renewal, proliferation and differentiation?
3. Why pluripotent stem cells are almost immortalized while adult stem cells like MSCs have limited lifespan?
4. How immuno-modulation, anti-inflammatory and homing properties of MSCs can be controlled?

The answers if discovered may pave the way for the following potential uses of stem cells:

1. Repair tissue/organ that is extensively damaged either by direct transplantation or through tissue engineering.
2. Stimulation of the intrinsic stem cell pool of an individual for effective repair.
3. Development of tissue/organ models for disease course study and their therapeutics.
4. Regulation of the developmental course of an individual.
5. Study inherited diseases like birth defects, Parkinson's disease, etc.
6. Understand cancer biology and, thus, control measures.

## References

Ab Kadir R, Hisham Zainal Ariffin S, Megat Abdul Wahab R, Senafi S (2012) Molecular characterisation of human peripheral blood stem cells. S Afr J Sci 108:1–7

Allsopp RC, Morin GB, DePinho R, Harley CB, Weissman IL (2003a) Telomerase is required to slow telomere shortening and extend replicative lifespan of HSCs during serial transplantation. Blood 102:517–520

Allsopp RC, Morin GB, Horner JW, DePinho R, Harley CB, Weissman IL (2003b) Effect of TERT over-expression on the long-term transplantation capacity of hematopoietic stem cells. Nat Med 9:369–371

Angelone M, Conti V, Biacca C, Battaglia B, Pecorari L, Piana F, Gnudi G, Leonardi F, Ramoni R, Basini G, Dotti S, Renzi S, Ferrari M, Grolli S (2017) The contribution of adipose tissue-derived Mesenchymal stem cells and platelet-rich plasma to the treatment of chronic equine laminitis: a proof of concept. Int J Mol Sci 18:2122

Ayala-Cuellar AP, Kim C-W, Hwang K-A, Kang J-H, Lee G, Cho J, Choi K-C (2018) Characterization of canine adipose tissue-derived mesenchymal stem cells immortalized by SV40-T retrovirus for therapeutic use. J Cell Physiol 234:9. https://doi.org/10.1002/jcp.28338

Baxter MA, Wynn RF, Jowitt SN, Wraith JE, Fairbairn LJ, Bellantuono I (2004) Study of telomere length reveals rapid aging of human marrow stromal cells following in vitro expansion. Stem Cells 22:675–682

Berner A, Reichert JC, Woodruff MA, Saifzadeh S, Morris AJ, Epari DR, Nerlich M, Schuetz MA, Hutmacher DW (2013) Autologous vs. allogenic mesenchymal progenitor cells for the reconstruction of critical sized segmental tibial bone defects in aged sheep. Acta Biomater 9(8):7874–7884

Bottagisio M, Lovati AB, Lopa S, Moretti M (2015) Osteogenic differentiation of human and ovine bone marrow stromal cells in response to b-glycerophosphate and monosodium phosphate. Cell Reprogram 17(4):235–242

Caplan AI (2012) MSCs as therapeutics. In: Hematti P, Keating A (eds) Stem cell biology and Reg medicine: Mesenchymal stromal cells: biology and clinical applications. Human Press, New York, pp 79–90

Caplan AI, Dennis JE (2006) Mesenchymal stem cells as trophic mediators. J Cell Biochem 98:1076–1084

Cardoso MT, Pinheiro AO, Vidane AS, Casals JB, de Oliveira VC, Gonçalves NJN, Martins DS, Ambrósio CE (2017) Characterization of teratogenic potential and gene expression in canine and feline amniotic membrane-derived stem cells. Reprod Domest Anim 52(s2):58–64

Carrade DD, Lame MW, Kent MS, Clark KC, Walker NJ, Borjesson DL (2012) Comparative analysis of the immunomodulatory properties of equine adult-derived mesenchymal stem cells. Cell Med 4:1–11

Cesarz Z, Tamama K (2016) Spheroid culture of Mesenchymal stem cells. Stem Cells Int 2016:9176357

Chavakis E, Hain A, Vinci M et al (2007) High-mobility group box 1 activates integrin-dependent homing of endothelial progenitor cells. Circ Res 100(2):204–212

DeCaprio JA, Ludlow JW, Figge J, Shew JY, Huang CM, Lee WH, Livingston DM (1988) SV40 large tumor antigen forms a specific complex with the product of the retinoblastoma susceptibility gene. Cell 54(2):275–283

Di Nicola M, Carlo-Stella C, Magni M, Milanesi M, Longoni PD, Matteucci P, Grisanti S, Gianni AM (2002) Human bone marrow stromal cells suppress T-lymphocyte proliferation induced by cellular or non-specific mitogenic stimuli. Blood 99(10):3838–3843

Drela K, Sarnowska A, Siedlecka P, Szablowska-Gadomska I, Wielgos M, Jurga M, Lukomska B, Domanska-Janik K (2014) Low oxygen atmosphere facilitates proliferation and maintains undifferentiated state of umbilical cord mesenchymal stem cells in an hypoxia inducible factor-dependent manner. Cytotherapy 16:881–892

Ferrari G, Cusella-De Angelis G, Coletta M, Paolucci E, Stornaiuolo A, Cossu G, Mavilio F (1998) Muscle regeneration by bone marrow-derived myogenic progenitors. Science 279:1528–1530
Fick LJ, Fick GH, Li Z, Cao E, Bao B, Heffelfinger D, Riabowol K (2012) Telomere length correlates with life span of dog breeds. Cell Rep 2(6):1530–1536
Fotia C, Massa A, Boriani F, Baldini N, Granchi D (2015) Hypoxia enhances proliferation and stemness of human adipose-derived mesenchymal stem cells. Cytotechnology 67:1073–1084
Fuchs E, Tumbar T, Guasch G (2004) Socializing with the neighbors: stem cells and their niche. Cell 116:769–778
Ghasemzadeh-Hasankolaei M, Sedighi-Gilani MA, Eslaminejad MB (2014) Induction of ram bone marrow mesenchymal stem cells into germ cell lineage using transforming growth factor-b superfamily growth factors. Reprod Domest Anim 49:588–598
Ghasemzadeh-Hasankolaei M, Eslaminejad BM, Sedighi-Gilani M (2015) Derivation of male germ cells from ram bone marrow mesenchymal stem cells by three different methods and evaluation of their fate after transplantation into the testis. In Vitro Cell Dev Biol Anim 52(1):49–61
Greenwood MJ, Lansdorp PM (2003) Telomeres, telomerase, and hematopoietic stem cell biology. Arch Med Res 34:489–495
Groh ME, Maitra B, Szekely E, Koc ON (2005) Human mesenchymal stem cells require monocyte-mediated activation to suppress alloreactive T cells. Exp Hematol 33:928–934
Guest DJ, Smith MR, Allen WR (2010) Equine embryonic stem-like cells and mesenchymal stromal cells have different survival rates and migration patterns following their injection into damaged superficial digital flexor tendon. Equine Vet J 42(7):636–642
Gugjoo MB, Amarpal (2018) Mesenchymal stem cell research in sheep: current status and future prospects. Small Rumin Res 169:46–56
Gugjoo MB, Amarpal, Kinjavdekar P, Aithal HP, Ansari MM, Pawde AM, Sharma GT (2015) Isolation, culturing and characterization of New Zealand white rabbit mesenchymal stem cells derived from bone marrow. Asian J Animal Vet Adv 10(10):537–548
Gugjoo MB, Amarpal, Abdelbaset-Ismail A, Aithal HP, Kinjavdekar P, Pawde AM, Kumar GS, Sharma GT (2017) Mesenchymal stem cells with IGF-1 and TGF-β1 in laminin gel for osteochondral defects in rabbits. Biomed Pharmacother 93:1165–1174
Gugjoo MB, Amarpal, Chandra V, Wani MY, Dhama K, Sharma GT (2018a) Mesenchymal stem cell research in veterinary medicine. Curr Stem Cell Res Ther 13(8):645–657
Gugjoo MB, Amarpal, Fazili MR, Shah RA, Sharma GT (2018b) Mesenchymal stem cell: basic research and potential applications in cattle and buffalo. J Cell Physiol 234(6):8618–8635
Gugjoo MB, Amarpal, Makhdoomi DM, Sharma GT (2019) Equine mesenchymal stem cells: properties, sources, characterization and potential therapeutic applications. J Equine Veter Sci 72:16–27
Gugjoo MB, Hussain S, Amarpal, Shah RA, Dhama K (2020) Mesenchymal stem cell mediated Immuno-modulatory and anti-inflammatory mechanisms in immune and allergic disorders. Recent Patents Inflamm Allergy Drug Discov 14:1. https://doi.org/10.2174/1872213X14666200130100236
Haider H, Jiang S, Idris NM et al (2008) IGF-1-overexpressing mesenchymal stem cells accelerate bone marrow stem cell mobilization via paracrine activation of SDF-1alpha/CXCR4 signaling to promote myocardial repair. Circ Res 103(11):1300–1308
He S, Nakada D, Morrison SJ (2009) Mechanisms of stem cell self-renewal. Annu Rev Cell Dev Biol 25:377–406
Hopper N, Wardale J, Brooks R, Power J, Rushton N, Henson F (2015) Peripheral blood mononuclear cells enhance cartilage repair in in vivo Osteochondral defect model. PLoS One 10(8):e0133937
Hristov M, Weber C (2009) Progenitor cell trafficking in vascular wall. J Thromb Haemost 7(suppl 1):31–34
Hu X, Wu R, Shehadeh LA, Zhou Q, Jiang C et al (2014) Severe hypoxia exerts parallel and cell-specifi c regulation of gene expression and alternative splicing in human mesenchymal stem cells. BMC Genom 15:303

Hu Y, Zhang Q, Zhang L, Tang X-X, He H-Y (2015) Basic fibroblast growth factor lentiviral vector-transfected sheep bone marrow mesenchymal stem cells and non-specific osteogenic gene expression. Mol Med Rep 12:267–272

Kang JW, Kang K-S, Koo HC, Park JR, Choi EW, Park YH (2008) Soluble factors-mediated immunomodulatory effects of canine adipose tissue-derived mesenchymal stem cells. Stem Cells Dev 17(4):681–693

Kim HS, Kim KH, Kim SH, Kim YS, Koo KT, Kim TI, Seol Y-J, Ku Y, Rhyu I-C, Chung C-P, Lee Y-M (2010) Immunomodulatory effect of canine periodontal ligament stem cells on allogenic and xenogenic peripheral blood mononuclear cells. J Periodontal Implant Sci 40:265–270

Klingelhutz AJ, Barber SA, Smith PP, Dyer K, McDougall JK (1994) Restoration of telomeres in human papillomavirus-immortalized human anogenital epithelial cells. Mol Cell Biol 14(2):961–969

Kode JA, Mukherjee S, Joglekar MV, Hardikar AA (2009) Mesenchymal stem cells: immunobiology and role in immunomodulation and tissue regeneration. Cytotherapy 11:377–791

Krampera M, Pizzolo G, Aprili G et al (2006) Mesenchymal stem cells for bone, cartilage, tendon and skeletal muscle repair. Bone 39(4):678–683

Kriston-Pál É, Czibula Á, Gyuris Z, Balka G, Seregi A, Sükösd F, Süth M, Kiss-Tóth E, Haracska L, Uher F, Monostori É (2017) Characterization and therapeutic application of canine adipose mesenchymal stem cells to treat elbow osteoarthritis. Can J Vet Res 81:73–78

Kubo H, Shimizu M, Taya Y, Kawamoto T, Michida M, Kaneko E, Igarashi A, Nishimura M, Segoshi K, Shimazu Y, Tsuji K, Aoba T, Kato Y (2009) Identification of mesenchymal stem cell (MSC)-transcription factors by microarray and knockdown analyses, and signature molecule marked MSC in bone marrow by immuno histochemistry. Genes Cells 14:407–424

Kyba M, Perlingeiro RC, Daley GQ (2002) HoxB4 confers definitive lymphoid-myeloid engraftment potential on embryonic stem cell and yolk sac hematopoietic progenitors. Cell 109:29–37

Le Blanc K, Tammik C, Rosendahl K, Zetterberg E, Ringden O (2003) HLA expression and immunologic properties of differentiated and undifferentiated mesenchymal stem cells. Exp Hematol 31:890–896

Lee WS, Suzuki Y, Graves SS, Iwata M, Venkataraman GM, Mielcarek M, Peterson LJ, Ikehara S, Torok-Storb B, Storb R (2011) Canine bone-marrow–derived mesenchymal stromal cells suppress alloreactive lymphocyte proliferation in vitro but fail to enhance engraftment in canine bone marrow transplantation. Biol Blood Marrow Transplant 17:465–475

Ley K, Laudanna C, Cybulsky MI, Nourshargh S (2007) Getting to the site of inflammation: leukocyte adhesion cascade updated. Nat Rev Immunol 7(9):678–689

Li L, Neaves WB (2006) Normal stem cells and cancer stem cells: the niche matters. Cancer Res 66:4553–4557

Li PZ, Yan GY, Han L, Pang J, Zhong BS, Zhang GM, Wang F, Zhang YL (2017) Overexpression of STRA8, BOULE, and DAZL genes promotes goat bone marrow-derived Mesenchymal stem cells in vitro Transdifferentiation toward putative male germ cells. Reprod Sci 24(2):300–312

Lilly AJ, Johnson WE, Bunce CM (2011) The haematopoietic stem cell niche: new insights into the mechanisms regulating haematopoietic stem cell behaviour. Stem Cells Int 2011:274564

Liu ZJ, Zhuge Y, Velazquez OC (2009) Trafficking and differentiation of Mesenchymal stem cells. Biochemist 106(6):984–991

Liu TM, Ng WM, Tan HS, Vinitha D, Yang Z, Fan JB, Zou Y, Hui JH, Lee EH, Lim B (2013) Molecular basis of immortalization of human mesenchymal stem cells by combination of p53 knockdown and human telomerase reverse transcriptase overexpression. Stem Cells 22:268–278

Lopez MJ, Jarazo J (2015) State of the art: stem cells in equine regenerative medicine. Equine Vet J 47:145–154

Luster AD, Alon R, von Adrian UH (2005) Immune cell migration in inflammation: present and future therapeutic targets. Nat Immunol 6(12):1182–1190

Mackenzie TC, Flake AW (2001) Human mesenchymal stem cells persist, demonstrate site-specific multipotential differentiation, and are present in sites of wound healing and tissue regeneration after transplantation into fetal sheep. Blood Cells Mol Dis 27(3):601–604

Mareddy S, Dhaliwal N, Crawford R, Xiao Y (2010) Stem cell-related gene expression in clonal populations of mesenchymal stromal cells from bone marrow. Tissue Eng Part A 16:749–758

McDuffee LA, Pack L, Lores M, Wright GM, Esparza-Gonzalez B, Masaoud E (2012) Osteoprogenitor cell therapy in an equine fracture model. Vet Surg 41:773–783

Menicanin D, Bartold PM, Zannettino AC, Gronthos S (2010) Identification of a common gene expression signature associated with immature clonal mesenchymal cell populations derived from bone marrow and dental tissues. Stem Cells Dev 19:1501–1510

Minguell JJ, Erices A, Conget P (2001) Mesenchymal stem cells. Exp Biol Med (Maywood) 226(6):507–520

Nedeau AE, Bauer RJ, Gallagher K et al (2008) A CXCL5- and bFGF-dependent effect of PDGF-B-activated fibroblasts in promoting trafficking and differentiation of bone marrow derived mesenchymal stem cells. Exp Cell Res 314(11–12):2176–2186

Obara C, Tomiyama K, Takizawa K, Islam R, Yasuda T, Gotoh T, Tajima K (2016) Characteristics of three-dimensional prospectively isolated mouse bone marrow mesenchymal stem/stromal cell aggregates on nanoculture plates. Cell Tissue Res 366:113–127

Oh HJ, Park JE, Kim MJ, Hong SG, Ra JC, Jo JY, Kang SK, Jang G, Lee BC (2011) Recloned dogs derived from adipose stem cells of a transgenic cloned beagle. Theriogenology 75:1221–1231

Ohlstein B, Kai T, Decotto E, Spradling A (2004) The stem cell niche: theme and variations. Curr Opin Cell Biol 16:693–699

Orlic D, Kajstura J, Chimenti S, Limana F, Jakoniuk I, Quaini F, Nadal-Ginard B, Bodine DM, Leri A, Anversa P (2001) Mobilized bone marrow cells repair the infarcted heart, improving function and survival. Proc Natl Acad Sci U S A 98:10344–10349

Palumbo R, Bianchi ME (2004) High mobility group box 1 protein, a cue for stem cell recruitment. Biochem Pharmacol 68(6):1165–1170

Pelagalli A, Nardelli A, Lucarelli E, Zannetti A, Brunetti A (2018) Autocrine signals increase ovine mesenchymal stem cells migration through Aquaporin-1 and CXCR4 overexpression. J Cell Physiol 233(8):6241–6249

Pipas JM (2009) SV40: cell transformation and tumorigenesis. Virology 384(2):294–303

Potten CS, Gandara R, Mahida YR, Loeffler M, Wright NA (2009) The stem cells of small intestinal crypts: where are they? Cell Prolif 42:731–750

Prockop DJ, Oh JY (2012) Mesenchymal stem/stromal cells (MSCs): role as guardians of inflammation. Mol Ther 20:14–20

Reesink HL, Sutton RM, Shurer CR, Peterson RP, Tan JS, Su J, Paszek MJ, Nixon AJ (2017) Galectin-1 and galectin-3 expression in equine mesenchymal stromal cells (MSCs), synovial fibroblasts and chondrocytes, and the effect of inflammation on MSC motility. Stem Cell Res Ther 8:243

Ruster B, Gottig S, Ludwig RJ, Bistrian R, Müller S, Seifried E, Gille J, Henschler R (2006) Mesenchymal stem cells display coordinated rolling and adhesion behavior on endothelial cells. Blood 108(12):3938–3944

Ryu H-H, Kang B-J, Park S-S, Kim Y, Sung G-J, Woo H-M, Kim WH, Kweon O-K (2012) Comparison of Mesenchymal stem cells derived from fat, bone marrow, Wharton's jelly, and umbilical cord blood for treating spinal cord injuries in dogs. J Vet Med Sci 74(12):1617–1630

Sackstein R, Merzaban JS, Cain DW, Dagia NM, Spencer JA, Lin CP, Wohlgemuth R (2008) Ex vivo glycan engineering of CD44 programs human multipotent mesenchymal stromal cell trafficking to bone. Nat Med 14(1):181–187

Singer NG, Caplan AI (2011) Mesenchymal stem cells: mechanisms of inflammation. Annu Rev Pathol 6:457–478

Sordi V, Malosio ML, Marchesi F, Mercalli A, Melzi R, Giordano T, Belmonte N, Ferrari G, Leone BE, Bertuzzi F, Zerbini G, Allavena P, Bonifacio E, Piemonti L (2005) Bone marrow mesenchymal stem cells express a restricted set of functionally active chemokine receptors capable of promoting migration to pancreatic islets. Blood 106(2):419–427

Spees JL, Lee RH, Gregory CA (2016) Mechanisms of mesenchymal stem/stromal cell function. Stem Cell Res Ther 7:125. https://doi.org/10.1186/s13287-016-0363-7

Spradling A, Drummond-Barbosa D, Kai T (2001) Stem cells find their niche. Nature 414:98–104
Stewart MC, Stewart AA (2011) Mesenchymal stem cells: characteristics, sources, mechanisms of action. Vet Clin North Am Equine Pract 27:243–261
Terraciano V, Hwang N, Moroni L, Park HB, Zhang Z, Mizrahi J, Seliktar D, Elisseeff J (2007) Differential response of adult and embryonic Mesenchymal progenitor cells to mechanical compression in hydrogels. Stem Cells 25:2730–2738
Till JE, McCulloch EA (1961) A direct measurement of the radiation sensitivity of normal mouse bone marrow cells. Radiat Res 14:213–222
Till JE, McCulloch EA, Siminovitch L (1964) A stochastic model of stem cell proliferation, based on the growth of spleen colony-forming cells. Proc Natl Acad Sci USA 51:29–36
Vieira NM, Brandalise V, Zucconi E, Secco M, Strauss BE, Zatz M (2010) Isolation, characterization, and differentiation potential of canine adipose-derived stem cells. Cell Transplant 19:279–289
Von Luttichau I, Notohamiprodjo M, Wechselberger A, Peters C, Henger A, Seliger C et al (2005) Human adult CD34- progenitor cells functionally express the chemokine receptors CCR1, CCR4, CCR7, CXCR5, and CCR10 but not CXCR4. Stem Cells Dev 14(3):329–336
Wagner W, Horn P, Castoldi M, Diehlmann A, Bork S, Saffrich R, Benes V, Blake J, Pfister S, Eckstein V, Ho AD (2008) Replicative senescence of mesenchymal stem cells: a continuous and organized process. PLoS One 3:2213
Wang L, Menendez P, Shojaei F, Li L, Mazurier F, Dick JE, Cerdan C, Levac K, Bhatia M (2005) Generation of hematopoietic repopulating cells from human embryonic stem cells independent of ectopic HOXB4 expression. J Exp Med 201:1603–1614
Wang M, Yang Y, Yang D, Luo F, Liang W, Guo S, Xu J (2009) The immunomodulatory activity of human umbilical cord blood-derived mesenchymal stem cells in vitro. Immunology 126(2):220–232
Wang Q, Wang X, Lai D, Deng J, Hou Z, Liang H, Liu D (2018) BIX-01294 promotes the differentiation of adipose mesenchymal stem cells into adipocytes and neural cells in Arbas cashmere goats. Res Vet Sci 119:9–18
Watt FM, Hogan BL (2000) Out of Eden: stem cells and their niches. Science 287:1427–1430
Willert K, Brown JD, Danenberg E, Duncan AW, Weissman IL, Reya T, Yates JR 3rd, Nusse R (2003) Wnt proteins are lipid-modified and can act as stem cell growth factors. Nature 423:448–452
Wilson A, Trumpp A (2006) Bone-marrow haematopoietic stem-cell niches. Nat Rev Immunol 6:93–106
Wong TY, Chang CH, Yu CH, Huang LLH (2017) Hyaluronan keeps mesenchymal stem cells quiescent and maintains the differentiation potential over time. Aging Cell 16:451–460
Wynn RF, Hart CA, Corradi-Perini C, O'Neill L, Evans CA, Wraith JE, Fairbairn LJ, Bellantuono I (2004) A small proportion of mesenchymal stem cells strongly expresses functionally active CXCR4 receptor capable of promoting migration to bone marrow. Blood 104(9):2643–2645
Xie T, Spradling AC (2000) A niche maintaining germ line stem cells in the *Drosophila* ovary. Science 290:328–330
Xiong Y, He J, Zhang W, Zhou G, Cao Y, Liu W (2015) Retention of the stemness of mouse adipose-derived stem cells by their expansion on human bone marrow stromal cell-derived extracellular matrix. Tissue Eng Part A 21:1886–1894
Yan G, Fan Y, Li P, Zhang Y, Wang F (2015) Ectopic expression of DAZL gene in goat bone marrow-derived mesenchymal stem cells enhances the trans-differentiation to putative germ cells compared to the exogenous treatment of retinoic acid or bone morphogenetic protein 4 signalling molecules. Cell Biol Int 39(1):74–83
Zhang YL, Li PZ, Pang J, Wan YJ, Zhang GM, Fan YX, Wang ZY, Tao NH, Wang F (2019) Induction of goat bone marrow mesenchymal stem cells into putative male germ cells using mRNA for STRA8, BOULE and DAZL. Cytotechnology 71(2):563–572

Zhuge Y, Liu Z-J, Velazquez OC (2010) Adult stem cell differentiation and trafficking and their implications in disease. In: Meshorer E, Plath K (eds) The cell biology of stem cells. Springer, New York, pp 169–183

Zuk P, Zhu M, Mizuno H, Huang J, Futrell JW, Katz AJ, Benhaim P, Lorenz HP, Hedrick MH (2001) Multilineage cells from human adipose tissue: implications for cell-based therapies. Tissue Eng 7:211–228

# Mesenchymal Stem Cell Isolation, Culture, Characterization and Cryopreservation

3

M. B. Gugjoo, Amar Pal, V. Chandra, and G. T. Sharma

**Abstract**

The characteristic features of mesenchymal stem cells (MSCs) have made them an important subject of the regenerative medicine. These cells are being harvested, isolated and characterized from numerous adult body tissues/organs, in addition to the foetal membranes. The cells in donor tissues are present in a very limited and variable concentrations making it imperative to culture expand them for effective utilization. The variations in tissue types and/in culture techniques of the cells make it difficult to understand their features and clinical applications. For ready to use application, the cells are being banked, mostly through cryopreservation. The details of isolation, culture and characterization, in addition to cryopreservation, of MSCs are detailed in this topic.

## 3.1 Introduction

Initially, mesenchymal stem cells (MSCs) were identified by Friedenstein and co-workers from bone marrow. The typical fibroblast-like morphology and ability to adhere to the plastic were basis of their characterization (Friedenstein et al. 1976;

M. B. Gugjoo (✉)
Division of Veterinary Clinical Complex, FVSc & AH, SKUAST-Kashmir, Jammu and Kashmir, India

A. Pal
Division of Surgery, Indian Veterinary Research Institute, Izatnagar, UP, India

V. Chandra · G. T. Sharma
Division of Physiology & Climatology, Indian Veterinary Research Institute, Izatnagar, UP, India





M. B. Gugjoo, A. Pal (eds.), *Mesenchymal Stem Cell in Veterinary Sciences*, https://doi.org/10.1007/978-981-15-6037-8_3

Phinney and Sensebe 2013). Subsequently, it had been shown that these cells are present in almost all the adult tissues of the body, in addition to the foetal membranes (Gugjoo and Amarpal 2018; Gugjoo et al. 2019). These cells represent very small population in a particular tissue and are mixed with other cell types. In bone marrow less than 0.1% of the mononuclear cells are MSCs and such a concentration decreases with ageing (Caplan 2007). The therapeutic applications of MSCs require much higher doses (Gugjoo et al. 2020a, b), making it imperative to harvest and culture expand them. The culture expansion of MSCs may change their phenotype making their regulations necessary. Bone marrow mononuclear cells and stromal vascular fraction that are minimally manipulated and applied directly are being used unrestrictedly (Gugjoo et al. 2018; Gugjoo et al. 2020a, b).

For development of any therapeutic agent comprehensive standardized procedures are mandatory in good manufacturing practices (GMPs). Likewise MSCs must be harvested, isolated, culture expanded and characterized as per the GMP norms for their successful utilization (Gugjoo et al. 2015; Akram et al. 2017). The techniques employed for MSCs processing, however, involve variable procedures. MSCs harvested from liquid source like bone marrow/peripheral blood are cultured from mononuclear cell (MNC) fraction being separated by density gradient method (Gugjoo et al. 2015). In contrast, MSCs harvested from solid tissue like adipose tissue or cord matrix, among others are harvested from stromal vascular fraction (SVF) or MNC fraction. MNC fraction conforms to the mixture of red blood cells, granulocytes, platelets and myeloid immature precursors, in addition to the MSCs obtained from peripheral blood or bone marrow. SVF refers to the nucleated cell fraction of heterogeneous mixture of endothelial, muscle, fibroblastic and mast cells, and pericytes, pre-adipocytes and MSCs (Mamidi et al. 2012; Salehinejad et al. 2012; Otsuru et al. 2013). The main difference in the procedure to harvest MSCs from these tissue types is that the solid tissues require an extra step of enzymatic digestion while such a step is not required for liquid sources (Fig. 3.1). For MSCs clinical applications, GMPs and good clinical practices (GCPs) must be ensured during the whole procedure of cell harvesting, isolation, culturing procedures and therapeutics (Gugjoo et al. 2019). Fortunately, GMP-compliant media and chemicals are available to process and culture expand MSCs (Bieback et al. 2008).

## 3.2 MSCs from Liquid Sources (Bone Marrow)

MSCs may be harvested from liquid sources by various methods involving classical and density gradient method. Density gradient solutions utilized are percoll (silica particle colloidal solution, density of 1.084–1.088) and ficoll (polysaccharide solution; density of 1.077) (Kadiyala et al. 1997; Smith et al. 2003; Hegewald et al. 2004; Vidal et al. 2006; Pacini et al. 2007; Wilke et al. 2007; Colleoni et al. 2009; Gugjoo et al. 2015). After bone marrow centrifugation, the supernatant is aspirated and pellet obtained therefrom is washed with phosphate buffer solution (PBS), followed by addition of growth medium in culture flask or plate. In case of density gradient methods, MNC fraction is obtained from bone marrow after addition of

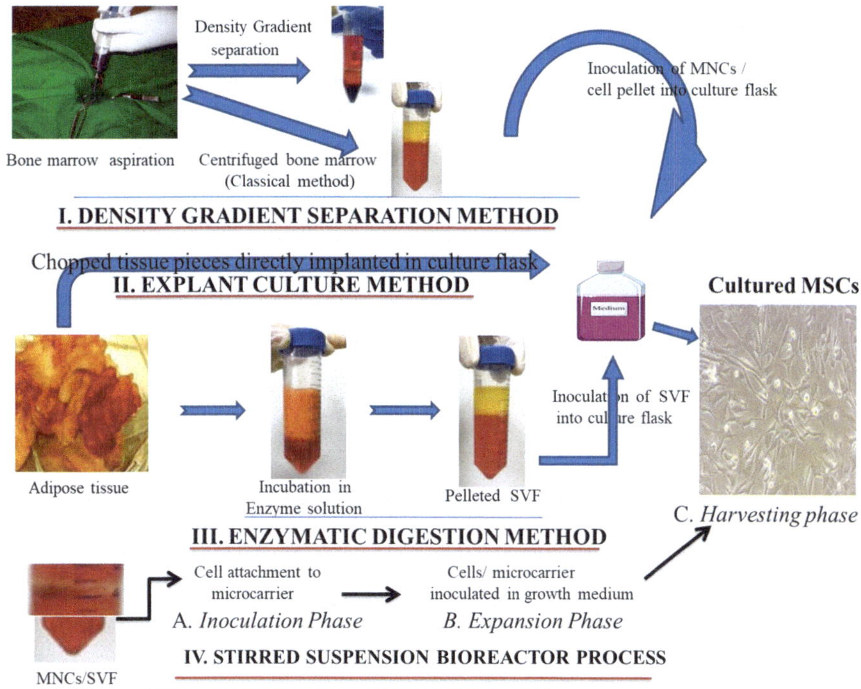


**Fig. 3.1** Mesenchymal stem cell isolation protocols and their culture expansion

density gradient solutions and centrifugation. Classical method utilizes MSCs intrinsic property of plastic adherence (plastic culture dishes) without any separation of the MNCs (Fortier et al. 1998; Arnhold et al. 2007). MNCs seeding densities are regarded as the most critical parameter for expanding MSCs (Bieback et al. 2008). Equine BM-MSCs' better yield had been achieved at MNCs implantation concentration of $4 \times 10^5$ to $1 \times 10^6$ (Zahedi et al. 2018). Most of the studies currently follow the protocol of density gradient centrifugation (Bieback et al. 2008).

## 3.3 MSCs from Solid Tissues (Adipose Tissue)

MSCs from solid tissues are usually derived either by explant culture or enzymatic protocol. Explant culture method is one of the earliest protocol, in which well rinsed and mechanically split tissue (<few millimetres tissue pieces have better gas and nutrient diffusion towards the cells) is directly placed into the culture flask containing the growth medium (Atala 2002). The smaller fraction of MSCs comes out from these smaller tissue pieces and adheres to the culture dish surface. The residual tissue pieces are washed off after the visible attachment of the cells. In enzymatic protocol, an extra step of adding the enzymatic solution like collagenase to degrade the extracellular matrix (ECM) is performed and the assembly is kept in shaking

incubator for proper digestion. The resultant SVF or MNC fraction obtained after centrifugation is transferred to culture dishes containing growth media (detailed below). The culture flasks/plates are maintained at 37 °C incubation temperature with 5% carbon dioxide and under humid conditions. The cells are harvested after attaining 70–80% confluency in flasks (Gugjoo et al. 2015).

There is a dearth of extensive literature comparing the isolation protocols. However, few studies have compared different density gradient media, and classical/explants methods to the enzymatic protocol. In an equine study, bone marrow MSCs (BM-MSCs) isolated by various density gradient solutions had shown matching differentiation potential, but yield and properties like self-renewal were more desirable in MSCs isolated by percoll solution (Bourzac et al. 2010). Comparison of the enzymatic method to the explant method had shown that the latter method provided little heterogeneous cell population that exhibited higher proliferation rates and better viability (Salehinejad et al. 2012; Hilkens et al. 2013; Yoon et al. 2013). The explant methods keep tissue pieces intact and their extracellular matrix undissociated and thus, the cells might remain shielded from proteolytic action and mechanical stress (Hynes 2009; Hendijani 2017). The release of the cytokines and growth factors from the local tissue may additionally provide cell growth promoters (Hynes 2009; Jing et al. 2011; Shah et al. 2013). However, the explant method is more cumbersome. In enzymatic methods, type of enzyme and its concentration play an important role in cell yield and their viability (Can and Karahuseyinoglu 2007; Mitchell et al. 2003; Wang et al. 2004). Dissociation of extracellular matrix (ECM) tends to give lower cell yield (Karahuseyinoglu et al. 2007; Seshareddy et al. 2008) and also increases time of cell adherence (Han et al. 2013). To improve the cell yield to about 70% or more, a combination of enzymatic method and mechanical dissociation can be performed (Tong et al. 2011).

## 3.4 Growth Medium

Contrary to embryonic stem cells (ESCs), the cell feeder layer remains dispensable in MSCs culture. MSCs are plated directly into the plastic culture dishes/flasks supplemented with growth media. This is due to their characteristic property of plastic adherence. MSCs expand and proliferate in growth medium in the culture flasks. Growth medium includes basal medium, foetal bovine serum (FBS) and antibiotics and/antimycotics. The commonly used basal culture media for MSCs culture is Dulbecco's Modified Eagle's Medium (DMEM) and/or alpha minimal essential medium (Sampaio et al. 2015). Other media like TCM199/DMEM (1:1) (Colleoni et al. 2005), Aminomax-II complete medium (da Silva et al. 2016) and Mesencult medium (Hepsibha et al. 2011) have also been used. One of the studies had demonstrated that mesencult medium had favoured better growth and proliferation of buffalo adipose tissue derived MSCs (AD-MSCs) as compared to the DMEM. The results however, were obvious only at passage 1 as cells at passage 3 had better colony formation even in DMEM (Hepsibha et al. 2011). In general, the cells are allowed to attach for 4–8 days depending upon the culture ingredients, cell density and species involved. Most of the cells in MNC or SVF fraction that lack plastic adherence properties (including haematopoietic cells) are removed after decanting the initial culture medium. After achieving 70–80% confluency in about 10–14 days, the cells are detached by various enzymes like trypsin, trypsin-ethylene diamine tetra acetic acid (EDTA), collagenase I or accutase (Fortier et al. 1998; Arnhold et al. 2007; Li et al. 2015a) and passaged for further proliferation. The buffalo amniotic membrane MSCs (bufAm-MSCs) extracted with the combined use of trypsin-EDTA and collagenase I had been reported to prove better than trypsin-EDTA alone (Deng et al. 2018). Between accutase and trypsin, immediate viability of human neural stem cells (NSCs) after trypsin disassociation had been lower as compared to that with accutase; however, subsequent apoptosis of NSCs caused by trypsin was demonstrated to be lower than that caused by accutase (Li et al. 2015a).

## 3.5 Foetal Bovine Serum (FBS) and Its Alternatives

FBS is one of the commonly used supplementation medium to culture MSCs (Vidal et al. 2008; Toupadakis et al. 2010; Gugjoo et al. 2015). FBS has been used at changeable concentration between 5 and 25%. Buffalo BM-MSCs had shown better growth at 15% FBS (Gade et al. 2013) and for cattle AD-MSCs and BM-MSCs 10% FBS concentration had been sufficient (Lu et al. 2014a, b). Although 20% FBS had also been used with good results (de Moraes et al. 2016; Deng et al. 2018). Goat- and equine-MSCs derived from adipose tissue, bone marrow and liver had accelerated cell confluency at higher FBS concentration (20%) than at its lower concentration (10%) (Heidari et al. 2013; Spaas et al. 2013; Martins et al. 2017). FBS furnishes nutrients, hormones/growth factors and cell adhesion plasma proteins for cellular growth and proliferation (Seo et al. 2013). GMP-compliant FBS is available and is being used to produce clinical-grade MSCs (Li et al. 2015a). However, there are associated limitations like potential threat of disease transmission, and xenogeneic protein induced change in cell behaviour and non-uniform cell preparations. FBS tends to hamper effective clinical applications of MSCs due to the potential immunogenic reactions that can arise from internalized FBS proteins (McIntosh et al. 2009).

To overcome these problems, FBS alternatives for culturing MSCs are being studied extensively. Among various FBS alternatives, autologous serum, allogeneic serum and platelet lysate (Del Bue et al. 2008; Toupadakis et al. 2010; Schwarz et al. 2012; Bieback 2013) and commercially available serum substitutes (Schwarz et al. 2012) have been evaluated. Overall, these alternatives have failed to provide consistent MSCs adhesion, morphology, growth patterns (Seo et al. 2013; Eydt et al. 2016; Schubert et al. 2018), proliferation rate (Toupadakis et al. 2010; Schwarz et al. 2012; Glynn et al. 2013) and immune-modulatory properties (Russell et al. 2015; Clark et al. 2016; Enciso et al. 2018) in animals. However, autologous serum had supported the replacement of foetal calf serum (FCS) to culture ovine BM-MSCs. The morphology, stem cell markers, osteogenic differentiation and protein expression of ovine BM-MSCs supplemented with autologous serum had been comparable to that obtained with FBS supplementation (Weigand et al. 2017). Similarly, a dog study had shown that plasma rich in growth factors was able to improve adipose tissue MSCs (AD-MSCs) survival and had promoted their proliferation, migration but had induced their differentiation (Mellado-López et al. 2017).

Serum-free media had failed to maintain equine MSCs consistency in morphology and immunophenotype (Schubert et al. 2018). Contrarily, a study had demonstrated that bovine umbilical cord MSCs (UC-MSCs) can be cultured in serum-free medium for longer periods. The normal karyotype cells had good proliferation with an ability to differentiate into neuron-like cells, in addition to tri-lineages up to passage 60 (Cardoso et al. 2012). In absence of the serum, application of various growth/ humoral factors like bovine fibroblast growth factor (bFGF), epidermal growth factor (EGF), platelet derived growth factor (PDGF), fibroblast growth factor-2 (FGF-2), insulin growth factor – 1 (IGF-1) and most significantly transforming growth factor-α (TGF-α) had supported MSCs proliferation especially of ovine and bovine sources (Mccarty et al. 2009; Lu et al. 2014a, b). Combined use of these growth factors may also affect MSCs proliferation, migration and invasion. Combined use of PDGF/TGF-β and PDGF/IGF-I had promoted proliferation; EGF/bFGF and EGF/bFGF/PDGF had promoted migration and EGF/bFGF/TGF-β had promoted BM-MSCs invasion (Somers et al. 2012). These growth factors however, tend to favour MSCs differentiation. However, further studies for better understanding in FBS alternatives are desired.

## 3.6 Modifications in Conventional MSCs Culture Protocol

The conventional culture expansion of MSCs follows static mechanism. This system is labour intensive, bears low reproducibility, poses contamination problem and above all is less economical for large-scale expansion. In order to make MSCs production economical and reduce chances of contamination various modifications are being tried. Among various modified techniques, the application of static magnetic field has proven to be useful to some extent. Application of static magnetic field to equine adipose tissue MSCs (eAD-MSCs) had accelerated their population doubling potential with high colony-forming potential. An enhanced secretory potential of growth factor rich microvesicles with no compromise on morphology was also demonstrated (Marędziak et al. 2015; Iacono et al. 2017).

### 3.6.1 Stirred Suspension Bioreactor

An alternative to the conventional system is the stirred suspension bioreactor for culture expansion of MSCs (Fig. 3.1), though it remains to be used extensively in veterinary practice. It can compensate the limitations normally associated with the conventional techniques. The cell population of clinical grade with effective numbers can be produced in shorter time at less input cost (Roberts et al. 2019). Among various factors selection of microcarrier (MC) is critical. Microcarrier (MC) acts as attachment surface for cellular growth. Commercially manufactured microcarriers for MSCs expansion are either porous or non-porous types with variable materials (Schop et al. 2010; Goh and Reuveny 2013; Petry et al. 2016; Rafiq et al. 2016). Various features of MC like composition, porosity, topography and surface charge determine cell attachment. MCs may also be cell line specific (Martin et al. 2011). Thus, selection of microcarrier is an important factor for appropriate cell expansion (Rafiq et al. 2014).

In stirred suspension bioreactor process, various phases involving inoculation, expansion and harvesting phases are being accomplished to achieve cultured MSCs. The first 24 h are of inoculation phase. The main aim remains to achieve maximum possible cellular attachment to the microcarriers (Balint et al. 2015). Among various factors, cell to microcarrier (MC) ratios remain important. The appropriate ratio of initial cell to microcarrier density is determined by microcarrier surface area. Usually low initial cell to MC density provides greater cell yield (Sun et al. 2010; Li et al. 2015).

The second expansion phase continues till the desired MC cell density is obtained and is affected by culture medium composition and frequency of medium change. Replacement medium removes undesirable growth inhibitors and renews exhausted nutrients. For equine cord blood MSCs (CB-MSCs), growth medium consisting of DMEM, along with FBS (30%), L-glutamine (1%) and antibiotics (1%) has been used (Koch et al. 2007). However, lower FBS concentrations may also be used. Apart from these factors, rate of agitation needs to be set right. The appropriate agitation rate is determined by the geometry of bioreactor and its impeller, viscosity and density of media, type of cell line and MC type. Lower agitation rate favours cell-MC aggregate formation (Jossen et al. 2016). The higher agitation rates develop higher shear in the bioreactors favouring cells to detach from MCs (Yuan et al. 2014; Takahashi et al. 2016).

In the harvesting stage, the cells are detached from the MCs by enzymatic action involving trypsin. The cells are then filtered to obtain MSCs suspension. Trypsin (porcine origin) however, is usually replaced by bacteria derived recombinant substitute, TrypLE. The enzyme activity is usually maintained for 5–15 min, though exposures up to 1 h have also been made (Roberts et al. 2019).

In an equine study, cord blood derived MSCs in stirred suspension bioreactors could achieve very large cell densities of about 75,000 per sq. cm which corresponds to $40 \times 10^6$ cells in 100 ml bioreactor. The harvesting cell efficiency had remained 80% approximately and the cells had shown comparable surface marker expression and differentiation potential to that of the conventional culture system (Roberts et al. 2019).

## 3.7 MSCs Characterization

Characterization of MSCs is mandatory to confirm their presence. These cells are characterized on the recommendations of the International Society for Cellular Therapy (ISCT). MSCs are plastic adherent cells that express specific surface markers, lack haematopoietic marker expression and are able to show tri-lineage (adipogenic, chondrogenic and osteogenic) differentiation. MSCs population must express ≥95% of the CD73 (5'-Nucleotidase), CD90 (Thy1) and CD105 (Endoglin) as measured by flow cytometry. There should be ≤2% of the CD11b (Integrin-α) or CD14, CD19, CD34, CD45, CD79α and HLA Class II. Multipotent tri-lineage differentiation potential is confirmed by staining these in vitro differentiated cells (Dominici et al. 2006).

These cells from various animal species generally meet the norm of plastic adherence and multipotency (Guest et al. 2008; Ranera et al. 2011; De Schauwer et al. 2012) but consensus on surface marker expression is debatable (Kato et al. 2004; Knippenberg et al. 2005; Kang et al. 2008; Vieira et al. 2010; Lyahyai et al. 2012; Corradetti et al. 2013; Caminal et al. 2014; Mediano et al. 2015; Khan et al. 2016; De Cesaris et al. 2017; Ghaneialvar et al. 2018). In some species like rabbit markers including CD90 have been demonstrated to be transcribed by MSCs but their protein translation has failed (Kovac et al. 2017). Variable surface marker expression of MSCs from various sources may be explained by donor tissue differences, variability in enzyme harvesting methods and specific reactivity of antibodies used (Ranera et al. 2011; De Schauwer et al. 2012; Colleoni et al. 2009; Radcliffe et al. 2010). The cell detaching agents like trypsin and accutase may vitiate cell surface receptors (Ranera et al. 2011; De Schauwer et al. 2012). All antibodies are not reactive to interspecies molecules (Radcliffe et al. 2010). MSCs dynamic immune-phenotype may additionally promulgate such differences (Huss et al. 2000; Mosna et al. 2010; Strioga et al. 2012). The differences in surface markers expression especially of CD44 and CD90 may impact MSCs ability to migrate (Cabezas et al. 2018). Recent mice study had demonstrated that MSCs chromatin accessibility signatures and gene expression programs may be useful modes to provide insights into tissue specific cell differences, though chromatin signature molecules had been more accurate. This demands extensive MSCs epigenome analysis studies to characterize them on the basis of tissue source (Ho et al. 2018).

In order to avoid such donor tissue specific cell characteristics, a uniform source like induced pluripotent stem cells (iPSCs) or embryonic stem cells (ESCs) are being evaluated. MSCs from these sources, although in preliminary studies, demand extensive studies so as to harvest sufficient and phenotypically stable cell number so as to utilize them clinically (Chan et al. 2018; Gugjoo et al. 2018; Steens and Klein 2018). Apart from basic characterization protocol, MSCs must be viable, free of microbes and/or endotoxin and pyrogen and finally dose titrated to conform to the standards of good-manufacturing and clinical practices (GMPs and GCPs) (Bieback et al. 2008).

## 3.8 Storage of MSCs

MSCs have the potential to be utilized as allogeneic therapy and as such may be stored for future use and for distant transport. Excipient medium (like PBS) can be retained along with the cells while ancillary medium (like bone marrow aspirate, plasma, serum and cryogenic medium) must be removed prior to the in vivo implantation of the cells. For shorter duration (9–12 h) storage, ancillary biological media have no observable benefit over the excipient in maintaining MSCs viability (Bronzini et al. 2012). Ancillary biological media had been demonstrated to keep cells equally viable for about 24 h that had decreased to 25–78% by 72 h. Cryogenic medium had maintained cell viability better than the biological fluids (Garvican et al. 2016). Cryogenic storage of MSCs in different cryogenic media like serum, dimethyl sulphoxide (DMSO), FBS and media had comparable and satisfactory post-thaw cell viability (80–90%) and growth evaluated for shorter duration (Mitchell et al. 2015; Garvican et al. 2016).

## 3.8.1 Cryopreservation of MSCs

### 3.8.1.1 Cryopreservation

Cryopreservation of cells is achieved either by fast or slow freezing rate. During this process, various steps are ensured to avoid cell dehydration and damage. During fast cell freezing rate solute and electrolyte imbalances can do the damage while during slow freezing process intra- and extracellular ice crystal formation can have adverse effects on the cell viability (Mazur 1977). During these processes, incorporation of cryoprotectant is an important step. To prevent cell damage during such a delicate process cryoprotectant is being added to reduce overall freezing point of the system. The combined mixture of the cells, medium and the cryoprotectant makes a eutectic system that has lower freezing temperature as compared to the individual components. It is important to note that freeze-thaw method directly affects cell viability. One of the studies had reported an unacceptable post-thaw viability of cells and might have been due to the faulty thaw process (Espina et al. 2016).

Another way of cryopreservation that involves an extremely rapid (>−1000 °C/s) cooling process known as vitrification has also been used for cell preservation (Fowler and Toner 2005). In this process, the cells and potentially cytotoxic concentration of cryoprotectant are immersed in open storage vessels. However, requirement of higher concentration of cryoprotectant, continuous maintenance of cells at cryogenic temperature and need of open containers (that act as a potential source of contamination) makes vitrification least opted technique for MSCs cryopreservation (Taylor et al. 2004; Fowler and Toner 2005).

### 3.8.1.2 Why to Cryopreserve MSCs

- MSCs, unlike that of embryonic stem cells, cannot be cultured indefinitely as they undergo senescence after certain culture passages. At temperature below −120 °C, most of the metabolic processes are at stasis and thus, cells can be preserved for long (Colosimo et al. 2013; Ji et al. 2016; Tian et al. 2016).
- MSCs are very limited in number under natural sources and thus, in vitro expansion remains a pre-requisite process. Such an expansion takes a minimum of 2 weeks and thus a delay in autologous therapy is imminent and can be sufficed by preserved autologous/ allogeneic MSCs (Gugjoo et al. 2016; Perez-Silos et al. 2016).

- MSCs concentration and characteristics vary with the source and are affected by the feeding, age and health status of the donor (Han et al. 2010; Pillai et al. 2016). Allogeneic therapy can avoid such variations and assure uniform applications in clinical cases and/or trials. Besides, the studies can effectively be studied and compared for the results.
- MSCs viability during long distance can be safeguarded in cryopreservation state.

### 3.8.1.3 Cryoprotectant

MSCs cryopreservation is done in cryogenic media containing cryoprotectant that prevents cell damage during freezing and thawing (Fuller 2004; Thirumala et al. 2010). Variable concentration of cryoprotectant has been used. The concentration of cryoprotectant varied among species and sources, in addition to the freezing technique used (Fuller 2004; Thirumala et al. 2005). Cryoprotectants are of two types including cell membrane permeable and impermeable (Fuller 2004), with highly permeable cryoprotectants tend to be more toxic (Gao and Critser 2000). To prevent such a toxicity permeable cryoprotectants like DMSO, ethylene glycol, etc. are combined with less permeable ones like hydroxyethyl starch (HES), dextran and polyethylene glycol (Gao and Critser 2000; Fuller 2004). In veterinary practice cryoprotectants like DMSO and FBS are most commonly used (Martinello et al. 2010; Marx et al. 2015). The ancillary cryoprotectants like DMSO exhibit carcinogenic properties (Quimby et al. 2013). The xenogeneic proteins in FBS may alter cellular behaviour (Duan and Lopez 2016) and need to be removed before cell implantation.

#### Dimethyl Sulphoxide (DMSO)

It is one of the most commonly used cryoprotectant in veterinary practice that works by stabilizing the cell proteins and equilibrating intracellular and extracellular electrolytes. Hydrophobic interactions between DMSO and proteins and/cell membrane phospholipids stabilize cellular proteins (Anchordoguy et al. 1991). Electrolyte equilibration occurs by displacing the intracellular fluids (Liu et al. 2010). In addition, DMSO forms high energy bonds with the water molecules that prevent ice crystal formation (Weng et al. 2011). Despite such features, DMSO can lead to potential chemical and osmotic shock to the MSCs (Scheinkonig et al. 2004; Liu et al. 2010). DMSO concentration has a significant effect on cell survival. An increase in DMSO concentration from 5% to 20% in the freezing medium tends to decrease cell survivability and enhance apoptotic gene expression in pig BM-MSCs (Ock and Rho 2011). Hydrophobic interactions that stabilize the cell membrane proteins may also lead to their denaturation and deactivation (Fahy et al. 1990). DMSO has been associated with the side effects like neurotoxicity and can lead to hypotension and anaphylactic shock upon intravenous injection along with the cells (Junior et al. 2008). Such side effects can be waived off by washing the cells, though that may result in lower colony-forming units (CFU) potential of the cells (Fry et al. 2015). Furthermore, DMSO complete removal is a complex and time-consuming process (Thirumala et al. 2009), demanding its alternatives.

**Foetal Bovine Serum (FBS)**

FBS apart from its utility as an important ingredient of culture medium, it acts as cryoprotectant. It protects the cell proteins and stabilizes the osmotic pressure. FBS provides nutrients and humoral/growth factors for cell proliferation. The post-thaw viability of the MSCs may increase with an increase in FBS concentration. Buffalo amniotic fluid MSCs (bufAF-MSCs) post-thaw viability had been demonstrated to improve from 28 to 85% with an increase in serum concentration (Yadav et al. 2011). Even the post-cryopreserved ovine amniotic-MSCs had failed to show any deleterious effect on their cartilage healing potential as the healing results were comparable to that of fresh amniotic-MSCs (Garcia et al. 2015).

FBS has issues of variable composition and can pose xenogeneic and/zoonotic threats, in addition to the ethical issues (Jochems et al. 2002; Liu et al. 2010; Verdanova et al. 2014). Even FBS may affect surface markers expression as an increased CD44 expression was demonstrated in dog cryopreserved AD-MSCs as compared to fresh ones (Duan and Lopez 2016). Considering all these impediments, FBS alternatives are being studied. Recently serum-free cryopreservation medium in human MSCs has been demonstrated to give similar or superior post-cryopreservation outcomes as compared to FBS-containing media (Freimark et al. 2011; Verdanova et al. 2014). However, such FBS alternatives are desired to be tested in animals given that such alternatives have failed to give satisfactory culture characteristics of animal MSCs (Schwarz et al. 2012; Seo et al. 2013; Russell et al. 2015; Clark et al. 2016; Enciso et al. 2018; Schubert et al. 2018).

For shorter duration (24–72 h) cryopreservation of MSCs during transit storage, allogeneic and/ autologous serum may replace FBS. Cryopreserved horse MSCs stored in allogeneic and/autologous serum had shown unaltered post-thaw growth and morphology (Mitchell et al. 2015). Conditioned medium may further support growth and viability of the cryopreserved equine umbilical cord MSCs (eUC-MSCs) (Rhodes et al. 2004; Maia et al. 2017). In veterinary species, MSCs are cryopreserved using cryoprotectant and/serum in the same way as are followed in human MSCs cryopreservation (Martinello et al. 2010; Ock and Rho 2011; Duan and Lopez 2016). The cells in our lab are cooled to about –80 °C in 1 °C cooler (containing isopropyl alcohol). The cells in 1 °C cooler are initially kept in –20 °C for few hours and transferred to –80 °C for over a night. Later the cells are transferred to the liquid nitrogen (Martinello et al. 2011). The cells are revitalized by thawing and then rinsed prior to culture or implantation (Gao and Critser 2000).

### 3.8.2 MSCs Characteristics After Cryopreservation

Post-cryopreservation viability of dog MSCs (dMSCs), equine MSCs (eMSCs), bovine MSCs (bMSCs), buffalo MSCs (bufMSCs) and sheep MSCs (sMSCs) had been demonstrated to remain about 60–70%, 64.0 ± 6.0%, 86.5%, 80% and >70%, respectively without any adverse effect upon their characteristics (Vidal et al. 2006; Laura et al. 2008; Martinello et al. 2011; de Moraes et al. 2016; Campos et al. 2017; Okamura et al. 2017; Korovina et al. 2019). One of the studies had an unacceptable post-thaw cellular viability at –20 and –80 °C, citing importance of freeze-thaw method (Espina et al. 2016).

The cryopreserved MSCs viability remains same irrespective of their sources (Knippenberg et al. 2005). Goat foetal adnexa derived MSCs had revived successfully upon thawing. The revived cells could express surface antigens and pluripotency markers and had shown tri-lineage differentiation (Somal et al. 2017). The population doubling time (PDT) too had remained the same (Somal et al. 2016, 2017) as had also been reported for ovine BM-MSCs (Rhodes et al. 2004). Storage of dog MSCs for a year had not affected fibroblast-like morphology, alkaline phosphatase (AP) activity, surface marker expression and tri-lineage differentiation potential. The cryopreserved cells however, had reduced proliferation ratio and telomerase activity in comparison to the fresh cells (Martinello et al. 2011).

Apart from the cells, cryopreservation of tissue engineered (TE) constructs may provide an off-the-shelf TE constructs. In one such study cryopreserved tissue engineered scaffold and goat BM-MSCs (gBM-MSCs) had good viability (Costa et al. 2012). Although other factors like scaffold design and structural properties might be affected by cryopreservation and should be studied in detail.

## 3.9 Conclusion

Culture expansion of MSCs is imperative for their effective utilization. The isolation of MNCs from liquid tissues can be achieved with conventional or density gradient separation methods while SVF from solid tissues can be separated either by direct explant method or by prior enzymatic digestion. MNCs and/SVF cells are culture expanded in growth media containing basal culture medium, FBS and antibiotic-antimycotic solution being kept under standard culture conditions. FBS poses various impediments for clinical applications and as such its alternatives are being studied. The plastic adhered cells are harvested by enzymatic digestion and characterized as per ISCT recommendations. The cells that are plastic adherent express specific surface markers while lacking haematopoietic markers and are able to differentiate into at least tri-lineages are termed as MSCs. For large-scale and cost-effective production of MSCs, stirred suspension bioreactor process is being tried and seems useful. The culture expanded cells are stored for longer period to be utilized as ready to use supply. The cells are preserved in cryopreservation media under liquid nitrogen.

## References

Akram T, Hakim M, Mir BA, Hussain SS, Shabir N, Andrabi M, Dar PA, Ganaia NA, Shabir N (2017) Comparative efficiency of goat mesenchymal stem cell isolation from bone marrow and bone chip. Small Rum Res 153:87–94

Anchordoguy TJ, Cecchini CA, Crowe JH, Crowe LM (1991) Insights into the cryoprotective mechanism of dimethyl sulfoxide for phospholipid bilayers. Cryobiology 28:467–473

Arnhold SJ, Goletz I, Klein H, Stumpf G, Beluche LA, Rohde C, Addicks K, Litzke LF (2007) Isolation and characterization of bone marrow-derived equine mesenchymal stem cells. Am J Vet Res 68:1095–1105

Atala A (2002) Methods of tissue engineering. Gulf Professional Publishing, Houston

Balint R, Richardson SM, Cartmell SH (2015) Low-density subculture: a technical note on the importance of avoiding cell-to-cell contact during mesenchymal stromal cell expansion. J Tissue Eng Regen Med 9:1200–1203

Bieback K (2013) Platelet lysate as replacement for fetal bovine serum in mesenchymal stromal cell cultures. Transf Med Hemother 40:326–335

Bieback K, Schallmoser K, Klüter H, Strunk D (2008) Clinical protocols for the isolation and expansion of mesenchymal stromal cells. Transf Med Hemother 35(4):286–294

Bourzac C, Smith LC, Vincent P, Beauchamp G, Lavoie JP, Laverty S (2010) Isolation of equine bone marrow-derived mesenchymal stem cells: a comparison between three protocols. Equine Vet J 42(6):519–527

Bronzini I, Patruno M, Iacopetti I, Martinello T (2012) Influence of temperature, time and different media on mesenchymal stromal cells shipped for clinical application. Vet J 194(1):121–123

Cabezas J, Rojas D, Navarrete F et al (2018) Equine mesenchymal stem cells derived from endometrial or adipose tissue share significant biological properties, but have distinctive pattern of surface markers and migration. Theriogenol 106:93–102

Caminal M, Fonseca C, Peris D, Moll X, Rabanal RM, Barrachina J, Codina D, García F, Cairó JJ, Gódia F, Pla A, Vives J (2014) Use of a chronic model of articular cartilage and meniscal injury for the assessment of long-term effects after autologous mesenchymal stromal cell treatment in sheep. New Biotechnol 31(5):492–498

Campos LL, Landim-Alvarenga FC, Ikeda TL, Monteiro BA, Maia L, Freitas-Dell'Aqua CP, De Vita B (2017) Isolation, culture, characterization and cryopreservation of stem cells derived from amniotic mesenchymal layer and umbilical cord tissue of bovine fetuses. Pesquisa Veterinária Brasileira 37(3):278–286

Can A, Karahuseyinoglu S (2007) Concise review: human umbilical cord stroma with regard to the source of fetus-derived stem cells. Stem Cells 25:2886–2895

Caplan AI (2007) Adult mesenchymal stem cells for tissue engineering versus regenerative medicine. J Cell Physiol 213:341–347

Cardoso TC, Ferrari HF, Garcia AF, Novais JB, Silva-Frade C, Ferrarezi MC, Andrade AL, Gamiero R (2012) Isolation and characterization of Whartons jelly-derived multipotent Mesenchymal stromal cells obtained from bovine umbilical cord and maintained in defined serum free three-dimensional system. BMC Biotechnol 12:18

Chan, CKF, Gulati GS, Sinha R, Tompkins JV, Lopez M, Carter AC, Ransom RC, Reinisch A, Wearda T, Murphy M, et al. (2018). Identification of the Human Skeletal Stem Cell. Cell 175(1): 43–56.e21.

Clark KC, Kol A, Shahbenderian S, Granick JL, Walker NJ, Borjesson DL (2016) Canine and equine mesenchymal stem cells grown in serum free media have altered immunophenotype. Stem Cell Rev Reports 12:245–256

Colleoni S, Donofrio G, Lagutina I, Duchi R, Galli C, Lazzari G (2005) Establishment, differentiation, electroporation, viral transduction, and nuclear transfer of bovine and porcine mesenchymal stem cells. Cloning Stem Cells 7(3):154–166

Colleoni S, Bottani E, Tessaro I et al (2009) Isolation, growth and differentiation of equine mesenchymal stem cells: effect of donor, source, amount of tissue and supplementation with basic fibroblast growth factor. Vet Res Commun 33:811–821

Colosimo A, Russo V, Mauro A, Curini V, Marchisio M, Bernabò N, Alfonsi M, Mattioli M, Barboni B (2013) Prolonged in vitro expansion partially affects phenotypic features and osteogenic potential of ovine amniotic fluid-derived mesenchymal stromal cells. Cytotherapy 15:930–950

Corradetti B, Meucci A, Bizzaro D, Cremonesi F, Lange Consiglio A (2013) Mesenchymal stem cells from amnion and amniotic fluid in the bovine. Reproduction 145:391–400

Costa PF, Dias AF, Reis RL, Gomes ME (2012) Cryopreservation of cell/scaffold tissue-engineered constructs. Tissue Engineering: Part C 18(11):852–858

da Silva CG, Martins CF, Cardoso TC, da Cunha ER, Bessler HC, MacManus CM, Pivato I, Bao SN (2016) Isolation and characterization of Mesenchymal stem cells derived from bovine Wharton's jelly and their potential use in cloning by nuclear transfer. Ciencia Rural, Santa Maria 46(10):1830–1837
De Cesaris V, Grolli S, Bresciani C, Conti V, Basini G, Parmigiani E, Bigliardi E (2017) Isolation, proliferation and characterization of endometrial canine stem cells. Reprod Domest Anim 52(2):235–242
de Moraes CN, Maia L, Dias MC, Dell'Aqua CP, da Mota LS, Chapwanya A, Landim-Alvarenga FD, Oba E (2016) Bovine endometrial cells: a source of Mesenchymal stem/progenitor cells. Cell Biol Int 40(12):1332–1339
De Schauwer C, Piepers S, Van de Walle GR et al (2012) In search for cross-reactivity to immunophenotype equine mesenchymal stromal cells by multicolor flow cytometry. Cytometry A 81(4):312–323
Del Bue M, Ricco S, Ramoni R, Conti V, Gnudi G, Grolli S (2008) Equine adipose-tissue derived mesenchymal stem cells and platelet concentrates: their association in vitro and in vivo. Vet Res Commun 32:S51–S55
Deng Y, Huang G, Zou L, Nong T, Yang X, Cui J, Wei Y, Yang S, Shi D (2018) Isolation and characterization of buffalo (Bubalus bubalis) amniotic mesenchymal stem cells derived from amnion from the first trimester pregnancy. Journal of Veterinary Medical Sciences 80(4):710–719
Dominici M, Le Blanc K, Mueller I et al (2006) Minimal criteria for defining multipotent mesenchymal stromal cells. The International Society for Cellular Therapy position statement. Cytotherarpy 8(4):315–317
Duan W, Lopez MJ (2016) Effects of cryopreservation on canine multipotent stromal cells from subcutaneous and infrapatellar adipose tissue. Stem Cell Rev Rep 12:257–268
Enciso N, Ostronoff LLK, Mejías G, León LG, Fermín ML, Merino E, Fragio C, Avedillo L, Tejero C (2018) Stem cell factor supports migration in canine mesenchymal stem cells. Vet Res Commun 42(1):29–38
Espina M, Jülke H, Brehm W, Ribitsch I, Winter K, Delling U (2016) Evaluation of transport conditions for autologous bone marrow-derived mesenchymal stromal cells for therapeutic application in horses. Peer J 4:e1773
Eydt C, Geburek F, Carmen S, Hambruch N, Rohn K, Pfarrer C, Staszyk C (2016) Sternal bone marrow derived equine multipotent mesenchymal stromal cells (MSCs): investigations considering the sampling site and the use of different culture media. Vet Med Sci 2:200
Fahy GM, Lilley TH, Linsdell H, Douglas MS, Meryman HT (1990) Cryoprotectant toxicity and cryoprotectant toxicity reduction: in search of molecular mechanisms. Cryobiology 27:247–268
Fortier LA, Nixon AJ, Williams J, Cable CS (1998) Isolation and chondrocytic differentiation of equine bone marrow-derived mesenchymal stem cells. Am J Vet Res 59:1182–1187
Fowler A, Toner M (2005) Cryo-injury and biopreservation. Ann N Y Acad Sci 1066:119–135
Freimark D, Sehl C, Weber C et al (2011) Systematic parameter optimization of a me 2 SO-and serum-free cryopreservation protocol for human mesenchymal stem cells. Cryobiology 63:67–75
Friedenstein AJ, Gorskaja JF, Kulagina NN (1976) Fibroblast precursors in normal and irradiated mouse hematopoietic organs. Exp Hematol 4(5):267–274
Fry L, Querol S, Gomez S, McArdle S, Rees R, Madrigal JA (2015) Assessing the toxic effects of DMSO on cord blood to determine exposure time limits and the optimum concentration for cryopreservation. Vox Sang 109:181–190
Fuller BJ (2004) Cryoprotectants: the essential antifreezes to protect life in the frozen state. Cryo Letters 25:375–388
Gade NE, Pratheesh MD, Nath A, Dubey PK, Amarpal, Sharma B, Saikumar G, Sharma GT (2013) Molecular and cellular characterization of buffalo bone marrow-derived mesenchymal stem cells. Reprod Domest Anim 48(3):358–367. https://doi.org/10.1111/j.1439-0531.2012.02156.x
Gao D, Critser J (2000) Mechanisms of cryoinjury in living cells. ILAR J 41:187–196

Garcia D, Longo UG, Vaquero J, Forriol F, Loppini M, Khan WS, Denaro V (2015) Amniotic membrane transplant for articular cartilage repair: an experimental study in sheep. Curr Stem Cell Res Therapy 10:77–83

Garvican ER, Cree S, Bull L, Smith RKW, Dudhia J (2016) Viability of equine mesenchymal stem cells during transport and implantation. Stem Cell Res Ther 5:1

Ghaneialvar H, Soltani L, Rahmani HR, Lotfi AS, Soleimani M (2018) Characterization and classification of mesenchymal stem cells in several species using surface markers for cell therapy purposes. Indian J Clin Biochem 33(1):46–52

Glynn ERA, Londono AS, Zinn SA, Hoagland TA, Govoni KE (2013) Culture conditions for equine bone marrow mesenchymal stem cells and expression of key transcription factors during their differentiation into osteoblasts. J Anim Sci Biotechnol 24:40

Goh TK, Reuveny S (2013) Microcarrier culture for efficient expansion and osteogenic differentiation of human fetal mesenchymal stem cells. Biores Open Access 2(2):84–97

Guest DJ, Smith MR, Allen WR (2008) Monitoring the fate of autologous and allogeneic mesenchymal progenitor cells injected into the superficial digital flexor tendon of horses: preliminary study. Equine Vet J 40:178–181

Gugjoo MB, Amarpal (2018) Mesenchymal stem cell research in sheep: current status and future prospects. Small Rumin Res 169:46–56

Gugjoo MB, Amarpal, Kinjavdekar P, Aithal HP, Ansari MM, Pawde AM, Sharma GT (2015) Isolation, culturing and characterization of New Zealand white rabbit mesenchymal stem cells derived from bone marrow. Asian J Anim Vet Adv 10(10):537–548

Gugjoo MB, Amarpal, Sharma GT, Aithal HP, Kinjavdekar P (2016) Cartilage tissue engineering: role of mesenchymal stem cells along with growth factors and scaffolds. Indian J Med Res 144:339–347

Gugjoo MB, Amarpal, Chandra V, Wani MY, Dhama K, Sharma GT (2018) Mesenchymal stem cell research in veterinary medicine. Current Stem Cell Research and Therapy 13(8):645–657

Gugjoo MB, Amarpal, Makhdoomi DM, Sharma GT (2019) Equine mesenchymal stem cells: properties, sources, characterization and potential therapeutic applications. Journal of Equine Veterinary Sciences 72:16–27

Gugjoo MB, Amarpal, Abdelbaset-Ismail A, Aithal HP, Kinjavdekar P, Kumar GS, Sharma GT (2020a) Allogeneic mesenchymal stem cells and growth factors in gel scaffold repair osteochondral defect in rabbits. Reg Med 15(2):1261–1275

Gugjoo MB, Amarpal, Fazili MR, Shah RA, Mir MS, Sharma GT (2020b) Goat mesenchymal stem cell basic research and potential applications. Small Rum Res 183:106045

Han J, Liu JY, Swartz DD, Andreadis ST (2010) Molecular and functional effects of organismal ageing on smooth muscle cells derived from bone marrow mesenchymal stem cells. Cardiovasc Res 87:147–155

Han YF, Tao R, Sun TJ, Chai JK, Xu G, Liu J (2013) Optimization of human umbilical cord mesenchymal stem cell isolation and culture methods. Cytotechnology 65:819–827

Hegewald AA, Ringe J, Bartel J, Krüger I, Notter M, Barnewitz D, Kaps C, Sittinger M (2004) Hyaluronic acid and autologous synovial fluid induce chondrogenic differentiation of equine mesenchymal stem cells: a preliminary study. Tissue Cell 36:431–438

Heidari B, Shirazi A, Akhondi MM, Hassanpour H, Behzadi B, Naderi MM, Sarvari A, Borjian S (2013) Comparison of proliferative and multilineage differentiation potential of sheep mesenchymal stem cells derived from bone marrow, liver, and adipose tissue. Avicenna J Med Biotechnol 5(2):104–117

Hendijani F (2017) Explant culture: an advantageous method for isolation of mesenchymal stem cells from human tissues. Cell Prolif 50:e12334

Hepsibha P, Meenambigai TV, Mangalagowri A, Palanisamy A, Stalin A, Nithya S, Kumanan K (2011) Multipotent differentiation potential of buffalo adipose tissue derived mesenchymal stem cells. Asian Journal of Animal Veterinary Advances 6:772–788

Hilkens P, Gervois P, Fanton Y, Vanormelingen J, Martens W, Struys T, Politis C, Lambrichts I, Bronckaers A (2013) Effect of isolation methodology on stem cell properties and multilineage differentiation potential of human dental pulp stem cells. Cell Tissue Res 353:65–78

Ho Y-T, Shimbo T, Wijaya E, Ouchi Y, Eiichi T, Yamamoto R, Kikuchi Y, Kaneda Y, Tama K (2018) Chromatin accessibility identifies diversity in mesenchymal stem cells from different tissue origins. Scientific Rep 8:17765
Huss R, Claudia L, Weissinger EM, Kolb HJ, Thalmeiler K (2000) Evidence of peripheral blood-derived, plastic-adherent CD34–/low hematopoietics stem cell clones with mesenchymal stem cell characteristics. Stem Cells 18:252–260
Hynes RO (2009) The extracellular matrix: not just pretty fibrils. Science 326:1216–1219
Iacono E, Pascuccib L, Rossia B et al (2017) Ultrastructural characteristics and immune profile of equine MSCs from fetal adnexa. Reprod 154(4):509–519
Ji M, Bai C, Li L, Fan Y, Ma C, Li X, Guan W (2016) Biological characterization of sheep kidney-derived mesenchymal stem cells. Exp Therap Med 12:3963–3971
Jing W, Xiao J, Xiong Z, Yang X, Huang Y, Zhou M, Chen S, Lin Y, Tian W (2011) Explant culture: an efficient method to isolate adipose-derived stromal cells for tissue engineering. Artif Organs 35:105–112
Jochems CE, van der Valk JB, Stafleu FR, Baumans V (2002) The use of fetal bovine serum: ethical or scientific problem? Altern Lab Anim 30:219–227
Jossen V, Schirmer C, Mostafa Sindi D et al (2016) Theoretical and practical issues that are relevant when scaling up hMSC microcarrier production processes. Stem Cells Int 2016:4760414. https://doi.org/10.1155/2016/4760414
Junior A, Arrais C, Saboya R, Velasques RD, Junqueira PL, Dulley FL (2008) Neurotoxicity associated with dimethyl sulfoxide preserved hematopoietic progenitor cell infusion. Bone Marrow Transplant 41:95–97
Kadiyala S, Young RG, Thiede MA, Bruder SP (1997) Culture expanded canine mesenchymal stem cells possess osteochondrogenic potential in vivo and in vitro. Cell Transplant 6(2):125–134
Kang JW, Kang K-S, Koo HC, Park JR, Choi EW, Park YH (2008) Soluble factors-mediated immunomodulatory effects of canine adipose tissue-derived mesenchymal stem cells. Stem Cells Development 17(4):681–693
Karahuseyinoglu S, Cinar O, Kilic E, Kara F, Akay GG, Demiralp DO, Tukun A, Uckan D, Can A (2007) Biology of stem cells in human umbilical cord stroma: in situ and in vitro surveys. Stem Cells 25:319–331
Kato Y, Imabayashi H, Mori T, Tani T, Taniguchi M, Higashi M, Matsumoto M, Umezawa A, Tsunoda Y (2004) Nuclear transfer of adult bone marrow mesenchymal stem cells: developmental totipotency of tissue-specific stem cells from an adult mammal. Biol Reprod 70(2):415–418
Khan MR, Chandrashekran A, Smith RKW, Dudhia J (2016) Immunophenotypic characterization of ovine mesenchymal stem cells. Cytometry A 89(5):443–450
Knippenberg M, Helder M, Doulabi BZ, Semeins CM, Wuisman PIJM, Klein-Nulend J (2005) Adipose tissue-derived mesenchymal stem cells acquire bone cell-like responsiveness to fluid shear stress on osteogenic stimulation. Tissue Eng 11(11/12):1780–1787
Koch TG, Heerkens T, Thomsen PD, Betts DH (2007) Isolation of mesenchymal stem cells from equine umbilical cord blood. BMC Biotechnol 7(1):26
Korovina DG, Volkova IM, Vasilieva SA, Gulukin MI, Savchenkova IP (2019) Multipotent mesenchymal stem cells derived from sheep bone marrow: isolation and cryopreservation. Cell and Tissue Biology 13(3):161–169
Kovac M, Vasicek J, Kulikova B, Bauer M, Curlej J, Balazi A, Chrenek P (2017) Different RNA and protein expression of surface markers in rabbit amniotic fluid derived mesenchymal stem cells. Biotechnol Prog 33(6):1601–1613
Laura C, Groza I, Oana L, Pall E, Pestean C, Cătană R, Cenariu M (2008) Canine mesenchymal stem cells isolation from bone marrow aspirates. Bulletin UASVM, Veterinary Medicine 65(2):96–101
Li P, Liu F, Wu C, Jiang W, Zhao G, Liu L, Bai T, Wang L, Jiang Y, Guo L, Qi X, Kou J, Fan R, Hao D, Lan S, Li Y, Liu JY (2015) Feasibility of human hair follicle-derived mesenchymal stem cells/CultiSpher®-G constructs in regenerative medicine. Cell Tissue Res 362(1):69–86
Li T, Li C, Zhang CY, Zhao J (2015a) Effect of accutase or trypsin dissociation on the apoptosis of human striatum-derived neural stem cells. Acta Academiae Medicinae Sinicae 37(2):185–194

Liu Y, Xu X, Ma X, Martin-Rendon E, Watt S, Cui Z (2010) Cryopreservation of human bone marrow-derived mesenchymal stem cells with reduced dimethylsulfoxide and well-defined freezing solutions. Biotechnol Prog 26:1635–1643

Lu T, Huang Y, Wang H, Ma Y, Guan W (2014a) Multi-lineage potential research of bone marrow-derived stromal cells (BMSCS) from cattle. Applied Biochemistry & Biotechnology 172:21–35

Lu T, Xiong H, Wang K, Wang S, Ma Y, Guan W (2014b) Isolation and characterization of adipose-derived mesenchymal stem cells (ADSCs) from cattle. Applied Biochemistry & Biotechnology 174(2):719–728

Lyahyai J, Mediano DR, Ranera B, Sanz A, Remacha AR, Bolea R, Zaragoza P, Rodellar C, Martín-Burriel I (2012) Isolation and characterization of ovine mesenchymal stem cells derived from peripheral blood. BMC Vet Res 8:169

Maia L, Dias MC, de Moraes CN et al (2017) Conditioned medium: a new alternative for cryopreservation of equine umbilical cord mesenchymal stem cells. Cell Biol Int 41(3):239–248

Mamidi MK, Nathan KG, Singh G, Thrichelvam ST, Yusof M, Nasim NA, Fakharuzi NA, Zakaria Z, Bhonde R, Das AK (2012) Comparative cellular and molecular analyses of pooled bone marrow multipotent mesenchymal stromal cells during continuous passaging and after successive cryopreservation. J Cell Biochem 113:3153–3164

Marędziak M, Marycz K, Lewandowski D, Siudzińska A, Śmieszek A (2015) Static magnetic field enhances synthesis and secretion of membrane-derived microvesicles (MVs) rich in VEGF and BMP-2 in equine adipose-derived stromal cells (EqASCs)—a new approach in veterinary regenerative medicine. In Vitro Cell Dev Biol Anim 51:230–240

Martin Y, Eldardiri MB, Lawrence-Watt DJ, Sharpe JR (2011) Microcarriers and their potential in tissue regeneration. Tissue Eng Part B Rev 17(1):71–81

Martinello T, Bronzini I, Maccatrozzo L et al (2010) Cryopreservation does not affect the stem characteristics of multipotent cells isolated from equine peripheral blood. Tissue Eng Part C Methods 16:771–781

Martinello T, Bronzini I, Maccatrozzo L et al (2011) Canine adipose derived- mesenchymal stem cells do not lose stem features after a long-term cryopreservation. Res Vet Sci 91:18–24

Martins GR, Marinho RC, Bezerra-Junior RQ, Câmara LMC, Albuquerque-Pinto LC, Teixeira MFS (2017) Isolation, culture and characterization of multipotent mesenchymal stem cells from goat umbilical cord blood. Pesquisa Vet Bras 37(6):643–649

Marx C, Silveira MD, Beyer NN (2015) Adipose-derived stem cells in veterinary medicine: characterization and therapeutic applications. Stem Cells Dev 24:803–813

Mazur P (1977) The role of intracellular freezing in the death of cells cooled at supraoptimal rates. Cryobiology 14:251–272

Mccarty R, Gronthos S, Zannettino AC, Foster BK, Xian CJ (2009) Characterisation and developmental potential of ovine bone marrow derived mesenchymal stem cells. J Cell Physiol 219:324–333

McIntosh KR, Lopez MJ, Borneman JN, Spencer ND, Anderson PA, Gimble JM (2009) Immunogenicity of allogeneic adipose-derived stem cells in a rat spinal fusion model. Tissue Eng Part A 15:2677–2686

Mediano DR, Sanz-Rubio D, Bolea R, Marín B, Vázquez FJ, Remacha AR, Pez-Pérez O, Fernández-Borges N, Castilla J, Zaragoza P, Badiola JJ, Rodellar C, Martín-Burriel I (2015) Characterization of mesenchymal stem cells in sheep naturally infected with scrapie. J Gen Virol 96:3715–3726

Mellado-López M, Griffeth RJ, Meseguer-Ripolles J, Cugat R, García M, Moreno Manzano V (2017) Plasma rich in growth factors induces cell proliferation, migration, differentiation, and cell survival of adipose-derived stem cells. Stem Cells International 2017:5946527

Mitchell KE, Weiss ML, Mitchell BM, Martin P, Davis D, Morales L, Helwig B, Beerenstrauch M, Abou-Easa K, Hildreth T et al (2003) Matrix cells from Wharton's jelly form neurons and glia. Stem Cells 21:50–60

Mitchell A, Rivas KA, Smith R III, Watts AE (2015) Cryopreservation of equine mesenchymal stem cells in 95% autologous serum and 5% DMSO does not alter post-thaw growth or mor-

phology in vitro compared to fetal bovine serum or allogeneic serum at 20 or 95% and DMSO at 10 or 5%. Stem Cell Res Ther 6:231
Mosna F, Sensebé L, Krampera M (2010) Human bone marrow and adipose tissue mesenchymal stem cells: a user's guide. Stem Cells Dev 19:1449–1470
Ock S-A, Rho G-J (2011) Effect of dimethyl sulfoxide (DMSO) on cryopreservation of porcine mesenchymal stem cells (pMSCs). Cell Transplant 20:1231–1239
Okamura LH, Cordero P, Palomino J, Parraguez VH, Torres CG, Peralta OA (2017) Myogenic differentiation potential of mesenchymal stem cells derived from fetal bovine bone marrow. Anim Biotechnol 29:1–11. https://doi.org/10.1080/10495398.2016.1276926
Otsuru S, Hofmann TJ, Olson TS, Dominici M, Horwitz EM (2013) Improved isolation and expansion of bone marrow mesenchymal stromal cells using a novel marrow filter device. Cytotherapy 15:146–153
Pacini S, Spinabella S, Trombi L et al (2007) Suspension of bone marrow-derived undifferentiated mesenchymal stromal cells for repair of superficial digital flexor tendon in race horses. Tissue Eng 13(12):2949–2955
Perez-Silos V, Camacho-Morales A, Fuentes-Mera L (2016) Mesenchymal stem cells subpopulations: application for orthopedic regenerative medicine. Stem Cells Int 2016:3187491
Petry F, Smith JR, Leber J, Salzig D, Czermak P, Weiss ML (2016) Manufacturing of human umbilical cord mesenchymal stromal cells on microcarriers in a dynamic system for clinical use. Stem Cells Int 2016:12
Phinney DG, Sensebe L (2013) Mesenchymal stromal cells: misconceptions and evolving concepts. Cytotherapy 15(2):140–145
Pillai SM, Sereda NH, Hoffman ML, Valley EV, Crenshaw TD, Park Y-K, Lee J-Y, Zinn SA, Govoni KE (2016) Effects of poor maternal nutrition during gestation on bone development and mesenchymal stem cell activity in offspring. PLoS One 11(12):e0168382
Quimby JM, Webb TL, Habenicht LM, Dow SW (2013) Safety and efficacy of intravenous infusion of allogeneic cryopreserved mesenchymal stem cells for treatment of chronic kidney disease in cats: results of three sequential pilot studies. Stem Cell Res Ther 4:48
Radcliffe CH, Flaminio JBF, Fortier LA (2010) Temporal analysis of equine bone marrow aspirate during establishment of putative mesenchymal progenitor cell populations. Stem Cells Dev 19:269–282
Rafiq QA, Coopman K, Hewitt CJ (2014) Scale-up of human mesenchymal stem cell culture: current technologies and future challenges. Curr Opin Chem Eng 2(1):8–16
Rafiq QA, Coopman K, Nienow AW, Hewitt CJ (2016) Systematic microcarrier screening and agitated culture conditions improve human mesenchymal stem cell yield in bioreactors. Biotechnol J 11:473–486
Ranera B, Lyahyai J, Romeroa A et al (2011) Immunophenotype and gene expression profiles of cell surface markers of mesenchymal stem cells derived from equine bone marrow and adipose tissue. Vet Immunol Immunopathol 144:147–154
Rhodes NP, Srivastava JK, Smith RF, Longinotti C (2004) Heterogeneity in proliferative potential of ovine mesenchymal stem cell colonies. J Mater Sci Mater Med 15(4):397–402
Roberts EL, Dang T, Lepage SIM, Alizadeh AH, Walsh T, Koch TG, Kallos MS (2019) Improved expansion of equine cord blood derived mesenchymal stromal cells by using microcarriers in stirred suspension bioreactors. Journal of Biological Engineering 13:25
Russell KA, Gibson TWG, Chong A, Co C, Koch TG (2015) Canine platelet lysate is inferior to fetal bovine serum for the isolation and propagation of canine adipose tissue- and bone marrow-derived mesenchymal stromal cells. Plos ONE 10(9):e0136621
Salehinejad P, Alitheen NB, Ali AM, Omar AR, Mohit M, Janzamin E, Samani FS, Torshizi Z, Nematollahi-Mahani SN (2012) Comparison of different methods for the isolation of mesenchymal stem cells from human umbilical cord Wharton's jelly. In Vitro Cell Dev Biol Anim 48:75–83
Sampaio RV, Chiaratti MR, Santos DCN, Bressan FF, Sangalli JR, Sá ALA, Silva TVG, Costa NN, Cordeiro MS, Santos SSD, Ambrosio CE, Adona PR, Meirelles FV, Miranda MS, Ohashi OM (2015) Generation of bovine (Bos indicus) and buffalo (Bubalus bubalis) adipose tissue derived

stem cells: isolation, characterization, and multipotentiality. Genetics & Molecular Research 14(1):53–62

Scheinkonig C, Kappicht S, Kolb HJ, Schleuning M (2004) Adoption of long-term cultures to evaluate the cryoprotective potential of trehalose for freezing hematopoietic stem cells. Bone Marrow Transplant. 34:531–536

Schop D, Dijkhuizen-Radersma R, Van Borgart E, Jansen FW, Rozemuller H, Prins H-J, de Bruijin JD (2010) Expansion of human mesenchymal stromal cells on microcarriers: growth and metabolism. J Tissue Eng Regen Med 4(7):131–140

Schubert S, Brehm W, Hillmann A, Burk J (2018) Serum-free human MSC medium supports consistency in human but not in equine adipose-derived multipotent mesenchymal stromal cell culture. Cytometry A 93(1):60–72

Schwarz C, Leicht U, Rothe C et al (2012) Effects of different media on proliferation and differentiation capacity of canine, equine and porcine adipose derived stem cells. Res Vet Sci 93:457–462

Seo JP, Tsuzuki N, Haneda S et al (2013) Comparison of allogeneic platelet lysate and fetal bovine serum for in vitro expansion of equine bone marrow-derived mesenchymal stem cells. Res Vet Sci 95:693–698

Seshareddy K, Troyer D, Weiss ML (2008) Method to isolate mesenchymal-like cells from Wharton's jelly of umbilical cord. Methods Cell Biol 86:101–119

Shah FS, Wu X, Dietrich M, Rood J, Gimble JM (2013) A non-enzymatic method for isolating human adipose tissue-derived stromal stem cells. Cytotherapy 15:979–985

Smith RKW, Korda M, Blunn GW, Goodship AE (2003) Isolation and implantation of autologous equine mesenchymal stem cells from bone marrow into the superficial digital flexor tendon as a potential novel treatment. Equine Vet J 35:99–102

Somal A, Bhat IA, Indu B, Pandey S, Panda BSK, Thakur N, Sarkar M et al (2016) A comparative study of growth kinetics, in vitro differentiation potential and molecular characterization of fetal adnexa derived caprine mesenchymal stem cells. PLoS ONE 11(6):e0156821

Somal A, Bhat IA, Baiju I, Singh AP, Panda BSK, Desingu PA, Pandey et al (2017) Impact of cryopreservation on caprine fetal adnexa derived stem cells and its evaluation for growth kinetics, phenotypic characterization and wound healing potential in xenogenic rat model. J Cellular Physiol 232:2186–2200

Somers P, Cornelissen R, Thierens H, Van Nooten G (2012) An optimized growth factor cocktail for ovine mesenchymal stem cells. Growth Factors 30(1):37–48

Spaas JH, Schauwer CD, Cornillie P, Meyer E, Soom AV, Van de Walle GR (2013) Culture and characterisation of equine peripheral blood mesenchymal stromal cells. Vet J 195(1):107–113

Steens J, Klein D. (2018). Current Strategies to Generate Human Mesenchymal Stem Cells In Vitro. Stem Cells International Art ID 6726185.

Strioga M, Viswanathan S, Darinskas A, Slaby O, Michalek J (2012) Same or not the same? Comparison of adipose tissue-derived versus bone marrow-derived mesenchymal stem and stromal cells. Stem Cells Dev 21(14):2724–2752

Sun LY, Hsieh DK, Syu WS, Li YS, Chiu HT, Chiou TW (2010) Cell proliferation of human bone marrow mesenchymal stem cells on biodegradable microcarriers enhances in vitro differentiation potential. Cell Prolif 43(5):445–456

Takahashi I, Sato K, Mera H, Wakitani S, Takagi M (2016) Effects of agitation rate on aggregation during beads-to-beads subcultivation of microcarrier culture of human mesenchymal stem cells. Cytotechnology 69(3):503–509

Taylor M, Song Y, Brockbank K (2004) Vitrification in tissue preservation: new developments. In: Fuller BJ, Lane N, Benson EE (eds) Life in the frozen state. CRC Press, Boca Raton, pp 604–641

Thirumala S, Zvonic S, Floyd E, Gimble JM, Devireddy RV (2005) Effect of various freezing parameters on the immediate post–thaw membrane integrity of adipose tissue derived adult stem cells. Biotechnol Prog 21:1511–1524

Thirumala S, Goebel WS, Woods EJ (2009) Clinical grade adult stem cell banking. Organogenesis 5:143–154

Thirumala S, Gimble JM, Devireddy RV (2010) Evaluation of methylcellulose and dimethyl sulfoxide as the cryoprotectants in a serum-free freezing media for cryopreservation of adipose derived adult stem cells. Stem Cells Dev 19:513–522

Tian Y, Tao L, Zhao S, Tai D, Liu D, Liu P (2016) Isolation and morphological characterization of ovine amniotic fluid mesenchymal stem cells. Exp Anim 65(2):125–134

Tong CK, Vellasamy S, Tan BC, Abdullah M, Vidyadaran S, Seow HF, Ramasamy R (2011) Generation of mesenchymal stem cell from human umbilical cord tissue using a combination enzymatic and mechanical disassociation method. Cell Biol Int 35:221–226

Toupadakis CA, Wong A, Genetos DC et al (2010) Comparison of the osteogenic potential of equine mesenchymal stem cells from bone marrow, adipose tissue, umbilical cord blood, and umbilical cord tissue. Am J Vet Res 71:1237–1245

Verdanova M, Pytlik R, Kalbacova MH (2014) Evaluation of sericin as a fetal bovine serum-replacing cryoprotectant during freezing of human mesenchymal stromal cells and human osteoblast-like cells. Biopreserv Biobank 12:99–105

Vidal M, Kilroy G, Johnson J, Lopez M, MoorGae R, Gimble J (2006) Cell growth characteristics and differentiation frequency of adherent equine bone marrow-derived mesenchymal stromal cells: adipogenic and osteogenic capacity. Vet Surg 35:601–610

Vidal MA, Robinson SO, Lopez MJ et al (2008) Comparison of chondrogenic potential in equine mesenchymal stromal cells derived from adipose tissue and bone marrow. Vet Surg 37:713–724

Vieira NM, Brandalise V, Zucconi E, Secco M, Strauss BE, Zatz M (2010) Isolation, characterization, and differentiation potential of canine adipose-derived stem cells. Cell Transplant 19:279–289

Wang HS, Hung SC, Peng ST, Huang CC, Wei HM, Guo YJ, Fu YS, Lai MC, Chen CC (2004) Mesenchymal stem cells in the Wharton's jelly of the human umbilical cord. Stem Cells 22:1330–1337

Weigand A, Beier JP, Schmid R, Knorr T, Kilian D, Götzl R, Gerber T, Horch RE, Boos AM (2017) Bone tissue engineering under xenogeneic-free conditions in a large animal model as a basis for early clinical applicability. Tissue Eng Part A 23(5–6):208–222

Weng L, Li W, Zuo J, Chen C (2011) Osmolality and unfrozen water content of aqueous solution of dimethyl sulfoxide. J Chem Eng Data 56:3175–3182

Wilke MM, Nydam DV, Nixon AJ (2007) Enhanced early chondrogenesis in articular defects following arthroscopic mesenchymal stem cell implantation in an equine model. J Orthop Res 25:913–925

Yadav P, Mann A, Singh V, Yashveer S, Sharma R, Singh I (2011) Expression of pluripotency genes in buffalo (Bubalus bubalis) amniotic fluid cells. Reprod Domest Anim 46:705–711

Yoon JH, Roh EY, Shin S, Jung NH, Song EY, Chang JY, Kim BJ, Jeon HW (2013) Comparison of explant-derived and enzymatic digestion-derived MSCs and the growth factors from Wharton's jelly. Biomed Res Int 2013:428726

Yuan Y, Kallos MS, Hunter C, Sen A (2014) Improved expansion of human bone marrow-derived mesenchymal stem cells in microcarrier-based suspension culture. J Tissue Eng Regen Med. 8(3):210–225

Zahedi M, Parham A, Dehghani H, Mehrjerdi HK (2018) Equine bone marrow-derived mesenchymal stem cells: optimization of cell density in primary culture. Stem Cell Investig 5:31

# Mesenchymal Stem Cell Immuno-Modulatory and/Anti-Inflammatory Properties

4

M. B. Gugjoo and Amar Pal

**Abstract**

Among various attributed features of mesenchymal stem cells (MSCs), immuno-compromise/immuno-evasive, and immuno-modulation and/anti-inflammatory properties have an important role to play in the therapeutics. The immuno-compromised feature makes them suitable for allogeneic and/xenogeneic application while immuno-modulation and/ anti-inflammatory properties facilitate the healing process. Immuno-compromised feature is attributed to their minimal ability to express MHC-II and co-stimulatory molecules. The immune regulation may be achieved through their direct contact to immune cells and/by paracrine effect. Currently, a very limited control can be exerted on these features of MSCs. An extensive available literature has demonstrated these properties of MSCs, although some studies have reported otherwise. The current chapter deliberates on the immuno-modulatory and/anti-inflammatory features of MSCs in veterinary species.

## Abbreviations

| | |
|---|---|
| ConA | Concanavalin A |
| COX2 | Cyclooxygenase 2 |
| FASL | Fas ligand |
| HGF | Hepatocyte growth factor |
| HMOX1 | Heme oxygenase 1 |

M. B. Gugjoo (✉)
Division of Veterinary Clinical Complex, FVSc & AH, SKUAST-Kashmir, Jammu and Kashmir, India

A. Pal
Division of Surgery, Indian Veterinary Research Institute, Izatnagar, UP, India



M. B. Gugjoo, A. Pal (eds.), *Mesenchymal Stem Cell in Veterinary Sciences*, https://doi.org/10.1007/978-981-15-6037-8_4

| | |
|---|---|
| IDOn | Indole-amine 2,3 dioxygenase |
| IL6 | Interleukin 6 |
| INFAC | Interferon alpha C (INF-alpha C) |
| INFB1 | Interferon, beta 1, INFG: Interferon gamma |
| iNOS | Inducible nitric oxide synthase |
| LPS | Lipopolysaccharide |
| LTBR | Lymphotoxin beta receptor (TNF superfamily member 3) |
| PDL1 | Programmed death ligand 1 |
| PMA | Phorbol 12-myristate 13-acetate |
| PTGES | Prostaglandin E synthase |
| RNA | Ribonucleic acid |
| STC-1 | Stanniocalcin 1 |
| TGFβ1 | Transforming growth factor beta 1 |
| TNF | Tumour necrosis factor |
| TNFα | Tumour necrosis factor |
| TSG | Tumour suppressor gene |
| VEGF | Vascular endothelial growth factor |

## 4.1 Introduction

MSCs arise from pericytes upon an injury/assault to a particular tissue. Activated MSCs prove doubly rewarding: they provide immuno-regulation on the blood vasculature end (preventing aggressive immune response and establishment of chronic autoimmune reaction) and on injured/damaged tissue side MSCs secrete trophic factors to help regenerate tissue (Caplan and Dennis 2006; Caplan 2012; Prockop and Oh 2012). These characteristic features of MSCs make them potential candidates for therapeutics. The trend of utilizing MSCs in medical and/veterinary therapeutics is increasing by each passing day and even appears prospective (Gugjoo and Amarpal 2018; Gugjoo et al. 2019).

MSCs may bring therapeutic effects either through differentiation into particular lineage, rescuing injured/dyeing cell via cell fusion or through secretome (Rehman et al. 2004). Current notion favours their therapeutic role mainly through the release of secretory factors. MSCs secretome contains pro-healing proteins/peptides, hormones, RNA and other chemicals. Such secretome is being mediated through extracellular vesicles like micro-vesicles and exosomes. Even organelle transfer is being ensued and may be mediated through tunnelling nanotubes (Spees et al. 2016; Phinney and Pittenger 2017; Gugjoo et al. 2018a, b). Based on these features MSCs conditioned medium (contains metabolites, growth factors and extracellular matrix proteins) is considered as potential therapeutic option (Bhang et al. 2014; Joseph et al. 2020). These factors show immune-modulation, prevent cell apoptosis and through chemotactic agents bring other pro-healing cells or factors to the area of injury (Koch et al. 2008; Fortier et al. 2010; Owens et al. 2016)

### 4.1.1 Restricted Therapeutic Applications of MSCs

MSCs definitive therapeutic applications are currently marred by limited comprehension in their cellular processes. Lack of infinite self-renewal properties under culture conditions, variability in properties with respect to donor tissue source, age and health condition and above all the control of gene expression by the available environment make their applications dubious. An in-depth understanding in stemness marker genes with a well control over their molecular mechanisms can make their definitive therapeutic applications a reality (Zheng 2018). Recent mice study could demonstrate that transcriptome and chromatin markers direct MSCs characteristics on the basis of their source (Ho et al. 2018). This demands extensive work over the epigenomic features that govern MSCs characteristics including their fate.

Apart from the limited understanding in the cellular features in vitro culture studies and in vivo experimental studies too have constraints. In vitro studies have frequently incorporated diverse culture ingredients like other cells, matrices and humoral or growth factors without any conclusion. In vivo studies conducted lack appropriate control groups, animal size and evaluation period (Park et al. 2012; Spencer et al. 2012; de Bakker et al. 2013; Saulnier et al. 2016; Kresic et al. 2017; Gugjoo et al. 2019). In veterinary practice, literature has frequently relied on study designs that have failed to conduct studies as evidence based medicine. Most studies lack blinded randomized control trials and the sample size has been too little to draw any definitive conclusion(s). It may be attributed to the machination and economical limitations leading to faulty/compromised study design (de Bakker et al. 2013; Gugjoo et al. 2019). The duration of studies has to be appropriate as per the study design and condition involved (Gugjoo et al. 2019). Additionally, MSC doses have been supplemented with other biological materials like bone marrow (BM) supernatant, autologous serum, platelet-rich plasma, and/or scaffolds and growth factors adding further misery to the understanding (Clegg and Pinchbeck 2011; De Schauwer et al. 2013; Gugjoo et al. 2017; Gugjoo et al. 2020a, b).

## 4.2 Immuno-Modulation and/Anti-Inflammatory Properties

Initially, autologous MSC applications were preferred but their immuno-compromised/ immuno-evasive nature has shifted the preference towards their allogeneic applications (Berner et al. 2015; Gugjoo et al. 2017). Allogeneic MSCs may be harvested at large scale with similar culture conditions. This makes their ready-to-use cell therapy quite possible as against their autologous therapy that has a lag period of 14–21 days for isolation and culture expansion. Most of the in vivo MSCs allogeneic studies have supported their safety and useful applicability (McDonald et al. 2015; Zorzi et al. 2015).

An extensive literature exists that support human and animal MSCs immuno-modulatory and anti-inflammatory activities. Various animal studies described in the current manuscript are of dog (Kang et al. 2008; Kim et al. 2010; Lee et al. 2011; Russell et al. 2016; Bearden et al. 2017; Chow et al. 2017; Wheat et al. 2017; Yang et al. 2018; Taguchi et al. 2019; Kafarnik et al. 2020), horse (Carrade and Borjesson 2011; Deuse et al. 2011; Peroni and Borjesson 2011; Carrade et al. 2012; Carrade et al. 2014; Colbath et al. 2017; Cassano et al. 2018; Cortés-Araya et al. 2018; Lepage et al. 2019; Longhini et al. 2019), sheep (Mrugala et al. 2008), chickens (Khatri et al. 2009), pig (Poncelet et al. 2007; Cho et al. 2008), rabbit (Liu et al. 2006; Moreno et al. 2010), goats (Somal et al. 2016), cattle (Cardoso et al. 2012; Jacca et al. 2014; Cardoso et al. 2017; de Moraes et al. 2017; Lara et al. 2017) and cat (Mumaw et al. 2015; Zajic et al. 2016; Chae et al. 2017; Clark et al. 2017; Parys et al. 2017). In general, MSCs immuno-modulation and/or anti-inflammatory activities occur through their interaction with immune cells. These activities are mediated through secretome or through cell-cell contact. MSCs lack direct cytotoxic or humoral defence activities unlike that of the immune cells. These cells fail to secrete granzymes or perforins, lack antibody production ability and are devoid of phagocytic activity (Tso et al. 2010).

### 4.2.1 MSCs Are Immuno-Privileged

Autologous and allogeneic MSCs are non-immunogenic and xenogeneic MSCs lack leukocyte proliferation in the absence of activation (Mrugala et al. 2008: Carrade and Borjesson 2011). The cells lack expression of major histocompatibility complex II (MHC-II), although variably and weakly express MHC-I (Krampera et al. 2006a, b; Carrade et al. 2012; De Schauwer et al. 2011; Schnabel et al. 2013; Berglund and Schnabel 2017; Cardoso et al. 2017; Merlo et al. 2019). Additionally, majority of the classical co-stimulatory molecules on MSCs have remained unexpressed (Fig. 4.1) (Tse et al. 2003; Carrade et al. 2012). The pro-inflammatory factors (IL2, IL6R, INFAC, INFB1, INFG, TNF and LTBR) too had remained downregulated in bovine Wharton's jelly MSCs (bWJ-MSCs) (Cardoso et al. 2017). Apart from the single in vivo parenteral implantation (Carrade et al. 2011), the repeated intra-articular or intradermal implantation of allogeneic equine MSCs too had failed to develop any hypersensitivity reaction (Ardanaz et al. 2016).

### 4.2.2 MSCs Immuno-Modulation

An initial study that had demonstrated immune-modulatory properties of MSCs was conducted by Bartholomew and colleagues in 2002. It was shown that MSCs suppressed mixed lymphocyte response and also prevented rejection of allograft in a baboon skin model (Bartholomew et al. 2002). Afterwards an extensive work in the field has been conducted piling up the huge literature on such properties (Fig. 4.1). MSCs are now known to bear anti-inflammatory properties, modulate immune response (Kang et al. 2008; Kim et al. 2010; Lee et al. 2011; Carrade and Borjesson 2011) and stimulate angiogenesis (Wu et al. 2007; El-Menoufy et al. 2010). These actions are undertaken when MSCs interact with immune cells. The inhibition may occur either by MSCs contact with immune cells and/or by their secretome. In

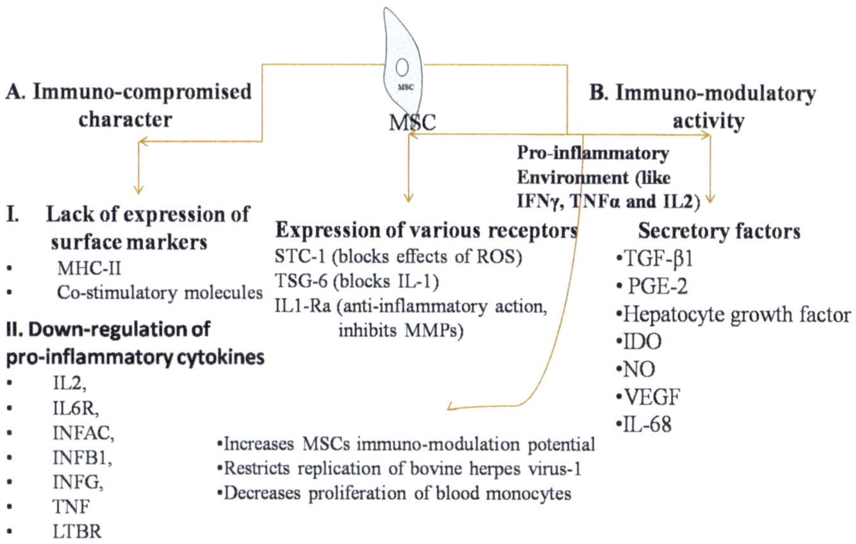


**Fig. 4.1** MSCs immuno-compromised and/ or immuno-modulatory properties. MSCs increased immuno-modulation potential in presence of inflammatory environment

humans, contact based immune-modulation may be mediated through toll-like receptors (TLRs), intracellular adhesion molecule-1 (ICAM) and vascular cell adhesion molecule-1 (VCAM-1) present on the surface of MSC (Tomchuck et al. 2008; DelaRosa and Lombardo 2010; Akiyama et al. 2012). In canine transwell assays, wherein cells are separated physically from lymphocytes, certain soluble factors had been released that had brought about the inhibition of immune response (Kang et al. 2008; Lee et al. 2011). The former way of immuno-suppression however seems to be stronger than later (Di Nicola et al. 2002).

MSCs express various receptors and/factors that tend to decrease inflammation and free radical damage. Equine MSCs had inhibited adverse reactions of reactive oxygen species (ROS) (through *STC-1*), pro-inflammatory cytokine IL-1 (through *TSG-6*) and inhibited matrix metallo-proteinases and brought anti-inflammatory actions (through *IL1-Ra*) (Angelone et al. 2017). Even unstimulated feline AD-MSCs had expressed various immune-modulatory genes like *IDO1, COX2, PTGES1, PTGES2, PTGES3, HGF, TGFβ1, IL6, PDL1 and HMOX1* barring *IL-10* (Parys et al. 2017).

In a recent advancement, MSCs are considered to secrete extracellular vesicles (exosomes) that carry modulatory paracrine actions (Katsuda et al. 2013; Yu et al. 2014), with some demonstrating immune-suppressive properties (Kordelas et al. 2014). MSCs derived exosomes are generated from plasma membrane lipid raft microdomains and after secretion accumulate in the culture medium (Tan et al. 2013; Perrini et al. 2016). Microvesicles secreted from equine MSCs had reduced apoptosis rate of equine endometrial cells, replenished their proliferation and prevented inflammation through inhibition of pro-inflammatory cytokines (Perrini et al. 2016). MSCs microvesicle production varies with respect to their source. Among various equine foetal adnexal sources, Wharton's jelly derived MSCs (WJ-MSCs) had copious micro-vesicles as compared to the other adnexal sources (Iacono et al. 2017).

### 4.2.3 Role of Inflammatory Environment

The pro-inflammatory mediators especially interferon-γ (IFNγ), tumour necrosis factor-α (TNFα) and interleukin 2 (IL2) available in inflammatory environment tend to variably stimulate the immuno-modulatory action of MSCs (Liu et al. 2006; Poncelet et al. 2007; Russell et al. 2016). Co-culture of feline adipose tissue-MSCs (AD-MSCs) with peripheral blood derived mononuclear cells and macrophages had significantly reduced pro-inflammatory cytokine production (Chae et al. 2017). MSCs challenge against the prostaglandin E2 ($PGE_2$) had rearranged its transcriptome profile. The obvious effect was downregulated immune response, and an enhanced cell proliferation and vasculogenesis. Such a challenge had additionally downregulated transforming growth factor β (TGF-β) and Wnt-β-catenin signalling pathways (Lara et al. 2017). Lipopolysaccharide (LPS) activated bovine endometrium-MSCs (End-MSCs) secretome analysis had revealed presence of anti-inflammatory and antibacterial proteins. These proteins had led to protective anti-oxidant activity and were related to tissue remodelling and angiogenesis (de Moraes et al. 2017).

MSCs mediated immuno-modulation occur either constitutively and/or after activation with inflammatory mediators. The secretome includes TGF-β1 and PGE2 (Liu et al. 2006; Poncelet et al. 2007; Kang et al. 2008; Berman et al. 2010; Lee et al. 2011; Carrade et al. 2012; Chow et al. 2017), hepatocyte growth factor (HGF) (Kang et al. 2008; Lee et al. 2011; Carrade et al. 2012) and indoleamine 2,3-dioxygenase (IDO) (Kang et al. 2008), nitric oxide (NO) (Khatri et al. 2009; Carrade et al. 2012), vascular endothelial growth factor (VEGF) (Berman et al. 2010; Lee et al. 2011), and interleukin-68 (IL-68) (Berman et al. 2010; Kang et al. 2008; Carrade et al. 2012; Bearden et al. 2017). The production of such factors, however, varies with species and tissue types though the functional basis of such differences are currently unknown (Carrade and Borjesson 2011).

The cell lines of various sources (adipose tissue-MSCs, Bone marrow-MSCs, Umbilical cord-MSCs, umbilical cord tissue-MSCs) may follow overlapping or varying pathways to exhibit immuno-modulatory effects. The source however, may not affect their comprehensive immune-modulatory effects under a particular inflammatory environment (Cassano et al. 2018). Dog bone marrow-MSCs (dBM-MSCs) and adipose tissue-MSCs (dAD-MSCs) (Russell et al. 2016; Chow et al. 2017) and horses cord blood MSCs (eCB-MSCs) and cord tissue-MSCs (eCT-MSCs) (Lepage et al. 2019) had shown comparable immuno-modulation, although variable pathways may be ensued (Fig. 4.2).

Apart from an immune-modulatory role, activated MSCs may restrict viral/bacterial growth. The antibacterial potential (against Escherichia coli) of conditioned

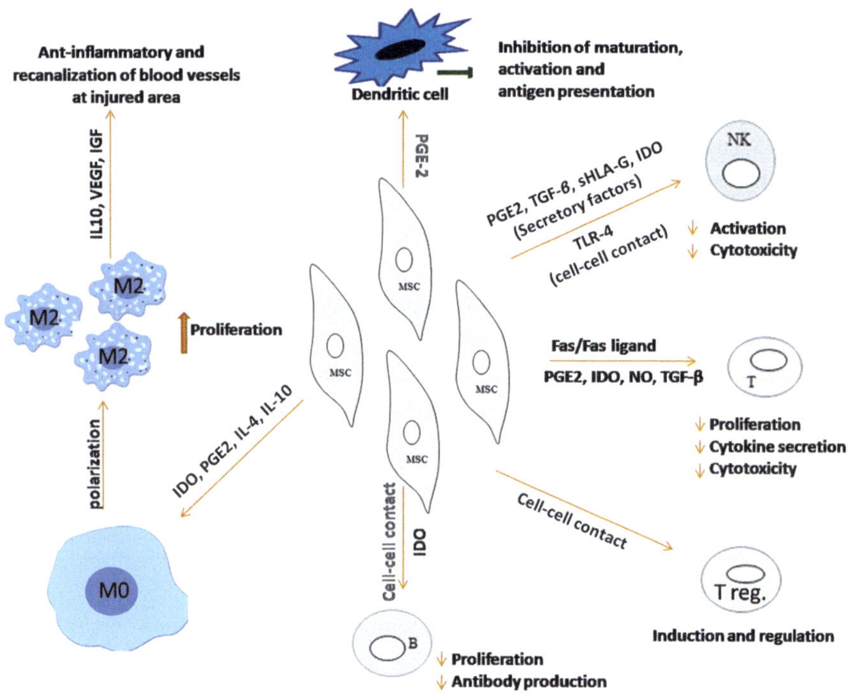


**Fig. 4.2** MSCs immuno-modulation mechanism through paracrine secretion. ↓: reduction

media of equine Endometrium-MSCs (eEnd-MSCs) and AD-MSCs had been more pronounced as compared to that of equine BM-MSCs. Lipocalin-2 (antimicrobial agent) was highly expressed in eEnd-MSCs in comparison to the eAD-MSCs and eBM-MSCs. Contrarily, lipopolysaccharide (LPS) had enhanced its comparable production in these three source-MSCs (Cortés-Araya et al. 2018). Likewise dog MSCs had shown antibacterial effects upon interaction with host innate immune response and applied antibiotics (Johnson et al. 2017). IFN-γ activation of bovine End-MSCs had induced replication restriction in bovine herpes virus-1, depicting their antiviral activity. The followed pathway might have been other than the indoleamine 2,3 dioxygenase (IDO-1) (Cardoso et al. 2012). Various factors secreted after MSCs activation are detailed below:

#### 4.2.3.1 Transforming Growth Factor-β1 (TGF-β1)

MSCs TGF-β1 production varies with respect to species and tissue type. Activated lymphocytes had failed to stimulate MSCs expression of TGF-β1 in dog BM-MSCs, horse BM-MSCs, AD-MSCs and cord tissue-MSCs (eCT-MSCs), and pig BM-MSCs (Poncelet et al. 2007; Lee et al. 2011; Carrade et al. 2012). Dog BM-MSC had failed to show improved production of TGF-β1 after exposure to activated lymphocytes (Lee et al. 2011) and even if produced levels were insufficient to inhibit lymphocyte proliferation. Contrarily, dog AD-MSC, and feline AD-MSCs and rabbit BM-MSCs had shown increased TGF-β1 production after exposure to lymphocytes (Kang et al. 2008) and IFN-γ pre-treatment (Liu et al. 2006; Kang et al. 2008; Clark et al. 2017), respectively. TGF-β1 production of feline AD-MSCs had also increased after stimulation with TNFα while combined stimulation (INFγ and TNFα) of feline AD-MSCs had resulted in decreased TGF-β1 production (Parys et al. 2017).

#### 4.2.3.2 Prostaglandin E2 and Indoleamine 2,3 Dioxygenase (IDO)

Dog MSCs (BM- and AD-), feline AD-MSCs secrete PGE2 (Kang et al. 2008; Lee et al. 2011; Chow et al. 2017; Clark et al. 2017; Parys et al. 2017; Taguchi et al. 2019) and IDO (Kang et al. 2008; Clark et al. 2017; Parys et al. 2017; Taguchi et al. 2019) and horse MSCs (AD-, BM-, CT- and CB-) secrete PGE2 (Carrade et al. 2012). The functional role of such factors has been shown to inhibit the lymphocyte proliferation as blocking of such factors restores the lymphocyte proliferation.

#### 4.2.3.3 Nitrous Oxide (NO)

Secretory factors like NO show variation with respect to the tissue source. Horse BM-MSCs and CB-MSCs (hemic sources) produce NO while CT-MSCs and AD-MSCs (solid tissues) do not (Carrade et al. 2012). This is considered to be important with respect to the angiogenesis but not immune-modulatory function. Anecdotal evidence had shown increased healing time of equine tendon upon injection of solid tissue derived MSCs as compared to CB-MSCs, and may be attributed to the increased angiogenesis with CB-MSCs (Carrade and Borjesson 2011).

Apart from above factors, INFγ had been demonstrated to upregulate various other factors of human and feline AD-MSCs including programmed death ligand (PDL-1), interleukin 6 (IL-6), cyclooxygenase 2 (COX2) and hepatocyte growth factor (HGF) (Clark et al. 2017; Parys et al. 2017). TNF-α stimulation of these cells had led to the increased expression of IL-6, prostaglandin E synthase 1 (PTGES1) and COX2 (Parys et al. 2017).

## 4.3 Contrasting Reports

Some of the recent studies have been unsuccessful to demonstrate immuno-privileged character of MSCs (Nauta et al. 2006; Gugjoo et al. 2018a, b; Gugjoo et al. 2020a, b). Various studies could demonstrate MSCs receptor expression. These receptors have the ability to commune with T-cells. Among notable ones are MHC-I, integrins and other adhesion molecules that have their analogue ligands on T-cells. These analogues are cell adhesion molecules: VCAM, intercellular adhesion molecule 1 (ICAM-1), activated leukocyte cell adhesion molecule (CD166;

ALCAM) and lymphocyte functional antigen-3 (LFA-3; CD58) (Chamberlain et al. 2007). Bovine End-MSCs had expressed IFN-γ receptors 1 and 2 without their ligand whose production was stimulated with an exposure to the exogenous IFNγ (Jacca et al. 2014). MSCs additionally bear intracellular proteins and inflammatory mediator (INFγ) that may induce them to express MHC-II alloantigen (Le Blanc et al. 2003; Klyushnenkova et al. 2005; Le Blanc and Ringdén 2007; Cassano et al. 2018; Gale et al. 2019). Further, few recent studies have demonstrated that allo-MSC antibodies may generate against the allogeneic MSCs (Pezzanite et al. 2015; Owens et al. 2016). Even MHC mismatched MSCs had been able to activate complement dependent cytotoxicity (Owens et al. 2016).

Apart from ex vivo studies, in vivo equine studies that had implanted allogeneic equine BM-MSCs/umbilical cord blood MSCs (eUCB-MSCs) either once or multiple times intra-articularly had reported an undesirable clinical response (Joswig et al. 2017; Bertoni et al. 2019). Comparing cell induced synovial effusions, equine BM-MSCs injected joints had more significant effusions as compared to equine UCB-MSCs injected ones. The inflammatory signs developed have been more evident with fewer doses of cells (ten million) as compared to higher cell doses (20 million) (Bertoni et al. 2019). Thus, it is important to note that allogeneic MSCs implantation might incite some adverse reactions rather than useful outcomes (Berglund and Schnabel 2017). However, such undesirable responses may also have resulted from foetal bovine serum (FBS) particles being internalized by the cultured MSCs. Modification of cell culture conditions for 48-hour duration may help to deplete FBS particles (Joswig et al. 2017). MSCs adaptation to autologous serum-supplementation may be an effective way to maintain cell viability and prevent any adverse immunological reactions (Haque et al. 2015).

## 4.4 Effect of MSCs on Body's Immune System

MSCs in general exhibit properties of immunomodulation and is achieved by secreting certain factors that inhibit immune cells proliferation and/or their secretory factors (Fig. 4.2). Below are given the interactions of MSCs and immune cells.

### 4.4.1 Innate Immunity and MSCs

This type of immunity plays a pivotal role in elimination of pathogens by adaptive immune responses (Yamane and Paul 2013). MSCs interaction with innate immune cells like dendritic cells (DCs), natural killer (NK) cells, neutrophils and macrophages inhibits inflammatory processes and promotes regeneration (Le Blanc and Mougiakakos 2012). Inhibition of peripheral blood mononuclear cells may be variable with respect to the MSCs from various sources (Cardoso et al. 2012). Goat WJ-MSCs had higher inhibitory potential over peripheral blood mononuclear cells in comparison to the amniotic sac MSCs (AS-MSCs). Former cells achieved it through indoleamine 2,3-dioxygenase (IDO) while later cell type had activated inducible nitric oxide synthase (iNOS) (Somal et al. 2016). The effect of MSCs on different immune cells is detailed below:

#### 4.4.1.1 Myeloid Dendritic Cells (MDCs)

MSCs immunosuppressive effects on dendritic cells (DCs) may be achieved through restriction of their maturation (differentiation from monocyte) (Jiang et al. 2005; Wheat et al. 2017), surface protein expression like MHC-II, CD86, CD83, CD40 and CD1-$\alpha$, capability to motivate lymphocytes (Gao et al. 2017), inhibiting their migration potential (Chiesa et al. 2011; Consentius et al. 2015) and by depressing their pro-inflammatory capacity (Aggarwal and Pittenger 2005). The maturation inhibition potential of MSCs increases in presence of inflammatory mediators like IFNγ and TNF-α (Wheat et al. 2017). The restricted DCs maturation may occur by PGE2 (Yañez et al. 2010). The decreased lymphocyte motivation of DCs occurs by downregulating IFN$\gamma$ and TNF-$\alpha$ expression and accelerating release of IL10 (Gao et al. 2017). These responses may be mediated through Notch pathway that relies on IFN$\gamma$-secretase (Xu et al. 2017).

#### 4.4.1.2 Natural Killer (NK) Cells

NK cells produce pro-inflammatory cytokines and carry cytolytic activity (Moretta 2002). MSCs tend to inhibit NK cell effects through secretory factors and/or direct cell interactions. MSCs secretome inhibiting NK cells are PGE2, TGF-β and sHLA-G (Sotiropoulou et al. 2006). MSCs suppress activated NK cell receptor expression being mediated by IDO and PGE2 (Cui et al. 2016). The direct cell-cell interaction may be achieved through MSCs Toll-like receptor-(TLR-) 4 (Michelo et al. 2016). These immune-suppressive effects of human occur in presence of high MSC:NK cell ratios (Sotiropoulou et al. 2006). Contrary to these findings, MSCs may be dissolved in presence of enough activating receptors on NK cells (Götherström et al. 2011). Thus, MSCs concentration and its surrounding environment has an important role to determine their immuno-modulatory properties.

#### 4.4.1.3 Neutrophils

MSCs cultured with neutrophils tend to decrease their reactive oxygen species (ROS) potential. The inhibition potential of MSCs varies with respect to the source of cell. Feline BM-MSCs had quite significantly inhibited neutrophilic ROS potential in comparison to the feline AD-MSCs (Mumaw et al. 2015).

#### 4.4.1.4 Macrophages

Macrophages based upon the available environment are polarized into two types: M1 possesses antimicrobial activity while M2 alleviates inflammation and expedites tissue repair. The anti-inflammatory activity and repair process of M2 may be ensured by secretion of IL10 and VEGF and IGF trophic factors (Mosser and Edwards 2008). MSCs induce production of M2 macrophages from M1. M2 increases the phagocytic activity and secretion of interleukin 10 and decreases inflammatory cytokines (Zhang et al. 2010; Selleri et al. 2016). Further, MSCs decrease immune disorders and hasten tissue regeneration by enhancing macrophage concentration locally (Liu et al. 2015; Chaturvedi et al. 2014).

### 4.4.2 Adaptive Immunity and MSCs

T and B lymphocytes are mainly involved in adaptive immune responses. T lymphocytes comprise CD4 T helper and CD8 T cytotoxic lymphocytes. These cells show antigen specific immune response and have immunological memory (Yamane and Paul 2013). B lymphocytes act by producing specific antibodies (De Silva and Klein 2015). Dog AD-MSC had decreased lymphocyte production of TNF-α while increasing their IFNγ production (Kang et al. 2008). Horse MSCs (AD-, BM-, CB- and CT-) had decreased TNFα and IFNγ production of lymphocytes (Peroni and Borjesson 2011). Pig BM-MSCs had decreased IFNγ and IL2 production of lymphocytes (Liu et al. 2004). Feline AD-MSCs lymphocyte inhibition tends to remain unaffected with donor age (Zajic et al. 2016). Below are the details of MSCs immune-modulation role against adaptive immunity.

#### 4.4.2.1 T Cells

MSCs are considered to repress T-cell proliferation either through cell based or nonspecific mitogenic stimuli (Di Nicola et al. 2002). The action is mediated through secretome being responsible for effective T-cell suppression and apoptosis (Ren et al. 2008; Ren et al. 2010). The secretome involved in proliferation inhibition include PGE2, IDO, NO and TGF-β (Gao et al. 2016). The cells may prevent T-cell activation and proliferation by decreasing the expression of activation markers like CD25, CD38 and CD69 (Le Blanc et al. 2003; Groh et al. 2005). Furthermore, MSCs may promote apoptosis of activated T-cells via the Fas/Fas ligand pathway (Akiyama et al. 2012). Allogeneic and autologous cells may comparably suppress proliferation and IFN-γ production of T-cells (Colbath et al. 2017).

Despite all these reports, it is worth mentioning here that MSCs immunosuppressive potential may not be activated every time. Such a response is affected and determined by the inflammatory mediators and their strength (Renner et al. 2009). Besides, MSCs activities to inhibit T-cell proliferation may not be sufficient in presence of pathogen associated molecules. Toll-like receptors (TLR3 and TLR4) which damage Notch signalling may help in T-cells recuperation (Rashedi et al. 2017). Blocking activity of human BM-MSC produced TGF-β1 had failed to decrease T-regulatory cell production (English et al. 2009).

#### 4.4.2.2 B Cells

MSCs suppress B cell proliferation, activation, differentiation, chemokine receptor expression either through cell-cell contact and/ or secretome (IDO) (Augello et al. 2005; O'Connor et al. 2006). However, MSCs ability to suppress B cell antibody production is dependent upon inflammatory stimulation strength, and MSCs:B cell ratio (Krampera et al. 2006a, b; Bernardo et al. 2009).

## 4.5 Conclusion(s)

Mesenchymal stem cells are immune-compromised/ immuno-evasive cells that achieve local immune-modulation and/anti-inflammatory role through secretion of paracrine factors and/or by cell-cell contact. The cells harvested from various sources may achieve comparable immune-modulatory role through different pathways. The cells act upon both the innate and acquired immunity without actually undertaking cytotoxic or humoral defence activities. The cells however, have failed to fully immune-suppress the in vivo immune responses and even some of the in vitro studies too have been unsuccessful to get the desired immuno-modulation. Thus, the mechanisms behind their immune-modulation and/anti-inflammatory action need further insights for their definitive clinical applications.

## References

Aggarwal S, Pittenger MF (2005) Human mesenchymal stem cells modulate allogeneic immune cell responses. Blood 105(4):1815–1822

Akiyama K, Chen C, Wang DD, Xu X, Qu C, Yamaza T, Cai T, Chen W, Sun L, Shi S (2012) Mesenchymal stem- cell-induced immune-regulation involves FAS-ligand-/FAS-mediated T cell apoptosis. Cell Stem Cell 10(5):544–555

Angelone M, Conti V, Biacca C, Battaglia B, Pecorari L, Piana F, Gnudi G, Leonardi F, Ramoni R, Basini G, Dotti S, Renzi S, Ferrari M, Grolli S (2017) The contribution of adipose tissue-derived Mesenchymal stem cells and platelet-rich plasma to the treatment of chronic equine laminitis: a proof of concept. Int J Mol Sci 18:2122

Ardanaz N, Vázquez FJ, Romero A, Remacha AR, Barrachina L, Sanz A, Ranera B, Vitoria A, Albareda J, Prades M, Zaragoza P, Martín-Burriel I, Rodellar C (2016) Inflammatory response to the administration of mesenchymal stem cells in an equine experimental model: effect of autologous, and single and repeat doses of pooled allogeneic cells in healthy joints. BMC Vet Res 12:65

Augello A, Tasso R, Negrini SM, Amateis A, Indiveri F, Cancedda R, Pennesi G (2005) Bone marrow mesenchymal progenitor cells inhibit lymphocyte proliferation by activation of the programmed death 1 pathway. Eur J Immunol 35(5):1482–1490

Bartholomew A, Sturgeon C, Siatskas M, Ferrer K, McIntosh K, Patil S, Hardy W, Devine S, Ucker D, Deans R, Moseley A, Hoffman R (2002) Mesenchymal stem cells suppress lymphocyte proliferation in vitro and prolong skin graft survival in vivo. Exp Hematol 30(1):42–48

Bearden RN, Huggins SS, Cummings KJ, Smith R, Gregory CA, Saunders WB (2017) In-vitro characterization of canine multipotent stromal cells isolated from synovium, bone marrow, and adipose tissue: a donor-matched comparative study. Stem Cell Res Ther 8:218

Berglund AK, Schnabel LV (2017) Allogeneic major histocompatibility complex-mismatched equine bone marrow-derived mesenchymal stem cells are targeted for death by cytotoxic anti-major histocompatibility complex antibodies. Equine Vet J 49(4):539–544

Berman DM, Willman MA, Han D, Kleiner G, Kenyon NM, Cabrera O, Karl JA, Wiseman RW, O'Connor DH, Bartholomew AM, Kenyon NS (2010) Mesenchymal stem cells enhance allogeneic islet engraftment in nonhuman primates. Diabetes 59:2558–2568

Bernardo ME, Locatelli F, Fibbe WE (2009) Mesenchymal stromal cells. Ann N Y Acad Sci 1176(1):101–117

Berner A, Henkel J, Woodruff MA, Steck R, Nerlich M, Schuetz MA, Hutmacher D, W. (2015) Delayed minimally invasive injection of Allogenic bone marrow stromal cell sheets regenerates large bone defects in an ovine preclinical animal model. Stem Cells Transl Med 4:503–512

Bertoni L, Branly T, Jacquet S, Desancé M, Desquilbet L, Rivory P, Hartmann D-J, Denoix J-M, Audigié F, Galéra P, Demoor M (2019) Intra-articular injection of 2 different dosages of autologous and allogeneic bone marrow- and umbilical cord-derived Mesenchymal stem cells triggers a variable inflammatory response of the fetlock joint on 12 sound experimental horses. Stem Cells Int 2019:9431894

Bhang SH, Lee S, Shin J-Y, Lee T-J, Jang H-K, Kim B-S (2014) Efficacious and clinically relevant conditioned medium of human adipose-derived stem cells for therapeutic angiogenesis. Mol Ther 22:862–872

Caplan AI (2012) MSCs as therapeutics. In: Hematti P, Keating A (eds) Stem cell biology and regenerative medicine: Mesenchymal stromal cells: biology and clinical applications. Human Press, New York, pp 79–90

Caplan AI, Dennis JE (2006) Mesenchymal stem cells as trophic mediators. J Cell Biochem 98:1076–1084

Cardoso TC, Ferrari HF, Garcia AF, Novais JB, Silva-Frade C, Ferrarezi MC, Andrade AL, Gamiero R (2012) Isolation and characterization of Wharton's jelly-derived multipotent Mesenchymal stromal cells obtained from bovine umbilical cord and maintained in defined serum free three-dimensional system. BMC Biotechnol 12:18

Cardoso TC, Okamura LH, Baptistella JC, Gameiro R, Ferreira HL, Marinho M, Flores EF (2017) Isolation, characterization and immunomodulatory-associated gene transcription of Wharton's jelly-derived multipotent mesenchymal stromal cells at different trimesters of cow pregnancy. Cell Tissue Res 367(2):243–256

Carrade DD, Borjesson DL (2011) Immunomodulation by Mesenchymal stem cells in veterinary species. Comp Med 63(3):207–2017

Carrade DD, Affolter VK, Outerbridge CA et al (2011) Intradermal injections of equine allogeneic umbilical cord-derived mesenchymal stem cells are well tolerated and do not elicit immediate or delayed hypersensitivity reactions. Cytotherapy 13(10):1180–1192

Carrade DD, Lame MW, Kent MS, Clark KC, Walker NJ, Borjesson DL (2012) Comparative analysis of the immunomodulatory properties of equine adult-derived mesenchymal stem cells. Cell Med 4:1–11

Carrade HDD, Wood JA, Granick JL, Walker NJ, Clark KC, Borjesson DL (2014) Equine Mesenchymal stem cells inhibit T cell proliferation through different mechanisms depending on tissue source. Stem Cells Dev 23(11):1258–1265

Cassano JM, Fortier LA, Hicks RB, Harman RM, Van de Walle GR (2018) Equine mesenchymal stem cells from different sources display comparable immune related gene expression profiles in response to interferon gamma (IFN-γ). Vet Immunol Immunopathol 202:25–30

Chae H-K, Song W-J, Ahn J-O, Li Q, Lee B-Y, Kweon K, Park S-C, Youn H-Y (2017) Immunomodulatory effects of soluble factors secreted by feline adipose tissue-derived mesenchymal stem cells. Vet Immunol Immunopathol. https://doi.org/10.1016/j.vetimm.2017.07.013

Chamberlain G, Fox J, Ashton B, Middleton J (2007) Concise review: mesenchymal stem cells: their phenotype, differentiation capacity, immunological features, and potential for homing. Stem Cells 25(11):2739–2749

Chaturvedi P, Gilkes DM, Takano N, Semenza GL (2014) Hypoxia-inducible factor-dependent signaling between triple-negative breast cancer cells and mesenchymal stem cells promotes macrophage recruitment. Proc Natl Acad Sci U S A 111(20):E2120–E2129

Chiesa S, Morbelli S, Morando S et al (2011) Mesenchymal stem cells impair in vivo T-cell priming by dendritic cells. Proc Natl Acad Sci U S A 108(42):17384–17389

Cho PS, Messina DJ, Hirsh EL, Chi N, Goldman SN, Lo DP, Harris IR, Popma SH, Sachs DH, Huang CA (2008) Immunogenicity of umbilical cord tissue derived cells. Blood 111:430–438

Chow L, Johnson V, Coy J, Regan D, Dow S (2017) Mechanisms of immune suppression utilized by canine adipose and bone marrow-derived Mesenchymal stem cells. Stem Cells Develop 26(5):374–389

Clark KC, Fierro FA, Ko EM, Walker NJ, Arzi B, Tepper CG, Dahlenburg H, Cicchetto A, Kol A, Marsh L, Murphy WJ, Fazel N, Borjesson DL (2017) Human and feline adipose-derived

mesenchymal stem cells have comparable phenotype, immunomodulatory functions, and transcriptome. Stem Cell Res Ther 8:69
Clegg PD, Pinchbeck GL (2011) Evidence-based medicine and stem cell therapy: how do we know such technologies are safe and efficacious? Vet Clin North Am Equine Pract 27:373–382
Colbath AC, Dow SW, Phillips JN, McIlwraith CW, Goodrich LR (2017) Autologous and allogeneic equine mesenchymal stem cells exhibit equivalent immunomodulatory properties in vitro. Stem Cells Dev 26(7):503–511
Consentius C, Akyüz L, Schmidt-Lucke JA, Tschöpe C, Pinzur L, Ofir R, Reinke P, Volk H-D, Juelke K (2015) Mesenchymal stromal cells prevent allostimulation in vivo and control checkpoints of Th1 priming: migration of human DC to lymph nodes and NK cell activation. Stem Cells 33(10):3087–3099
Cortés-Araya Y, Amilon K, Rink BE, Black G, Lisowski Z, Donadeu FX, Esteves CL (2018) Comparison of antibacterial and immunological properties of mesenchymal stem/stromal cells from equine bone marrow, endometrium and adipose tissue. Stem Cells Dev 27(21):1518–1525
Cui R, Rekasi H, Hepner-Schefczyk M, Kai F, Petri RM, Bruderek K, Brandau S, Jäger M, Flohé SB (2016) Human mesenchymal stromal/stem cells acquire immunostimulatory capacity upon cross-talk with natural killer cells and might improve the NK cell function of immunocompromised patients. Stem Cell Res Ther 7(1):88
de Bakker E, Van Ryssen B, De Schauwer C, Meyer E (2013) Canine mesenchymal stem cells: state of the art, perspectives as therapy for dogs and as a model for man. Vet Q 33(4):225–233
de Moraes CN, Maia L, de Oliveira E, de Paula Freitas Dell'Aqua C, Chapwanya A, da Cruz Landim-Alvarenga F, Oba E (2017) Shotgun proteomic analysis of the secretome of bovine endometrial mesenchymal progenitor/stem cells challenged or not with bacterial lipopolysaccharide. Vet Immunol Immunopathol 187:42–47
De Schauwer C, Meyer E, Van de Walle GR, Van Soom A (2011) Markers of stemness in equine mesenchymal stem cells: a plea for uniformity. Theriogenology 8(5):1431–1443
De Schauwer C, Van de Walle GR, Soom AV, Meyer E (2013) Mesenchymal stem cell therapy in horses: useful beyond orthopedic injuries? Vet Q 33(4):234–241
De Silva NS, Klein U (2015) Dynamics of B cells in germinal centres. Nat Rev Immunol 15(3):137–148
DelaRosa O, Lombardo E (2010) Modulation of adult mesenchymal stem cells activity by toll-like receptors: implications on therapeutic potential. Mediat Inflamm 2010:865601
Deuse T, Stubbendorff M, Tang-Quan K, Phillips N, Kay MA, Eiermann T (2011) Immunogenicity and immunomodulatory properties of umbilical cord lining mesenchymal stem cells. Cell Transplant 20(5):655–667
Di Nicola M, Carlo-Stella C, Magni M, Milanesi M, Longoni PD, Matteucci P, Grisanti S, Gianni AM (2002) Human bone marrow stromal cells suppress T-lymphocyte proliferation induced by cellular or nonspecific mitogenic stimuli. Blood 99:3838–3843
El-Menoufy H, Aly LA, Aziz MT, Atta HM, Roshdy NK, Rashed LA, Sabry D (2010) The role of bone marrow-derived mesenchymal stem cells in treating formocresol induced oral ulcers in dogs. J Oral Pathol Med 39:281–289
English K, Ryan JM, Tobin L, Murphy MJ, Barry FP, Mahon BP (2009) Cell contact, prostaglandin E(2) and transforming growth factor β 1 play nonredundant roles in human mesenchymal stem cell induction of CD4+CD25highforkhead box P3+ regulatory T cells. Clin Exp Immunol 156:149–160
Fortier LA, Potter HG, Rickey EJ et al (2010) Concentrated bone marrow aspirate improves full-thickness cartilage repair compared with microfracture in the quine model. J Bone Joint Surg 92:1927–1937
Gale AL, Linardi RL, McClung G, Mammone RM, Ortved KF (2019) Comparison of the Chondrogenic differentiation potential of equine synovial membrane-derived and bone marrow-derived Mesenchymal stem cells. Front Vet Sci 6:178
Gao F, Chiu SM, Motan DAL, Zhang Z, Chen L, Ji H-L, Tse H-F, Fu Q-L, Lian Q (2016) Mesenchymal stem cells and immunomodulation: current status and future prospects. Cell Death Dis 7:e2062

Gao WX, Sun YQ, Shi J et al (2017) Effects of mesenchymal stem cells from human induced pluripotent stem cells on differentiation, maturation, and function of dendritic cells. Stem Cell Res Ther 8(1):48

Götherström C, Lundqvist A, Duprez IR, Childs R, Berg L, le Blanc K (2011) Fetal and adult multipotent mesenchymal stromal cells are killed by different pathways. Cytotherapy 13(3):269–278

Groh ME, Maitra B, Szekely E, Koc ON (2005) Human mesenchymal stem cells require monocyte-mediated activation to suppress alloreactive T cells. Exp Hematol 33:928–934

Gugjoo MB, Amarpal (2018) Mesenchymal stem cell research in sheep: current status and future prospects. Small Rumin Res 169:46–56

Gugjoo MB, Amarpal, Abdelbaset-Ismail A, Aithal HP, Kinjavdekar P, Pawde AM, Kumar GS, Sharma GT (2017) Mesenchymal stem cells with IGF-1 and TGF- β1 in laminin gel for osteochondral defects in rabbits. Biomed Pharmacother 93:1165–1174

Gugjoo MB, Amarpal, Chandra V, Wani MY, Dhama K, Sharma GT (2018a) Mesenchymal stem cell research in veterinary medicine. Curr Stem Cell Res Ther 13(8):645–657

Gugjoo MB, Amarpal, Fazili MR, Shah RA, Sharma GT (2018b) Mesenchymal stem cell: basic research and potential applications in cattle and buffalo. J Cell Physiol 234(6):1–18

Gugjoo MB, Amarpal, Makhdoomi DM, Sharma GT (2019) Equine mesenchymal stem cells: properties, sources, characterization and potential therapeutic applications. J Equine Vet Sci 72:16–27

Gugjoo MB, Hussain S, Amarpal SRA, Dhama K (2020a) Mesenchymal stem cell mediated Immuno-modulatory and anti-inflammatory mechanisms in immune and allergic disorders. Recent Patents Inflamm Allergy Drug Discov 14(1):3–14. https://doi.org/10.2174/1872213X14666200130100236

Gugjoo MB, Amarpal, Ismail AA, Aithal HP, Kinjavdekar P, Saikumar GS, Sharma GT (2020b) Allogeneic mesenchymal stem cells and growth factors in gel scaffold repair osteochondral defect in rabbit. Regen Med 14:1261–1275. https://doi.org/10.2217/rme-2018-0138

Haque N, Kasim NHA, Rahman MT (2015) Optimization of pre-transplantation conditions to enhance the efficacy of mesenchymal stem cells. Int J Biol Sci 11:324–334

Ho Y-T, Shimbo T, Wijaya E, Ouchi Y, Takaki E, Yamamoto R, Kikuchi Y, Kaneda Y, Tamai K (2018) Chromatin accessibility identifies diversity in mesenchymal stem cells from different tissue origins. Sci Rep 8:1776

Iacono E, Pascuccib L, Rossia B et al (2017) Ultrastructural characteristics and immune profile of equine MSCs from fetal adnexa. Reproduction 154(4):509–519

Jacca S, Franceschi V, Agosti M, Cavirani S, Mistretta F, Donofrio G (2014) Interferon gamma-mediated BoHV-4 replication restriction in bovine endometrial stromal cells is host IDO1 gene expression independent and BoHV-4 IE2 gene expression dependent. Biol Reprod 91(5):112, 1–112,14

Jiang XX, Zhang Y, Liu B et al (2005) Human mesenchymal stem cells inhibit differentiation and function of monocyte derived dendritic cells. Blood 105(10):4120–4126

Johnson V, Webb T, Norman A, Coy J, Kurihara J, Regan D, Dow S (2017) Activated Mesenchymal stem cells interact with antibiotics and host innate immune responses to control chronic bacterial infections. Sci Rep 7(1):9575

Joseph A, Baiju I, Bhat IA, Pandey S, Bharti M, Verma M, Pratap Singh A, Ansari MM, Chandra V, Saikumar G, Amarpal, Sharma GT (2020) Mesenchymal stem cell-conditioned media: a novel alternative of stem cell therapy for quality wound healing. J Cell Physiol 235(7–8):5555–5569

Joswig A-J, Mitchell A, Cummings KJ et al (2017) Repeated intra-articular injection of allogeneic mesenchymal stem cells causes an adverse response compared to autologous cells in equine model. Stem Cell Res Ther 8:42

Kafarnik C, McClellan A, Dziasko M, Daniels JT, Guest DJ (2020) Canine corneal stromal cells have multipotent mesenchymal stromal cell properties in vitro. Stem Cells Dev 29(7):429–435

Kang JW, Kang K-S, Koo HC, Park JR, Choi EW, Park YH (2008) Soluble factors-mediated immunomodulatory effects of canine adipose tissue-derived mesenchymal stem cells. Stem Cells Develop 17(4):681–693

Katsuda T, Tsuchiya R, Kosaka N, Yoshioka Y, Takagaki K, Oki K, Takeshita F, Sakai Y, Kuroda M, Ochiya T (2013) Human adipose tissue-derived mesenchymal stem cells secrete functional neprilysin-bound exosomes. Sci Rep 3:1197

Khatri M, O'Brien TD, Sharma JM (2009) Isolation and differentiation of chicken mesenchymal stem cells from bone marrow. Stem Cells Dev 18:1485–1492

Kim HS, Kim KH, Kim SH, Kim YS, Koo KT, Kim TI, Seol Y-J, Ku Y, Rhyu I-C, Chung C-P et al (2010) Immunomodulatory effect of canine periodontal ligament stem cells on allogenic and xenogenic peripheral blood mononuclear cells. J Periodont Implant Sci 40:265–270

Klyushnenkova E, Mosca JD, Zernetkina V, Majumdar MK, Beggs KJ, Simonetti DW, Deans RJ, McIntosh KR (2005) T cell responses to allogeneic human mesenchymal stem cells: immunogenicity, tolerance, and suppression. J Biomed Sci 12(1):47–57

Koch TG, Berg LC, Betts DH (2008) Concepts for the clinical use of stem cells in equine medicine. Canadian Vet J 49:1009–1017

Kordelas L, Rebmann V, Ludwig AK, Radtke S, Ruesing J, Doeppner TR, Epple M, Horn PA, Beelen DW, Giebel B (2014) MSC-derived exosomes: a novel tool to treat therapy-refractory graft-versus-host disease. Leukemia 28:970–973

Krampera M, Pasini A, Pizzolo G, Cosmi L, Romagnani S, Annunziato F (2006a) Regenerative and immunomodulatory potential of mesenchymal stem cells. Curr Opin Pharmacol 6(4):435–441

Krampera M, Cosmi L, Angeli R et al (2006b) Role for interferon gamma in the immunomodulatory activity of human bone marrow mesenchymal stem cells. Stem Cells 24(2):386–398

Kresic N, Simic I, Lojkic I, Bedekovic T (2017) Canine adipose derived mesenchymal stem cells transcriptome composition alterations: a step towards standardizing therapeutic. Stem Cells Int 2017:12. Art ID:4176292

Lara E, Velásquez A, Cabezas J, Rivera N, Pacha P, Rodríguez-Alvarez L, Saravia F, Castro FO (2017) Endometritis and in vitro PGE2 challenge modify properties of cattle endometrial Mesenchymal stem cells and their Transcriptomic profile. Stem Cells Int 2017:4297639

Le Blanc K, Mougiakakos D (2012) Multipotent mesenchymal stromal cells and the innate immune system. Nat Rev Immunol 12(5):383–396

Le Blanc K, Ringdén O (2007) Immunomodulation by mesenchymal stem cells and clinical experience. J Intern Med 262(5):509–525

Le Blanc K, Tammik C, Rosendahl K, Zetterberg E, Ringden O (2003) HLA expression and immunologic properties of differentiated and undifferentiated mesenchymal stem cells. Exp Hematol 31:890–896

Lee WS, Suzuki Y, Graves SS, Iwata M, Venkataraman GM, Mielcarek M, Peterson LJ, Ikehara S, Torok-Storb B, Storb R (2011) Canine bone-marrow–derived mesenchymal stromal cells suppress alloreactive lymphocyte proliferation in vitro but fail to enhance engraftment in canine bone marrow transplantation. Biol Blood Marrow Transplant 17:465–475

Lepage SIM, Lee OJ, Koch TG (2019) Equine cord blood Mesenchymal stromal cells have greater differentiation and similar immunosuppressive potential to cord tissue Mesenchymal stromal cells. Stem Cells Dev 28(3):227–237

Liu J, Lu XF, Wan L, Li YP, Li SF, Zeng LY, Zeng YZ, Cheng LH, Lu YR, Cheng JQ (2004) Suppression of human peripheral blood lymphocyte proliferation by immortalized mesenchymal stem cells derived from bone marrow of Banna minipig inbred line. Transplant Proc 36:3272–3275

Liu H, Kemeny DM, Heng BC, Ouyang HW, Melendez AJ, Cao T (2006) The immunogenicity and immunomodulatory function of osteogenic cells differentiated from mesenchymal stem cells. J Immunol 176:2864–2871

Liu W, Zhang S, Gu S, Sang L, Dai C (2015) Mesenchymal stem cells recruit macrophages to alleviate experimental colitis through TGF$\beta$1. Cell Physiol Biochem 35(3):858–865

Longhini ALF, Salazar TE, Vieira C, Trinh T, Duan Y, Pay LM, Li Calzi S, Losh M, Johnston NA, Xie H, Kim M, Hunt RJ, Yoder MC, Santoro D, McCarrel TM, Grant MB (2019) Peripheral blood-derived mesenchymal stem cells demonstrate immunomodulatory potential for therapeutic use in horses. PLoS One 14(3):e0212642

McDonald CA, Oehme D, Pham Y, Kelly K, Itescu S, Gibbon A, Jenkin R (2015) Evaluation of the safety and tolerability of a high-dose intravenous infusion of allogeneic mesenchymal precursor cells. Cytotherapy 17(9):1178–1187
Merlo B, Teti G, Lanci A, Burk J, Mazzotti E, Falconi M, Iacono E (2019) Comparison between adult and foetal adnexa derived equine post-natal mesenchymal stem cells. BMC Vet Res 15:277
Michelo CM, Fasse E, van Cranenbroek B et al (2016) Added effects of dexamethasone and mesenchymal stem cells on early natural killer cell activation. Transpl Immunol 37:1–9
Moreno R, Martinez-Gonzalez I, Rosal M, Farwati A, Gratacos E, Aran JM (2010) Characterization of mesenchymal stem cells isolated from the rabbit fetal liver. Stem Cells Dev 19:1579–1588
Moretta A (2002) Natural killer cells and dendritic cells: rendezvous in abused tissues. Nat Rev Immunol 2(12):957–965
Mosser DM, Edwards JP (2008) Exploring the full spectrum of macrophage activation. Nat Rev Immunol 8(12):958–969
Mrugala D, Bony C, Neves N, Caillot L, Fabre S, Moukoko D, Jorgensen C, Noel D (2008) Phenotypic and functional characterisation of ovine mesenchymal stem cells: application to a cartilage defect model. Ann Rheum Dis 67:288–295
Mumaw JL, Schmiedt CW, Breidling S, Sigmund A, Norton NA, Thoreson M, Peroni JF, Hurley DJ (2015) Feline mesenchymal stem cells and supernatant inhibit reactive oxygen species production in cultured feline neutrophils. Res Vet Sci 103:60–69
Nauta AJ, Westerhuis G, Kruisselbrink AB, Lurvink EG, Willemze R, Fibbe WE (2006) Donor-derived mesenchymal stem cells are immunogenic in an allogeneic host and stimulate donor graft rejection in a non-myeloablative setting. Blood 108:2114–2120
O'Connor BP, Vogel LA, Zhang W, Loo W, Shnider D, Lind FE, Ratliff M, Noelle RJ, Erickson LD (2006) Imprinting the fate of antigen-reactive B cells through the affinity of the B cell receptor. J Immunol 177(11):7723–7732
Owens SD, Kol A, Walker NJ, Borjesson DL (2016) Allogeneic Mesenchymal stem cell treatment induces specific alloantibodies in horses. Stem Cells Int 2016:830103
Park SS, Lee YJ, Lee SH, Lee D, Choi K, Kim WH, Kweon OK, Han HJ (2012) Functional recovery after spinal cord injury in dogs treated with a combination of Matrigel and neural-induced adipose-derived mesenchymal stem cells. Cytotherapy 14:584–597
Parys M, Kruger JM, Yuzbasiyan-Gurkan V (2017) Evaluation of immunomodulatory properties of feline mesenchymal stem cells. Stem Cells Dev 26(10):776–785
Peroni JF, Borjesson DL (2011) Anti-inflammatory and immunomodulatory activities of stem cells. Vet Clin North Am Equine Pract 27:351–362
Perrini C, Strillacci MG, Bagnato A et al (2016) Microvesicles secreted from equine amniotic-derived cells and their potential role in reducing inflammation in endometrial cells in an in-vitro model. Stem Cell Res Ther 7:169
Pezzanite LM, Fortier LA, Antczak DF, Cassano JM, Brosnahan MM, Miller D et al (2015) Equine allogeneic bone marrow-derived mesenchymal stromal cells elicit antibody responses in vivo. Stem Cell Res Ther 6:54
Phinney DG, Pittenger MF (2017) Concise review: MSC-derived Exosomes for cell-free therapy. Stem Cells 35(4):851–858
Poncelet AJ, Vercruysse J, Saliez A, Gianello P (2007) Although pig allogeneic mesenchymal stem cells are not immunogenic in vitro, intracardiac injection elicits an immune response in vivo. Transplantation 83:783–790
Prockop DJ, Oh JY (2012) Mesenchymal stem/stromal cells (MSCs): role as guardians of inflammation. Mol Ther 20:14–20
Rashedi I, Gómez-Aristizábal A, Wang XH, Viswanathan S, Keating A (2017) TLR3 or TLR4 activation enhances mesenchymal stromal cell-mediated Treg induction via notch signaling. Stem Cells 35(1):265–275
Rehman J, Traktuev D, Li J, Merfeld-Clauss S et al (2004) Secretion of angiogenic and antiapoptotic factors by human adipose stromal cells. Circulation 109:1292–1298

Ren G, Zhang L, Zhao X, Xu G, Zhang Y, Roberts AI, Zhao ZR, Shi Y (2008) Mesenchymal stem cell mediated immunosuppression occurs via concerted action of chemokines and nitric oxide. Cell Stem Cell 2(2):141–150

Ren G, Zhao X, Zhang L, Zhang J, L'Huillier A, Ling W, Roberts AI, Le AD, Shi S, Shao C, Shi Y (2010) Inflammatory cytokine-induced intercellular adhesion molecule-1 and vascular cell adhesion molecule-1 in mesenchymal stem cells are critical for immunosuppression. J Immunol 184(5):2321–2328

Renner P, Eggenhofer E, Rosenauer A, Popp FC, Steinmann JF, Slowik P, Geissler EK, Piso P, Schlitt HJ, Dahlke MH (2009) Mesenchymal stem cells require a sufficient, ongoing immune response to exert their immunosuppressive function. Transplant Proc 41(6):2607–2611

Russell KA, Chow NHC, Dukoff D, Gibson TWG, LaMarre J, Betts DH, Koch TG (2016) Characterization and Immunomodulatory effects of canine adipose tissue- and bone marrow-derived Mesenchymal stromal cells. PLoS One 11(12):e0167442

Saulnier N, Loriau J, Febre M, Robert C, Rakic R, Bonte T, Buff S, Maddens S (2016) Canine placenta: a promising potential source of highly proliferative and immunomodulatory mesenchymal stromal cells? Vet Immunol Immunopathol 171:47–55

Schnabel LV, Fortier LA, McIlwraith CW, Nobert KM (2013) Therapeutic use of stem cells in horses: which type, how, and when? Vet J 197:570–577

Selleri S, Bifsha P, Civini S et al (2016) Human mesenchymal stromal cell-secreted lactate induces M2-macrophage differentiation by metabolic reprogramming. Oncotarget 7(21):30193–30210

Somal A, Bhat IA, Indu B, Pandey S, BSK P, Bharti MK et al (2016) Comparative immunomodulatory analysis of caprine fetal adnexa derived mesenchymal stem cells. Am J Reprod Immunol 76(S1):1–65

Sotiropoulou PA, Perez SA, Gritzapis AD, Baxevanis CN, Papamichail M (2006) Interactions between human mesenchymal stem cells and natural killer cells. Stem Cells 24(1):74–85

Spees JL, Lee RH, Gregory CA (2016) Mechanisms of mesenchymal stem/stromal cell function. Stem Cell Res Ther 7:125. https://doi.org/10.1186/s13287-016-0363-7

Spencer ND, Chun R, Vidal MA, Gimble JM, Lopez MJ (2012) In vitro expansion and differentiation of fresh and revitalized adult canine bone marrow-derived and adipose tissue-derived stromal cells. Vet J 191(2):231–239

Taguchi T, Borjesson DL, Osmond C, Griffon DJ (2019) Influence of Donor's age on Immunomodulatory properties of canine adipose tissue-derived Mesenchymal stem cells. Stem Cells Dev 28(23):1562–1571

Tan SS, Yin Y, Lee T, Lai RC, Yeo RW, Zhang B, Choo A, Lim SK (2013) Therapeutic MSC exosomes are derived from lipid raft microdomains in the plasma membrane. J Extracell Vesicles 2:22614. https://doi.org/10.3402/jev.v2i0.22614

Tomchuck SL, Zwezdaryk KJ, Coffelt SB, Waterman RS, Danka ES, Scandurro AB (2008) Toll-like receptors on human mesenchymal stem cells drive their migration and immunomodulating responses. Stem Cells 26:99–107

Tse WT, Pendleton JD, Beyer WM, Egalka MC, Guinan EC (2003) Suppression of allogeneic T-cell proliferation by human marrow stromal cells: implications in transplantation. Transplantation 75:389–397

Tso GH, Law HK, Tu W, Chan GC, Lau YL (2010) Phagocytosis of apoptotic cells modulates mesenchymal stem cells osteogenic differentiation to enhance IL-17 and RANKL expression on CD4+ T cells. Stem Cells 28:939–954

Wheat WH, Chow L, Kurihara JN, Regan DP, Coy JW, Webb TL, Dow SW (2017) Suppression of canine dendritic cell activation/maturation and inflammatory cytokine release by Mesenchymal stem cells occurs via multiple distinct biochemical pathways. Stem Cells Development 26(4):249–262

Wu Y, Chen L, Scott PG, Tredget EE (2007) Mesenchymal stem cells enhance wound healing through differentiation and angiogenesis. Stem Cells 25:2648–2659

Xu LL, Fu HX, Zhang JM et al (2017) Impaired function of bone marrow mesenchymal stem cells from immune thrombocytopenia patients in inducing regulatory dendritic cell differentiation through the Notch-1/Jagged-1 signaling pathway. Stem Cells Dev 26(22):1648–1661

Yamane H, Paul WE (2013) Early signaling events that underlie fate decisions of naive CD4+ T cells toward distinct T-helper cell subsets. Immunol Rev 252(1):12–23

Yañez R, Oviedo A, Aldea M, Bueren JA, Lamana ML (2010) Prostaglandin E2 plays a key role in the immunosuppressive properties of adipose and bone marrow tissue derived mesenchymal stromal cells. Exp Cell Res 316(19):3109–3123

Yang H-M, Song W-J, Qiang L, Kim S-Y, Kim H-J, Ryu M-O, Ahn J-O, Youn H-Y (2018) Canine mesenchymal stem cells treated with TNF-α and IFN-γ enhance anti-inflammatory effects through the COX-2/PGE2 pathway. Res Vet Sci 119:19–26

Yu B, Zhang X, Li X (2014) Exosomes derived from mesenchymal stem cells. Int J Mol Sci 15:4142–4157

Zajic LB, Webb TL, Webb P, Coy JW, Dow SW, Quimby JM (2016) Comparison of proliferative and immune-modulatory potential of adipose-derived mesenchymal stem cells from young and geriatric cats. J Feline Med Surg. https://doi.org/10.1177/1098612X16680703

Zhang QZ, Su WR, Shi SH et al (2010) Human gingiva derived mesenchymal stem cells elicit polarization of m2 macrophages and enhance cutaneous wound healing. Stem Cells 28(10):1856–1868

Zheng L (2018) Epigenetic regulation on stem cell fate and regeneration. Curr Stem Cell Res Ther 13(1):3

Zorzi AR, Amstalden EMI, Plepis AMG, Martins VCA, Ferretti M, Antonioli E, Duarte ASS, Luzo ACM, Miranda JB (2015) Effect of human adipose tissue Mesenchymal stem cells on the regeneration of ovine articular cartilage. Int J Mol Sci 16:26813–26831

# Mesenchymal Stem Cell Differentiation Properties and Available Microenvironment

5

M. B. Gugjoo and Amar Pal

**Abstract**

Mesenchymal stem cells (MSCs) have characteristic properties, the expression of which is being determined by the available microenvironment/niche. Extensive ex vivo studies have been conducted, aimed at understanding the effects of various niche based signals on MSCs. Numerous microenvironment cues studied include the cell, chemical and mechanical, and of the topography. The available conducted studies have mostly focused on limited and isolated cues. An ex vivo system is less extensive with more controllability. In contrast, in vivo system exposes them to an uncontrolled environment of numerous simultaneous cues. The ex vivo results of MSCs therefore may not be recapitulated under in vivo environment unless many simaltaneous cues aping in vivo environment are studied. The current chapter focuses on the MSCs fate under various cues in ex vivo system.

## 5.1 Introduction

Mesenchymal stem cells (MSCs) due to their characteristic features are considered to have an all-in-one solution for diverse ailments (Gugjoo and Amarpal 2018; Gugjoo et al. 2019a). The characteristics of MSCs are believed to be controlled by the available microenvironment/or niche. The "microenvironment or niche" comprises cells, various chemical/mechanical and topographical cues. Such an

M. B. Gugjoo (✉)
Division of Veterinary Clinical Complex, FVSc & AH, SKUAST-Kashmir, Jammu and Kashmir, India

A. Pal
Division of Surgery, Indian Veterinary Research Institute, Izatnagar, UP, India


M. B. Gugjoo, A. Pal (eds.), *Mesenchymal Stem Cell in Veterinary Sciences*, https://doi.org/10.1007/978-981-15-6037-8_5

environment serves as source of cues that determine the cell migration and homing, local proliferation, secretome, and ultimately their fate towards particular lineage (Place et al. 2009). Understanding the mechanisms involved in MSCs fate based on these signals or cues becomes imperative for their fruitful applications. To confirm proof of principle, in vitro studies are initially being conducted which allows a wide range of permutations and combinations without any kind of involved risk. Such a system, however, has limitations of being less extensive and is more controlled. In vivo system in contrast presents an uncontrolled system subjected to numerous simultaneous cues from various quarters (Hwang et al. 2011).

The scientific field that studies such cell–microenvironment/niche interactions is tissue engineering. The *tissue engineering* (TE) concept has been introduced by Professor Robert Nerem in 1988 and was defined as "*the application of life sciences and engineering to develop a basic understanding of the functional and structural relationships of natural and pathologic mammalian tissues and the development of bio-substitutes that can be utilized to restore, maintain, or improve tissues damaged or lost by various disease conditions*" (Meyer et al. 2009; Lanza et al. 2011). To utilize tissue engineering and stem cells to the service of medicine, William Haseltine in 1999 has coined the term *regenerative medicine* (Atala and Lanza 2001). It is a translational research branch of tissue engineering and molecular biology and is aimed to develop technologies/therapeutics that can remove/provide relief in sufferings of diseased/malfunctional/non-functional tissues/organs (Gregor et al. 2017; Gugjoo et al. 2019a).

TE employs various ingredients like cells, signaling factors, and the biofabricated scaffold to develop the tissue graft (Gugjoo et al. 2018). In vivo implantation exposes cells to the unfavorable environment. To prevent such adverse effects and keep the cells viable and effective for a prolonged period, the biocompatible scaffold is being incorporated. To promote the cell specific expression various signaling factors and scaffolds play an important role. The role played by these critical factors in relation to the MSCs characteristics is of utmost importance for latter's effective clinical applications. The current chapter provides an overview of the ex vivo effects of such cues on the MSCs activities including their differentiation (Fig. 5.1).

## 5.2 Effect of In Vitro Culture Environment on MSCs Viability and Activity

MSCs availability in donor tissues is very limited and as such their culture expansion becomes imperative. These cells are cultured in a specific medium being renewed after certain intervals. An extensive culturing exhausts the culture ingredients and needs to be replenished. Accumulation of higher lactate and ammonia concentration affects goat BM-MSCs growth and proliferation under in vitro conditions. To prevent the adverse effects of these accumulated noxious ingredients, the culture medium is being replaced after some intervals. Prolonged proliferation of goat MSCs had been maintained by replacing about 30% of medium after every 3 days.

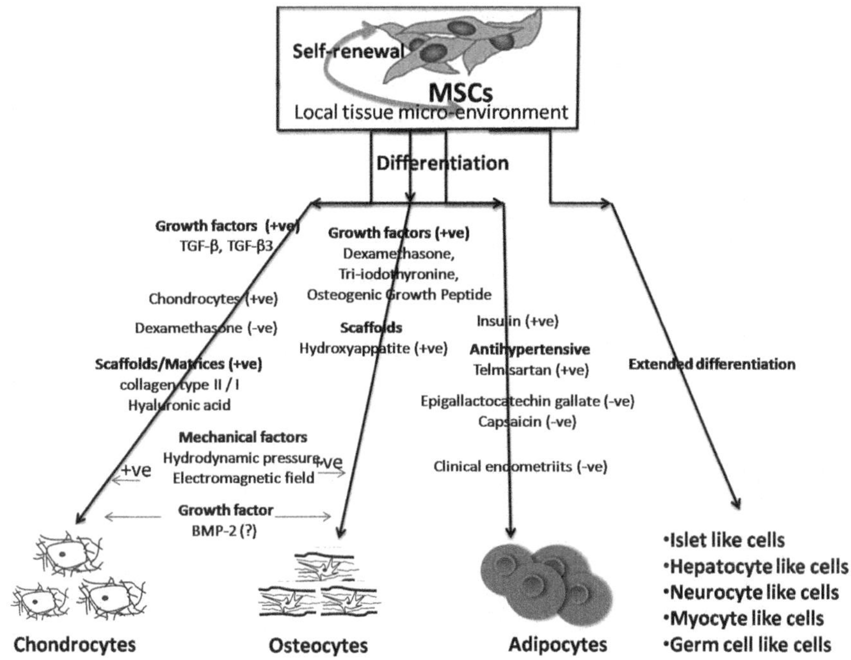


**Fig. 5.1** Pictorial representation of role of different in vitro cues in determining MSCs fate

This helps to dilute noxious metabolites (ammonia), provides energy metabolites, and increases surface area to the cells (Schop et al. 2008). As such it becomes necessary to offer safe in vivo heaven to the implanted cells.

To aid MSCs proliferation autologous cellular therapy products like stromal vascular fraction (SVF), bone marrow (BM)- and cord blood (CB)-mononuclear cells (MNCs), and platelet lysate (PL) have been added to the culture systems. These products have promoted equine MSCs (eMSCs) proliferation, migration, and interleukin secretion, although variably. The proliferation and chemotaxis of eMSCs had been demonstrated to be promoted by all these products. The effect of stromal vascular fraction (SVF) had been more pronounced. Equine BM-MNCs in the presence of SVF had more potential of chemo-invasion. Equine BM-MSCs (eBM-MSCs) under the influence of CB-MNCs had secreted interleukin 6 (IL-6), transforming growth factor beta 1 (TGF-β1) and prostaglandin E2 ($PGE_2$) while such effects were not demonstrated with SVF and PL (Kol et al. 2012). These findings may be cited as the prime reasons to supplement MSCs with autologous therapy products for in vivo applications. It is worth mentioning here that MSCs cultured under ex vivo system may undergo significant alterations in their phenotype and plasticity (Colosimo et al. 2013) and result in undesirable effects. Due to the associated risk factors, innovative MSCs have been put under the list of "Advanced Therapy Medicinal Products (ATMPs)" by European Medicines Agency (Gugjoo et al. 2019a).

MSCs may release various secretory factors and differentiate or express genes of particular lineages based on the availability of microenvironment/niche cues. The paracrine secretion of MSCs especially in relation to the immune-modulation and/or anti-inflammatory properties has been evaluated in a separate chapter. This chapter focuses on the MSCs in vitro differentiation potential with respect to the available environment and is summarized in Fig. 5.1. Animal MSCs have been studied in detail for their tri-lineage differentiation with respect to the growth factors, scaffolds, tissue specific cells, mechanical cues, and other miscellaneous factors. The extended differentiation potential of animal MSCs has also been evaluated, although not extensively. The mechanisms involved in such trans-differentiation of MSCs are not fully understood as the cells from different sources undergo variable differentiation pathways towards particular cell lineage.

## 5.3 Tri-Lineage Differentiation of MSCs

### 5.3.1 Chondrogenesis of MSCs

MSCs chondrogenesis has generally been studied under pellet or micromass system, although 3D culture system is now increasingly being studied. MSCs in pellet culture system had undergone spontaneous chondrogenesis without the addition of any chondrogenic factor. At the center of bovine BM-MSCs pellet, cartilage specific matrix (collagen type II) had accumulated with its progressive increase as culturing continued (Bosnakovski et al. 2004). To achieve desired and specific cartilaginous matrix formation, chondrogenic factors like growth factors, scaffolds, cells, or mechanical factors are being evaluated (Gugjoo et al. 2019b, c, d). There are various types of commercially available chondrogenic differentiation media that contains basic ingredients required for chondrogenesis.

#### 5.3.1.1 Growth Factors and MSCs Chondrogenesis

Culture addition of growth factors like TGF-β1, bone morphogenetic protein 2 (BMP-2), and BMP-7 promote MSCs chondrogenesis (Bosnakovski et al. 2005; Park et al. 2005; Jäger et al. 2006; Knippenberg et al. 2006; Zeiter et al. 2009; Kalaszczynska et al. 2013; Bottagisio et al. 2015; Hu et al. 2015a, b; Goldman and Barabino 2016; Desancé et al. 2018). Cattle BM-MSCs (cBM-MSCs) had expressed chondrogenic genes upon culture exposure to transforming growth factor β1 (TGF-β1) (10 ng/mL) (Zeiter et al. 2009) or TGF-β3 (1, 10, and 100 ng/mL) (Goldman and Barabino 2016) after 14 days. TGF-β had increased aggrecan expression but the collagen II expression had remained lower than the non-treated cells (Park et al. 2005). Bone morphogenetic protein 2 (BMP-2) has variably supported MSCs chondrogenesis. BMP-2 had promoted chondro-specific MSCs expression when loaded into the alginate beads. Its concentration, however, provided a ceiling effect on MSCs chondrogeneic potential. At 50 ng/mL BMP-2 concentration a plateau phase of MSCs chondrogenesis was developed (Park et al. 2005). Contrarily, one of the studies had demonstrated that BMP-2 (50 ng/mL) had no effect on MSCs (Zeiter et al. 2009), while other studies had shown that culture addition of BMP-2 (1, 10, and 100 ng/mL) yielded osteogenic lineage rather than the chondrogenic

(Knippenberg et al. 2006; Zhang and Jiang 2006; Goldman and Barabino 2016). This shows that combined utilization of growth factor (BMP-2) and scaffold (alginate beads) may promote MSCs chondrogenesis, while BMP-2 alone may favor their osteogenesis, though further studies are desired. The role of BMP-7 in MSCs differentiation is again controversial. BMP-7 may promote chondrogenic (Knippenberg et al. 2006), osteogenic (Elkhenany et al. 2016) differentiation of goat adipose tissue derived MSCs (gAD-MSCs) and adipogenic differentiation of human BM-MSCs (Katja et al. 2007). gAD-MSCs short (15 min) incubation under BMP-7 (10 ng/mL) had been sufficient to induce their chondrogenic differentiation while long term incubation had no additional effect on such differentiation (Knippenberg et al. 2006). Thus, MSCs may or may not undergo chondrogenesis under the influence of these above mentioned humoral factors. It is worth mentioning here that biologic activities of these recombinant growth factors are much lower than their analogues being synthesized endogenously and as such ectopic incorporation of these factors may be required in higher concentrations under in vivo conditions (Wang et al. 2010; Gugjoo et al. 2017; Gugjoo et al. 2020).

#### 5.3.1.2 Scaffolds and MSCs Chondrogenesis

MSCs supplementation with specific matrices predisposes them towards specific lineage. Equine and bovine MSCs chondrogenic potential enhances in the presence of natural and synthetic joint and cartilage ingredients like synovial fluid, collagen type II, I, hyaluronic acid hydrogel, polyethylene glycol hydrogels, alginate hydrogel microcapsules, and chondroitin sulfate (Hegewald et al. 2004; Varghese et al. 2008; Hwang et al. 2011; Toh et al. 2012; Toh et al. 2013; Yin et al. 2016; Gugjoo et al. 2016; Santos et al. 2018). The proteoglycan expression in the presence of these tissue matrices had remained quite lower as compared to that stimulated by TGF-β1 (Hegewald et al. 2004). However, addition of TGF-β1 had further promoted such a differentiation potential of bovine MSCs (Bosnakovski et al. 2006). The hypertrophic chondrocyte formation, usually seen under in vivo MSCs studies, had been prevented by the addition of polyethylene glycol (PEG) to the chondroitin sulfate (Varghese et al. 2008). Similarly, incorporation of collagen type I in sheep BM-MSCs 2D-cultures (5–10 × $10^6$/cm$^3$ collagen type I) and 3D cultures (50 × $10^6$/cm$^3$ collagen I) had improved their potential to synthesize hyaline cartilage (Bornes et al. 2016). In dog study, pentosan polysulfate (PPS) in micromass culture had significantly enhanced dog MSCs (dMSCs) chondrogenesis. This was unlike the polysulfated glycosaminoglycans (PSGAG) that had inhibited dMSCs chondrogenesis while promoting their fibrocartilage-like phenotype. In alginate culture, MSCs differentiation potential had remained unaffected with these scaffolds (PPS and PSGAG) (Bwalya et al. 2017). Thus, culture technique in addition to the scaffold influences MSCs chondrogenesis.

The scaffold matrix is composed of intricate design with its nature and composition determining the spatial distribution and secretory action of the cells (Prins et al. 2016). As such impact of scaffold matrix design (comprising fiber alignment and number of layers) on MSCs secretory properties must be evaluated. Cattle bone marrow MSCs (cBM-MSCs) had been grown on a single layer scaffold (poly(ε-caprolactone)) for 3–6 weeks. Cell/scaffold assembly bearing fiber alignment of 0° had better stiffness as compared to that of 90° aligned scaffold fibers. These results were further pronounced in multilayer 0° aligned fiber scaffold/cell assembly. The specific components like collagen concentration and cell infiltration however may remain neutral with respect to scaffold fiber alignment. Collagen deposition of MSCs although unenhanced with scaffold design but had followed the fiber alignment (Fisher et al. 2015). Cross-linking of hydrogel scaffolds like hyaluronic acid and tyramine (HA-Ty) conjugate tends to have a considerable impact on MSCs localization, condensation, matrix biosynthesis, and on their fate (Vickers et al. 2010; Toh et al. 2012). Cross-linking of cartilage specific scaffolds (collagen type II and glycosaminoglycan) that had allowed cell mediated contraction had promoted goat BM-MSCs chondrogenic expression. Further increase in such cross-linking had resulted in decreased cell mediated contraction and thereby inhibited their chondrogenic expression (Vickers et al. 2010).

#### 5.3.1.3 Various Combined Factors and MSCs Chondrogenesis

Mechanical factor like hydraulic pressure may not have any observable effect on cattle MSCs (cMSCs) differentiation (Zeiter et al. 2009). Contrarily hydrodynamic pressure had differentiated them towards musculoskeletal lineage (chondrogenic and/or osteogenic lineages) (Goldman and Barabino 2016). Even the adverse effects of inflammatory mediator (interleukin 1β) on cMSCs chondrogenesis had been prevented by combined action of electromagnetism and transforming growth factor β3 (TGF-β3) (Ongaro et al. 2012). The shear stress may prone MSCs to undergo hypertrophic pathway under insufficient TGF-β3 concentration (Goldman and Barabino 2016). Mechanical compression on hydrogel embedded goat BM-MSCs carrying TGF-β1 had upregulated their chondrogenic expression. However, in the absence of TGF-β1 mechanical compression had inhibited such a differentiation. It may be inferred that growth factor induced MSCs chondrogenesis may be hastened in the presence of mechanical compression (Terraciano et al. 2007). Chondrogenic differentiation of cattle bone marrow MSCs (cBM-MSCs) through mechanical loading may occur by rapid chromatin condensation that remains coincident with the fibrochondrogenic marker upregulation. This has been observed with the blockade of histone methyltransferase EZH2, a promoter of chromatin condensation and gene silencing (Heo et al. 2015).

#### 5.3.1.4 Chondrocytes and MSCs Chondrogenesis

Resident chondrocytes too favor MSCs chondrogenic differentiation. Fresh chondrocytes undergo superior chondrogenesis than cultured MSCs. There expression of chondrogenic matrices like aggrecan and collagen II remains high. Chondrocyte secretion of aggrecan and collagen II had been two and 30 fold higher as compared to the alginate loaded MSCs in the presence of BMP-2 (Park et al. 2005). Even agarose gel (4% w/v) laden cattle BM-MSCs (cBM-MSCs) had secreted mechanically inferior cartilage matrix as compared to that of chondrocytes. cBM-MSCs laden hydrogel constructs had secreted matrices like glycosaminoglycan (GAG) and its equilibrium modulus had attained plateau with the passage of time, inferring that delayed differentiation may not be the reason behind their diminished chondrogenic capacity (Mauck et al. 2006). Despite these properties of chondrocytes, their culture expansion deleteriously affects their chondrogenic potential. Culture expanded chondrocytes had secreted mechanically inferior (more Col I and less Col II) matrices that lacked integration with native tissue. Contrarily, cBM-MSCs had secreted cartilage specific matrices of superior mechanical properties (Rackwitz et al. 2014). MSCs chondrogenic potential may be further promoted in the presence of chondrocytes. Co-culture of cBM-MSCs and fibrochondrocytes had favored latter's chondrogenic expression. The cBM-MSCs:fibrochondrocytes ratio in co-culture, however, may affect MSCs differentiation. In monoculture a significant GAG content was produced in a cell ratio of 100:0, while highest secretory activity had been demonstrated at equal cellular concentrations (50:50). Such a combined co-culture chondrogenesis may also prevent hypertrophic chondrocyte formation (type X collagen) (McCorry et al. 2016). Similarly, infrapatellar fat pad derived gMSCs in the presence of chondrocytes had shown improved secretion of cartilage specific matrices (Arora et al. 2017). In vitro mono and co-cultures too had supported these results (McCorry and Bonassar 2017). It may be inferred that MSCs chondrogenesis improves in the presence of chondrocytes and as such results may be more encouraging in less damaged cartilage than extensively damaging one. To enhance MSCs in vitro chondrogenic matrices synthesis, co-culture may be an effective option (McCorry and Bonassar 2017).

#### 5.3.1.5 Hypoxia and MSCs Chondrogenesis

Sustained hypoxia tends to enhance MSCs chondrogenesis (Zscharnack et al. 2009; Bornes et al. 2015). cBM-MSCs laden in collagen/hyaluronic acid scaffold incubated under hypoxic conditions had expressed significantly superior chondrogenic matrices in comparison to those cultured under normoxia. The enhanced expression of collagen X in the presence of hypoxia however needs to be checked (Bornes et al. 2015). Hypoxia may enhance sheep BM-MSCs population doubling (Bornes et al. 2015) but might induce sheep MSCs apoptosis (Zhao et al. 2014). Such an adverse effect may be reduced by culture incorporation of proto-oncogenic serine/threonine kinase, Pim-1 (Zhao et al. 2014). Acid ceramidase incorporation to the culture medium had led to the two fold increase in equine and feline MSCs growth. On addition to the chondrogenic media, the chondrogenic potential of these cells had increased and was synergistic with TGF-β1 (Simonaro et al. 2013).

#### 5.3.1.6 Miscellaneous Factors Affecting MSCs Chondrogenesis

Growth factor- and scaffold-induced MSCs chondrogenesis may also be affected by the donor age and source location. cBM-MSCs harvested from different age donors had variable chondrogenic potential. MSCs from immature donors had greater chondrogenic expression in the presence of TGF-β3 and hyaluronic acid hydrogels as compared to the mature MSCs (Erickson et al. 2011). MSCs from the same tissue source but at different locations of an individual may or may not have comparable in vitro chondrogenic potential. Intra-articular fat pad derived MSCs had been demonstrated to exhibit superior chondrogenic expression potential in comparison to the non-articular AD-MSCs (Stewart 2011). Synovial fluid and synovial membrane MSCs of osteoarthritic and osteochondritis dissecans horses and healthy tissues tend to show comparable phenotypic and multipotency potentials. However, the chondrogenic differentiation potential of MSCs from healthy joints had been better (Fulber et al. 2016). Breed wise variation also ensues as chondrogenic differentiation potential of BM-MSCs from Labrador retriever and Hovawart dog breed had been superior to that from Border collie and German shepherd breeds (Bertolo et al. 2015). Thus, allogeneic sources may show variable results.

In general, MSCs chondrogenesis may be promoted through Sox9 expression. Pyruvate dehydrogenase kinase isoform 2 (PDK2) is possibly upregulating Sox9 gene from upregulated Sox6 (Wang et al. 2017). Enhanced Sox9 expression tends to repress chondrocyte hypertrophy possibly through inhibition of Runx2 (Zhou et al. 2006), Wnt (Topol et al. 2009), Col10a1, and VEGFA (Hattori et al. 2010; Leung et al. 2011) pathways.

## 5.3.2 Osteogenesis of MSCs

MSCs osteogenic lineage differentiation may occur in the presence of relevant cues eliciting from available environment. In vitro MSCs osteogenic studies have utilized commercially available medium (containing growth factors). However, various other factors, either alone or in combination, including matrices or growth factors or cells, in addition to mechanical forces also direct such differentiation. It may be emphasized that β-glycerolphosphate appears to be unsuitable phosphate ion source for sheep adipose tissue MSCs (sAD-MSCs) and sBM-MSCs. These cells had failed to show alkaline phosphatase activity (Kalaszczynska et al. 2013). Human MSCs although do show such activity and thus, MSCs mineralization status may sometimes be misleading.

### 5.3.2.1 Growth Factors and MSCs Osteogenesis

Growth factors like dexamethasone, ascorbic-acid-2-phosphate, monosodium phosphate, β-glycerolphosphate, 1,25-dihydroxyvitamin-D3 (1,25(OH)2), basic fibroblast growth factor (bFGF), BMP-2, BMP-7 and osteogenic growth peptide (OGP) favor MSCs osteogenesis (Knippenberg et al. 2006; Jäger et al. 2006; Zhang and Jiang 2006; Tjabringa et al. 2008; Zhu et al. 2008; Huang et al. 2013; Kalaszczynska et al. 2013; Randau et al. 2013; Bottagisio et al. 2015; Hu et al. 2015b; Elkhenany et al. 2016). As mentioned above the growth factors like BMP-2 and BMP-7 may have variable effects on MSCs differentiation especially in relation to source. Goat BM-MSCs (gBM-MSCs) osteogenesis may potentially be mediated through p38 mitogen-activated protein kinase (MAPK) pathway and in the goat AD-MSCs p44/42 MAPK pathway might be involved (Elkhenany et al. 2016). OGP may favor cBM-MSCs osteogenic differentiation through endogenous nitric oxide synthases (eNOS) (Zhu et al. 2008). gAD-MSCs osteogenesis under the influence of 1,25-dihydroxyvitamin-D3 may be mediated through polyamine metabolism (Tjabringa et al. 2008). Thus, specific growth factors may promote MSCs osteogenesis and the involved mechanism may vary.

#### 5.3.2.2 Scaffold and MSCs Osteogenesis

Osteogenic differentiation potential of goat, sheep and dog BM-MSCs and horse AD-MSCs had been demonstrated to enhance in the presence of various scaffolds like chitosan, chitosan and poly ε-caprolactone, demineralized bone matrix, hydroxyapatite and hyaluronic acid, methacrylate-endcapped caprolactone, nano-bioactive glass ceramic coated PLLA, oxidized graphene and porous tantalum, silica-coated bioactive ceramic, starch and poly-caprolactone, zinc silicate mineral (Fischer et al. 2004; Nair et al. 2009a, b, Ivirico et al. 2009; García Cruz et al. 2010; Hwang et al. 2011; Pereira-Junior et al. 2013; Rodrigues et al. 2014; Elkhenany et al. 2015; Lee et al. 2015a, b; Fang et al. 2017; Bwalya et al. 2017; Das et al. 2017; Jo et al. 2017; Mahdavia et al. 2017; Trindade et al. 2017; Bageshlooyafshar et al. 2019; Wei et al. 2019).

sBM-MSCs had differentiated into osteoblasts and chondrocytes when placed on porous calcium polyphosphate along with tri-iodothyronine (T3). The assembly had led to the formation of cartilaginous structure at the top and osteogenic tissue at bottom near osteo-conductive calcium polyphosphate (Lee et al. 2015a, b). The cells may remain aligned or scattered depending upon the scaffold type. gMSCs cultured on tricalcium phosphate and hydroxyapatite had undergone osteogenic differentiation; however, the cellular interconnection and alignment were maintained in former scaffold and scattered in the latter (Prins et al. 2016). Scaffold or implant biodegradability needs special emphasis as degraded products tend to alter local pH and thereby MSCs viability (Liu 2011; Johnson et al. 2012). Completely demineralized bone matrix may act as more favorable osteogenic agent for MSCs as compared to partial or intact bone matrix (Jo et al. 2017). Comparing hydroxyapatite and silica-coated bioactive ceramic scaffolds gBM-MSCs osteogenic potential was more enhanced under the latter scaffold (Nair et al. 2009a, b). Scaffold matrices like bovine teeth hydroxyapatite had been demonstrated to be biocompatible to cBM-MSCs with the ability to improve MSCs osteogenesis. The scaffold concentration, however, had affected the cell viability being better at 10% than at 50% and 100% (Setiawatie et al. 2017). Even cross-linked natural biological scaffolds like small intestine submucosa and goat acellular lung scaffold had favored gBM-MSCs osteogenesis (Li et al. 2006; Gupta et al. 2017).

Scaffold synthesis technique and composition have direct effect on scaffold design that in turn affects MSCs growth and differentiation. Freeze gelation and particle leaching out prepared chitosan had supported gBM-MSCs osteogenic differentiation (García Cruz et al. 2010). Scaffold hydrophilicity too affects MSCs differentiation. Higher hydration content (50%) had decreased osteogenic activity of gBM-MSCs as depicted by their compromised alkaline phosphatase activity (Ivirico et al. 2009). Dog BM-MSCs (dBM-MSCs) on nanomaterial based thin film scaffolds of carbon nanotubes (CNT), chitosan, and poly ε-caprolactone had shown slow proliferation but without decrease in overall number. Although initially cells had shown slow proliferation on CNT but that had improved as the time progressed. This depicted adaptation of dBM-MSCs (Das et al. 2017).

#### 5.3.2.3 Miscellaneous Factors and MSCs Osteogenesis

Mechanical factors may have direct effect upon MSCs differentiation (Versari et al. 2007). Pulsatile fluid flow and extracorporeal shock wave therapy had favored gMSCs osteogenic differentiation (Knippenberg et al. 2005; Tan et al. 2017). This might probably have been through adenosine release regulation and A2B receptor activation (Tan et al. 2017). Thus, the cells may show better clinical response in the case of a specific mechanical force.

Some of the ailments and aging may also affect MSCs osteogenic potential. BM-MSCs in an osteoporotic goat had shown restricted osteogenesis and had been compensated with the addition of beta tricalcium phosphate (β-TCP) (Cao et al. 2012). Platelet plasma although promotes gBM-MSCs proliferation but may inhibit their osteogenesis (Cheng et al. 2007). An age related reduction in the proliferation (40% reduction) and osteogenic potential of dBM-MSCs had been demonstrated (Volk et al. 2012).

### 5.3.3 Adipogenesis of MSCs

Adipogenesis is achieved with commercial adipogenic differentiation media preparations. The culture ingredients that favor MSCs adipogenesis include indomethacine, insulin, 3-isobutyl-1-methylxanthine, dexamethasone (Jäger et al. 2006; Martinello et al. 2011; Oh et al. 2011; Kalaszczynska et al. 2013; Bottagisio et al. 2015; Hu et al. 2015a, b). One of the studies had been unsuccessful to obtain optimal cBM-MSCs adipogenesis but could be achieved with the addition of insulin (Lee et al. 2015a, b). These differentiated cells express peroxisome proliferator-activated receptor gamma (PPAR-γ, preadipocytic commitment) and lipoprotein lipase (LPL) genes in the early adipogenic differentiation stage (Jiang et al. 2002; Martinello et al. 2011; Oh et al. 2011).

Various conventional chemicals/drugs have been demonstrated to affect MSCs adipogenesis. Among notable ones are telmisartan (antihypertensive drug), capsaicin, and epigallocatechin gallate. The former drug had promoted cBM-MSCs (Ramirez-Espinosa et al. 2015), while latter two components had inhibited such a differentiation (Jeong et al. 2014; Jeong et al. 2015). The decreased adipogenic expression may occur by decreased expression of PPAR-γ, cytosine-cytosine-adenosine-adenosine thymidine/enhancer binding protein alpha, fatty acid binding protein-4, and stearoyl CoA desaturase (Jeong et al. 2014; Jeong et al. 2015). In clinical endometritis adipogenesis of cMSCs may also reduce (Lara et al. 2017).

## 5.4 Extended Differentiation Potential of MSCs

MSCs in addition to tri-lineage differentiation show extended differentiation potential as depicted by their ability to express pluripotency markers.

### 5.4.1 Neural Cell like Differentiation

MSCs neural cell like trans-differentiation has been achieved by various factors, although the actual differentiation is questioned. Cattle umbilical cord MSCs (cUC-MSCs) had differentiated into the neural like cells under low glucose Dulbecco's modified Eagle's medium (L-DMEM), fetal bovine serum (FBS), and basal eagle medium (BME) (Xiong et al. 2014). Neural induction medium for dog AD-MSCs included FBS, dimethyl sulfoxide, butylated hydroxyanisole, hydrocortisone, insulin, 3-isobutyl-1-metylxanthine, and adenosine 3′5- cyclic monophosphate sodium salt monohydrate (Oh et al. 2011). Likewise, other specific medium had promoted neurogenesis in cattle Wharton's jelly (cWJ-MSCs). Neural differentiation was characterized by their ability to express neuronal/glial markers like neurofilament 200 (N200), neuronal microtubule associated protein 2 (MAP2), neurotrophin 3 (NT3), Tau, glial fibrillary acidic protein (GFAP) (Cardoso et al. 2012). Neural like cell trans-differentiation of gAD-MSCs had also been achieved with BIX-01294, a specific inhibitor of methlytransferase G9a. gAD-MSCs trans-differentiation achieved with methlytransferase inhibitor relies on Nanog regulatory network (Wang et al. 2018). The neural cell like differentiation of bufAm-MSCs had been promoted by glycogen synthase kinase 3 inhibitor (kenpaullone) (Deng et al. 2018). The neurogenic markers are variably expressed at different time intervals. Nestin and Map-2 expression mostly occurs at 24 hr. culture period while tropomyosin receptor kinase A (TRKA) and PRion Protein Cellular ($PrP^c$) expression is seen at 144 hrs (Dueñas et al. 2014). Therefore, at different time intervals MSCs neural expression profile is variable making it important to study their differentiation as per the developmental phase of an individual.

### 5.4.2 Differentiation into Islet Like Cells

Bovine umbilical cord MSCs (cUC-MSCs) differentiate into the islet like cells upon supplementation with media like H-DMEM, HGF, mercaptoethanol, and niacinamide. The trans-differentiation had been demonstrated through pancreatic and duodenal homeobox 1 (PDX-1) and insulin secretion (Xiong et al. 2014). Similarly, a culture medium containing nicotinamide, extendin-4, glucose, and poly-D-lysine had promoted cattle placental MSCs islet like cell differentiation. The differentiated cells had exhibited special characteristics including zinc ion staining, secretion of insulin, glucagon, paired box protein (Pax-4), Pax-6, homeobox protein Nkx6.1 and Forkhead box protein (Fox) (Peng et al. 2017). During MSCs islet-like cell differentiation, glucose promotes growth of pancreatic β-cells through phosphatidylinositol 3-kinase pathway (Wu et al. 1997). During islet like cells trans-differentiation of MSCs poly-D-lysine acts an important ingredient. It provides an intercellular connection and support, water retention, compression resistance, cell migration, cell physiology, and finally cell differentiation (Peng et al. 2017). These in vitro studies need to be translated under in vivo applications to evaluate actual feasibility and utility.

### 5.4.3 Differentiation into Hepatocyte like Cells

cUC-MSCs tend to differentiate into the hepatocyte like cells upon culturing in L-DMEM, FBS, hepatocyte growth factor (HGF), fibroblast growth factor 4 (FGF-4), dexamethasone, and ITS (Xiong et al. 2014). FGF-4 acts as an important growth and development factor for hepatocytes, while HGF increases their mitosis and resists their apoptosis (Jiang et al. 2002). cBM-MSCs upon culture supplementation with α-fetoprotein, albumin, α1-antitrypsin, connexin 32, tyrosine aminotransferase, and cytochrome P450 had differentiated to hepatogenic lineage. MSCs hepatogenesis was characterized on the basis of their significant increase in the expression of genes like cytochrome P34A4 (CYP3A4), ALB, α1-AT, CNX32, and TAT. CYP3A4 mRNA expression was higher as early as day 7 up to day 28 (Scott and Halpert 2005; Dueñas et al. 2014). α1AT and TAT, late markers of the hepatocyte lineage, play the role of tyrosine degradation and inhibit proteases (Rehman et al. 2004; Jedicke et al. 2014). CNX32 accounts for majority of liver connexin proteins (90%) (Nakashima et al. 2004) whose expression occurs at the later stage of differentiation. It points out that MSCs must be exposed to hepatogenic factors for longer time to trigger them towards the hepatic line.

### 5.4.4 Differentiation into Myocyte like Cells

Myogenic differentiation of bovine fetal BM-MSCs has been achieved with various protocols like incorporation of 5-Aza-2′-deoxycytidine (5-Aza), myoblast-secreted factor Galectin-1 (Gal-1), and/or myoblast culture medium SkGM-2 BulletKit (Oh et al. 2011; Okamura et al. 2017). Variable concentration of 5-Aza has bearing on cell viability. Higher concentration (100 μM) may increase cellular differentiation but may be more toxic to cells. It had been concluded that cBM-MSCs were myogenically differentiated to some extent (Okamura et al. 2017). Similarly, bezafibrate had promoted myogenesis of cBM-MSCs (Ramirez-Espinosa et al. 2015). Muscle derived cells (MDCs) had differentiated gBM-MSCs through their fusion producing myotubes (Kulesza et al. 2016). Sheep fetal derived BM-MSCs supplemented with reversine and 5-azacytidine had differentiated into cardiomyocyte like cells (Soltani et al. 2016). Arterial matrix supplementation to culture had directed sheep AD-MSCs differentiation towards endothelial like cells (Zhang et al. 2016).

### 5.4.5 Differentiation into Germ Cell like Cells

In vitro cultured sheep MSCs and bovine MSCs under the influence of TGF-β1, BMP-4, BMP8b, and retinoic acid transdifferentiate into the germ cells (Ghasemzadeh-Hasankolaei et al. 2014; Ghasemzadeh-Hasankolaei et al. 2015). Dog adipose tissue and ovarian derived MSCs were transdifferentiated into primordial germ cell like cells and male germ cell like cells under the influence of insulin, epidermal growth factor (EGF), glial derived nerve factor (GDNF), basic fibroblast growth factor (bFGF), and BMP-4 (Wei et al. 2016; Trindade et al. 2017).

TGF-β1 supplementation had differentiated sheep BM-MSC to spermatogonia stem cell like cells and spermatogonia-like cells. These cells were differentiated to primordial germ cell like cells through BMP-4 and BMP-8 supplementation (Ghasemzadeh-Hasankolaei et al. 2014). Zinc supplementation had changed gene expression of sheep BM-MSCs approximating germinal cell genes but had failed to promote germinal cell like trans-differentiation (Ghasemzadeh-Hasankolaei et al. 2012). Under the influence of above mentioned factors cBM-MSCs had expressed higher levels of pluripotency markers and deleted in azoospermia-like (Dazl) but failed to transcribe mRNA of germinal cell genes like Vasa, Fragilis, Stella male germinal cell genes (Stra8 and Piwil2). TGF-β1 and BMP-4 had transiently increased expression of Nanog and Dazl, while retinoic acid had effectively promoted Dazl expression. Such expression pattern had signified early germ cell like differentiation of cBM-MSCs (Cortez et al. 2018). gBM-MSCs had been trans-differentiated into the germ cell-like cells exhibited by expression of both early as well as late stage germ cell markers. Even limited number of differentiated cells had a potential to enter meiosis (Zhang et al. 2019). Dazl gene ectopic expression in gBM-MSCs had improved their ability to transdifferentiate to the putative germ cells as compared to the effect of retinoic acid which in turn had better effect than BMP-4 (Yan et al. 2015).

Apart from ectopic expression of germ cell like cells, follicular fluid may also favor MSCs germinal cell trans-differentiation. Higher concentration (20%) of follicular fluid had been able to promote gUC-MSCs differentiation towards oocyte like cells (Qiu et al. 2012). However, these transdifferentiated cells actual identity and utility can be evaluated under in vivo studies.

Goat mammary fat pad adipose stem cells had been transdifferentiated towards epithelial lineage with an initial cocktail of insulin, hydrocortisone, and epidermal growth factor and subsequently with keratinocyte growth factor (Reza et al. 2014). All these studies are promising but in infancy and need extensive studies before any conclusion can be drawn.

## 5.5 Conclusion(s)

MSCs characteristic features are controlled by the available microenvironment/niche. An extensive work on MSCs differentiation has been undertaken. These cells may undergo differentiation towards particular lineage based on the available cues in the form of available matrix, cells, humoral/growth factors, and mechanical factors. Most of the studies in available literature have mainly focused on individual cues, while in vivo cues are diverse and can act simultaneously leading to diverse results. Variable mechanisms have been demonstrated in differentiation of MSCs and have varied even on the basis of cell source. It is important to understand MSCs differentiation with respect to the developmental stages. Commonly used antihypertensive drugs may have effect on adipogenicity of MSCs. Effect of these drugs on the possible obesity in such individuals may be evaluated. A little evidence on hepatocyte, islet like cell, neural and germinal differentiation could be seen though preliminary studies support such cell type differentiations of MSCs. Thus, further studies are desired to be conducted in order to understand the particular lineage differentiation. Besides, multiple factorial studies may be conducted to understand combined effect of these factors on MSCs expression.

## References

Arora A, Sriram M, Kothari A, Katti DS (2017) Co-culture of infrapatellar fat pad–derived mesenchymal stromal cells and articular chondrocytes in plasma clot for cartilage tissue engineering. Cytotherapy 19(7):881–894

Atala A, Lanza RP (2001) Preface. In: Atala A, Lanza RP (eds) Methods of tissue engineering. Academic Press, San Diego

Bageshlooyafshar B, Isangcheshmeh AM, Vakilian S et al (2019) Zinc silicate mineral-coated scaffold improved in vitro osteogenic differentiation of equine adipose-derived mesenchymal stem cells. Res Vet Sci 124:444–451

Bertolo A, Steffen F, Malonzo-Marty C, Stoyanov J (2015) Canine mesenchymal stem cell potential and the importance of dog breed: implication for cell-based therapies. Cell Transplant 24(10):1969–1980

Bornes TD, Jomha NM, Mulet-Sierra A, Adesida AB (2015) Hypoxic culture of bone marrow-derived mesenchymal stromal stem cells differentially enhances in vitro chondrogenesis within cell-seeded collagen and hyaluronic acid porous scaffolds. Stem Cell Res Ther 6:84

Bornes TD, Jomha NM, Mulet-Sierra A, Adesida AB (2016) Optimal seeding densities for *In Vitro* Chondrogenesis of two- and three-dimensional-isolated and -expanded bone marrow-derived Mesenchymal stromal stem cells within a porous collagen scaffold. Tissue Eng Part C Methods 22(3):208–220

Bosnakovski D, Mizuno M, Kim G, Ishiguro T, Okumura M, Iwanaga T, Kadosawa T, Fujinaga T (2004) (2004) Chondrogenic differentiation of bovine bone marrow mesenchymal stem cells in pellet cultural system. Exp Hematol 32(5):502–509

Bosnakovski D, Mizuno M, Kim G, Takagi S, Okumura M, Fujinaga T (2005) Isolation and multilineage differentiation of bovine bone marrow mesenchymal stem cells. Cell Tissue Research 319(2):243–253

Bosnakovski D, Mizuno M, Kim G, Takagi S, Okumura M, Fujinaga T (2006) Chondrogenic differentiation of bovine bone marrow mesenchymal stem cells (MSCs) in different hydrogels: influence of collagen type II extracellular matrix on MSC chondrogenesis. Biotechnol Bioeng 93(6):1152–1163

Bottagisio M, Lovati AB, Lopa S, Moretti M (2015) Osteogenic differentiation of human and ovine bone marrow stromal cells in response to b-Glycerophosphate and monosodium phosphate. Cell Reprogram 17(4):235–242

Bwalya EC, Kim S, Fang J, Wijekoon HMS, Hosoya K, Okumura M (2017) Effects of pentosan polysulfate and polysulfated glycosaminoglycan on chondrogenesis of canine bone marrow-derived mesenchymal stem cells in alginate and micromass culture. J Vet Med Sci 79(7):1182–1190

Cao L, Liu G, Gan Y, Fan Q, Yang F, Zhang X, Tang T, Dai K (2012) The use of autologous enriched bone marrow MSCs to enhance osteoporotic bone defect repair in long-term estrogen deficient goats. Biomaterials 33(20):5076–5084

Cardoso TC, Novais JB, Antello TF, Silva-Frade C, Ferrarezi MC, Ferrari HF, Gameiro R, Flores EF (2012) Susceptibility of neuron-like cells derived from bovine Wharton's jelly to bovine herpesvirus type 5 infections. BMC Veterinary Research 10(8):242

Cheng W, Jin D, Zhao Y (2007) Effect of platelet-rich plasma on proliferation and osteogenic differentiation of bone marrow stem cells in China goats. Zhongguo Xiu Fu Chong Jian Wai Ke Za Zhi 21(4):386–389

Colosimo A, Russo V, Mauro A, Curini V, Marchisio M, Bernabò N, Alfonsi M, Mattioli M, Barboni B (2013) Prolonged in vitro expansion partially affects phenotypic features and osteogenic potential of ovine amniotic fluid-derived mesenchymal stromal cells. Cytotherapy 15:930–950
Cortez J, Bahamonde J, De los Reyes M, Palomino J, Torres CG, Peralta OA (2018) In vitro differentiation of bovine bone marrow-derived mesenchymal stem cells into male germ cells by exposure to exogenous bioactive factors. Reprod Domest Anim 53(3):700–709
Das K, Mili B, Madhusoodan AP, Saxena AC, Kumar A, Singh P, Vermae MR, Sarkara M, Bag S (2017) Proliferation of canine bone marrow derived mesenchymal stem cells on different nanomaterial based thin film scaffolds. Tissue Cell 49(2):270–274
Deng Y, Huang G, Zou L, Nong T, Yang X, Cui J, Wei Y, Yang S, Shi D (2018) Isolation and characterization of buffalo (*bubalus bubalis*) amniotic mesenchymal stem cells derived from amnion from the first trimester pregnancy. J Vet Med Sci 80(4):710–719
Desancé M, Contentin R, Bertoni L, Gomez-Leduc T, Branly T, Jacquet S, Betsch J-M, Batho A, Legendre F, Audigié F, Galéra P, Demoor M (2018) Chondrogenic differentiation of defined equine Mesenchymal stem cells derived from umbilical cord blood for use in cartilage repair therapy. Int J Mol Sci 19:537
Dueñas F, Becerra V, Cortes Y, Vidal S, Sáenz L, Palomino J, De Los Reyes M, Peralta OA (2014) Hepatogenic and neurogenic differentiation of bone marrow mesenchymal stem cells from abattoir-derived bovine fetuses. BMC Vet Res 10:154
Elkhenany H, Amelse L, Lafont A, Bourdo S, Caldwell M, Neilsen N, Dervishi E, Derek O, Biris AS, Anderson D, Dhar M (2015) Graphene supports in vitro proliferation and osteogenic differentiation of goat adult mesenchymal stem cells: potential for bone tissue engineering. J Appl Toxicol 35(4):367–374
Elkhenany H, Amelse L, Caldwell M, Abdelwahed R, Dhar M (2016) Impact of the source and serial passaging of goat mesenchymal stem cells on osteogenic differentiation potential: implications for bone tissue engineering. J Anim Sci Biotechnol 7:16
Erickson IE, van Veen SC, Sengupta S, Kestle SR, Mauck RL (2011) Cartilage matrix formation by bovine Mesenchymal stem cells in three-dimensional culture is age-dependent. Clin Orthop Relat Res 469:2744–2753
Fang J, Wei Y, Teng X, Zhao S, Hua J (2017) Immortalization of canine adipose-derived mesenchymal stem cells and their seminiferous tubule transplantation. J Cell Biochem 119(4):3663–3670
Fischer EM, Layrolle P, Van Blitterswijk CA, De Bruijn JD (2004) Bone formation by Mesenchymal progenitor cells cultured on dense and microporous hydroxyapatite particles. Tissue Eng 9(6):1179–1188
Fisher MB, Henning EA, Söegaard N, Bostrom M, Esterhai JL, Mauck RL (2015) Engineering meniscus structure and function via multi-layered Mesenchymal stem cell-seeded Nanofibrous scaffolds. J Biomech 48(8):1412–1419
Fulber J, Maria DA, Silva LCLCD, Massoco CO, Agreste F, Baccarin RYA (2016) Comparative study of equine mesenchymal stem cells from healthy and injured synovial tissues: an in vitro assessment. Stem Cell Res Ther 7(1):35
García Cruz DM, Gomes M, Reis RL, Moratal D, Salmerón-Sánchez M, Gómez Ribelles JL, Mano JF (2010) Differentiation of mesenchymal stem cells in chitosan scaffolds with double micro and macroporosity. J Biomed Mat Res Part A. 95A:1182–1193
Ghasemzadeh-Hasankolaei M, Sedighi-Gilani MA, Eslaminejad MB (2014) Induction of ram bone marrow Mesenchymal stem cells into germ cell lineage using transforming growth factor-b superfamily growth factors. Reprod Dom Anim 49:588–598
Ghasemzadeh-Hasankolaei M, Eslaminejad BM, Sedighi-Gilani M (2015) Derivation of male germ cells from ram bone marrow mesenchymal stem cells by three different methods and evaluation of their fate after transplantation into the testis. In Vitro Cell Dev Biol Anim 52(1):49–61
Ghasemzadeh-Hasankolai M, Batavani R, Eslaminejad BM, Sedighi-Gilani M (2012) Effect of zinc ions on differentiation of bone marrow-derived Mesenchymal stem cells to male germ cells and some germ cell-specific gene expression in rams. Biol Trace Elem Res 150:137–146

Goldman SM, Barabino GA (2016) Hydrodynamic loading in concomitance with exogenous cytokine stimulation modulates differentiation of bovine mesenchymal stem cells towards osteochondral lineages. BMC Biotechnol 16:10
Gregor A, Filová E, Novák M et al (2017) Designing of PLA scaffolds for bone tissue replacement fabricated by ordinary commercial 3D printer. J Biol Eng 11(1):31
Gugjoo MB, Amarpal (2018) Mesenchymal stem cell research in sheep: current status and future prospects. Small Rum Res 169:46–56
Gugjoo MB, Amarpal, Sharma GT, Kinjavdekar P, Aithal HP, Pawde AM (2016) Cartilage tissue engineering: role of mesenchymal stem cells along with growth factors and scaffolds. Indian J Med Res 144(3):339–347
Gugjoo MB, Amarpal, Abdelbaset-Ismail A, Aithal HP, Kinjavdekar P, Pawde AM, SaiKumar G, Sharma GT (2017) Mesenchymal stem cells with IGF-1 and TGF- β1 in laminin gel for osteochondral defects in rabbits. Biomed Pharmacother 93:1165–1174
Gugjoo MB, Amarpal, Chandra V, Wani MY, Dhama K, Sharma GT (2018) Mesenchymal stem cell research in veterinary medicine. Curr Stem Cell Res Ther 13(8):645–657
Gugjoo MB, Amarpal, Makhdoomi DM, Sharma GT (2019a) Equine mesenchymal stem cells: properties, sources, characterization and potential therapeutic applications. J Equine Vet Sci 72:16–27
Gugjoo MB, Amarpal, Fazili MR, Shah RA, Sharma GT (2019b) Mesenchymal stem cell: basic research and potential applications in cattle and buffalo. J Cell Physiol 234(6):8618–8635
Gugjoo MB, Amarpal, Sharma GT (2019c) Mesenchymal stem cell basic research and applications in dog medicine. J Cell Physiol 234(10):16779–16811
Gugjoo MB, Amarpal, Fazili MR, Gayas MA, Ahmad RA, Dhama K (2019d) Animal mesenchymal stem cell research in cartilage regenerative medicine – a review. Vet Q 39(1):95–120
Gugjoo MB, Amarpal AAI, Aithal HP, Kinjavdekar P, Saikumar GS, Sharma GT (2020) Allogeneic Mesenchymal stem cells and growth factors in gel scaffold repair Osteochondral defect in rabbit. Regen Med 14. https://doi.org/10.2217/rme-2018-0138
Gupta SK, Kumar R, Mishra NC (2017) Influence of quercetin and nanohydroxyapatite modifications of decellularized goat-lung scaffold for bone regeneration. Mat Sci Eng C 71:919–928
Hattori T, Muller C, Gebhard S, Bauer E, Pausch F, Schlund B, Bosl MR, Hess A, Surmann-Schmitt C, von der Mark H et al (2010) SOX9 is a major negative regulator of cartilage vascularization, bone marrow formation and endochondral ossification. Development 137(6):901–911
Hegewald AA, Ringe J, Bartel J (2004) Kr¨uger I, Notter M, Barnewitz D. Kaps C, Sittinger M. hyaluronic acid and autologous synovial fluid induce chondrogenic differentiation of equine mesenchymal stem cells: a preliminary study. Tissue Cell 36:431–438
Heo S-J, Thorpe SD, Driscoll TP, Duncan RL, Lee DA, Mauck RL (2015) Biophysical regulation of chromatin architecture instills a mechanical memory in Mesenchymal stem cells. Sci Rep 5:16895
Hu P, Pu Y, Li X, Zhu Z, Zhao Y, Guan W, Ma Y (2015a) Isolation, *in vitro* culture and identification of a new type of mesenchymal stem cell derived from fetal bovine lung tissues. Mol Med Rep 12:3331–3338
Hu Y, Zhang Q, Zhang L, Tang X-X, He H-Y (2015b) Basic fibroblast growth factor lentiviral vector-transfected sheep bone marrow mesenchymal stem cells and non-specific osteogenic gene expression. Mol Med Rep 12:267–272
Huang Y-Z, Cai J-Q, Lv F-J, Xie H-L, Yang Z-M, Huang Y-C, Deng L (2013) Species variation in the spontaneous calcification of bone marrow-derived mesenchymal stem cells. Cytotherapy 15:323e329
Hwang NS, Varghese S, Li H, Elisseeff J (2011) Regulation of osteogenic and chondrogenic differentiation of mesenchymal stem cells in PEG-ECM hydrogels. Cell Tissue Res 344:499–509
Ivirico JL, Salmerón-Sánchez M, Ribelles JL, Pradas MM, Soria JM, Gomes ME, Reis RL, Mano JF (2009) Proliferation and differentiation of goat bone marrow stromal cells in 3D scaffolds with tunable hydrophilicity. J Biomed Mat Res B Appl Biomater 91(1):277–286
Jäger M, Bachmann R, Scharfstädt A, Krauspe R (2006) Ovine cord blood accommodates multipotent Mesenchymal progenitor cells. In Vivo 20:205–214

Jedicke N, Struever N, Aggrawal N, Welte T, Manns MP, Malek NP, Zender L, Janciauskiene S, Wuestefeld T (2014) Alpha-1-antitrypsin inhibits acute liver failure in mice. Hepatology 59:2299–2308

Jeong JY, Suresh S, Park MN, Jang M, Park S, Gobianand K, You S, Yeon S-H, Lee H-J (2014) Effects of capsaicin on adipogenic differentiation of bovine bone marrow Mesenchymal stem cell. Asian Australas J Anim Sci 27(12):1783–1793

Jeong JY, Suresh S, Jang M, Park MN, Gobianand K, Seungkwon Y, Yeon S-H, Lee H-J (2015) Epigallocatechin-3-gallate suppresses the lipid deposition through the apoptosis during differentiation in bovine bone marrow mesenchymal stem cells. Cell Biol Int 39:52–64

Jiang Y, Jahagirdar BN, Reinhardt RL, Schwartz RE, Keene CD, OrtizGonzalez XR, Reyes M, Lenvik T, Lund T, Blackstad M (2002) Pluripotency of mesenchymal stem cells derived from adult marrow. Nature 418:41–49

Jo K, Kim Y, Lee SH, Yoon YS, Kim WH, Kweon O-K (2017) Effect of canine cortical bone demineralization on osteogenic differentiation of adipose- derived mesenchymal stromal cells. Heliyon 3(8):e00383

Johnson I, Perchy D, Liu H (2012) In vitro evaluation of the surface effects on magnesium-yttrium alloy degradation and mesenchymal stem cell adhesion. J Biomed Mat Res Part A 100A:477–485

Kalaszczynska I, Ruminski S, Platek AE, Bissenik I, Zakrzewski P, Noszczyk M, Lewandowska-Szumiel M (2013) Substantial differences between human and ovine mesenchymal stem cells in response to osteogenic media: how to explain and how to manage? Biores Open Access 2(5):356–363

Katja N, Michaela E, Jochen R, Bernd F, Rudi M, Thomas H et al (2007) BMP7 promotes adipogenic but not osteo−/chondrogenic differentiation of adult human bone marrow-derived stem cells in high-density micro-mass culture. J Cell Biochem 102:3,626–3,637

Knippenberg M, Helder M, Doulabi BZ, Semeins CM, Wuisman PIJM, Klein-Nulend J (2005) Adipose tissue-derived Mesenchymal stem cells acquire bone cell-like responsiveness to fluid shear stress on Osteogenic stimulation. Tissue Eng 11(11/12):1780–1787

Knippenberg M, Helder M, Doulabi BZ, Semeins CM, Wuisman PIJM, Klein-Nulend J (2006) Osteogenesis versus chondrogenesis by BMP-2 and BMP-7 in adipose stem cells. Biochem Biophys Res Commun 342:902–908

Kol A, Walker NJ, Galuppo LD, Clark KC, Buerchler S, Bernanke A, Borjesson DL (2012) Autologous point-of-care cellular therapies variably induce equine mesenchymal stem cell migration, proliferation and cytokine expression. Equine Vet J 45(2):193–198

Kulesza A, Burdzinska A, Szczepanska I, Zarychta-Wisniewska W, Pajak B, Bojarczuk K, Dybowski B, Paczek L (2016) The mutual interactions between Mesenchymal stem cells and myoblasts in an autologous co-culture model. PLoS One 11(8):e0161693

Lanza R, Langer R, Vacanti JP (2011) Principles of tissue engineering. Academic Press, Cambridge

Lara E, Velásquez A, Cabezas J, Rivera N, Pacha P, Rodríguez-Alvarez L, Saravia F, Castro FO (2017) Endometritis and in vitro PGE2 challenge modify properties of cattle endometrial Mesenchymal stem cells and their Transcriptomic profile. Stem Cells Int 2017:4297639

Lee SH, Cha S-H, Kim C-L, Lillehoj HS, Song J-Y, Lee K-W (2015a) Enhanced adipogenic differentiation of bovine bone marrow derived mesenchymal stem cells. J Appl Anim Res 43(1):15–21

Lee WD, Hurtig MB, Pilliar RM, Stanford WL, Kandel RA (2015b) Engineering of hyaline cartilage with a calcified zone using bone marrow stromal cells. Osteoarthr Cartil 23(8):1307–1315

Leung VYL, Gao B, Leung KKH, Melhado IG, Wynn SL, Au TYK, Dung NWF, Lau JYB, Mak ACY, Chan D et al (2011) SOX9 governs differentiation stage-specific gene expression in growth plate chondrocytes via direct concomitant transactivation and repression. PLoS Genet 7(11):e1002356

Li H, Ji T, Xu LQ, Hu YJ, Li SY, Zhang CP (2006) Study on the osteogenesis ability of co-culturing bone marrow stromal cells (BMSCs) and small intestinal submucosa. Shanghai Kou Qiang Yi Xue 15(2):167–171

Liu H (2011) The effects of surface and biomolecules on magnesium degradation and mesenchymal stem cell adhesion. J Biomed Mat Res Part A 99A:249–260

Mahdavia FS, Salehia A, Seyedjafarib E, Mohammadi-Sangcheshmeha A, Ardeshirylajimi A (2017) Bioactive glass ceramic nanoparticles-coated poly(L-lactic acid) scaffold improved osteogenic differentiation of adipose stem cells in equine. Tissue Cell 49(5):565–572

Martinello T, Bronzini I, Maccatrozzo L, Mollo A, Sampaolesi M, Mascarello F, Decaminada M, Patruno M (2011) Canine adipose-derived-mesenchymal stem cells do not lose stem features after a long-term cryopreservation. Res Vet Sci 91(1):18–24

Mauck RL, Yuan X, Tuan RS (2006) Chondrogenic differentiation and functional maturation of bovine mesenchymal stem cells in long-term agarose culture. Osteoarthr Cartil 14(2):179–189

McCorry MC, Bonassar LJ (2017) Fiber development and matrix production in tissue engineered menisci using bovine Mesenchymal stem cells and Fibrochondrocytes. Connect Tissue Res 58(3–4):329–341

McCorry MC, Puetzer JL, Bonassar LJ (2016) Characterization of mesenchymal stem cells and fibrochondrocytes in three-dimensional co-culture: analysis of cell shape, matrix production, and mechanical performance. Stem Cell Res Ther 12(7):39

Meyer U, Meyer T, Handschel J, Wiesmann HP (2009) Fundamentals of tissue engineering and regenerative medicine. Springer, Berlin

Nair MB, Varma HK, John A (2009a) Triphasic ceramic coated hydroxyapatite as a niche for goat stem cell-derived osteoblasts for bone regeneration and repair. J Mat Sci Mat Med 20:S251–S258

Nair MB, Varma HK, Menon KV, Shenoy SJ, John A (2009b) Tissue regeneration and repair of goat segmental femur defect with bioactive triphasic ceramic-coated hydroxyapatite scaffold. Biomed Mat Res 91A:855–865

Nakashima Y, Ono T, Yamanoi A, El-Assal ON, Kohno H, Nagasue N (2004) Expression of gap junction protein connexin32 in chronic hepatitis, liver cirrhosis, and hepatocellular carcinoma. J Gastroenterol 39:763–768

Oh HJ, Park JE, Kim MJ, Hong SG, Ra JC, Jo JY et al (2011) Recloned dogs derived from adipose stem cells of a transgenic cloned beagle. Theriogenology 75(7):1221–1231

Okamura LH, Cordero P, Palomino J, Parraguez VH, Torres CG, Peralta OA (2017) Myogenic differentiation potential of Mesenchymal stem cells derived from fetal bovine bone marrow. Anim Biotechnol 29(1):1–11

Ongaro A, Pellati A, Setti S, Masieri FF, Aquila G, Fini M, Caruso A, De Mattei M (2012) Electromagnetic fields counteract IL-1b activity during chondrogenesis of bovine mesenchymal stem cells. J Tissue Eng Regen Med 9(12):E229–E238

Park Y, Sugimoto M, Watrin A, Chiquet M, Hunziker EB (2005) BMP-2 induces the expression of chondrocyte-specific genes bovine synovium-derived progenitor cells cultured in three-dimensional alginate hydrogel. Osteoarthr Cartil 13:527e536

Peng S-Y, Chou C-W, Kuo Y-H, Shen P-C, Steven Shaw S-W (2017) Potential differentiation of islet like cells from pregnant cow derived placental stem cells. Taiwan J Obstet Gynaecol 56:306–311

Pereira-Junior OCM, RahalII SCJ, Lima-Neto F, da Cruz Landim-Alvarenga F, Monteiro FOB (2013) In vitro evaluation of three different biomaterials as scaffolds for canine mesenchymal stem cells. Acta Cir Bras 28(5):353–360

Place ES, Evans ND, Stevens MM (2009) Complexity in biomaterials for tissue engineering. Nat Mater 8:457–470

Prins H-J, Fernandes H, Rozemuller H, van Blitterswijk C, de Boer J, Martens ACM (2016) Spatial distribution and survival of human and goat mesenchymal stromal cells on hydroxyapatite and β-tricalcium phosphate. J Tissue Eng Reg Med 10:233–244

Qiu P, Bai Y, Liu C, He X, Cao H, Li M, Zhu H, Hua J (2012) A dose-dependent function of follicular fluid on the proliferation and differentiation of umbilical cord mesenchymal stem cells (MSCs) of goat. Histochem Cell Biol 138:593–603

Rackwitz L, Djouad F, Janjanin S, Nöth U, Tuan RS (2014) Functional cartilage repair capacity of de-differentiated, chondrocyte- and mesenchymal stem cell-laden hydrogels in vitro. Osteoarthr Cartil 22(8):1148–1157

Ramirez-Espinosa JJ, Gonzalez-Davalos L, Shimada A, Pina E, Varela-Echavarria A, Mora O (2015) Bovine *(Bos taurus)* bone marrow mesenchymal cell differentiation to Adipogenic and myogenic lineages. Cells Tissues Organs 201:51–64

Randau TM, Schildberg FA, Alini M, Wimmer MD, Haddouti E-M, Gravius S, Ito KJ, Stoddart M (2013) The effect of dexamethasone and Triiodothyronine on terminal differentiation of primary bovine chondrocytes and Chondrogenically differentiated Mesenchymal stem cells. PLoS One 8(8):e72973

Rehman KK, Ayesha Q, Khan AA, Ahmed N, Habibullah CM (2004) Tyrosine aminotransferase and gamma-glutamyl transferase activity in human fetal hepatocyte primary cultures under proliferative conditions. Cell Biochem Funct 22:89–96

Reza AMMT, Shiwani S, Singh NK, Lohakare JD, Lee SJ, Jeong DK, Han JY, Rengaraj D, Lee BW (2014) Keratinocyte growth factor and thiazolidinediones and linolenic acid differentiate characterized mammary fat pad adipose stem cells isolated from prepubertal Korean black goat to epithelial and adipogenic lineage. In Vitro Cell Dev Biol–Anim 50:194–206

Rodrigues MT, Leonor IB, Gröen N, Viegas CA, Dias IR, Caridade SG, Mano JF, Gomes ME, Reis RL (2014) Bone marrow stromal cells on a 3D bioactive fiber mesh undergo osteogenic differentiation in the absence of osteogenic media supplements: the effect of silanol groups. Acta Biomater 10(10):4175–4185

Santos VH, Pfeifer JPH, de Souza JB, Milani BHG, de Oliveira RA, Assis MG, Deffune E, Moroz A, Garcia Alves AL (2018) Culture of mesenchymal stem cells derived from equine synovial membrane in alginate hydrogel microcapsules. BMC Vet Res 14:114

Schop D, Janssen FW, Borgart E, de Bruijn JD, van Dijkhuizen-Radersma R (2008) Expansion of mesenchymal stem cells using a microcarrier-based cultivation system: growth and metabolism. J Tissue Eng Reg Med 2:126–135

Scott EE, Halpert JR (2005) Structure of cytochrome P450 3A4. Trends Biochem Sci 30:5–7

Setiawatie EM, Ulfah N, Wahjuningrum DA, Sari DS, Rubianto M (2017) Viability bovine tooth Hydroxyapatite on bone marrow Mesenchymal stem cells. In: International medical device and technology conference. Universiti Teknologi Malaysia, Johor Bahru, pp 78–81

Simonaro CM, Sachot S, Ge Y, He X, DeAngelis VA, Eliyahu E, Leong D, Sun HB, Mason JB, Haskins ME, Richardson DW, Schuchman EH (2013) Acid Ceramidase maintains the Chondrogenic phenotype of expanded primary chondrocytes and improves the Chondrogenic differentiation of bone marrow-derived Mesenchymal stem cells. PLoS One 8(4):e62715

Soltani L, Rahmani HR, Joupari MD, Ghaneialvar H, Mahdavi AH, Shamsara M (2016) Ovine fetal mesenchymal stem cell differentiation to cardiomyocytes, effects of co-culture, role of small molecules; reversine and 5-azacytidine. Cell Biochem Funct 34:250–261

Stewart MC (2011) Cell-based therapies: current issues and future directions. Vet Clin North Am Equine Pract 27(2):393–399

Tan L, Zhao B, Ge FT, Sun DH, Yu T (2017) Shockwaves inhibit chondrogenic differentiation of human mesenchymal stem cells in association with adenosine and A2B receptors. Sci Rep 7:1437

Terraciano V, Hwang N, Moroni L, Park HB, Zhang Z, Mizrahi J, Seliktar D, Elisseeff J (2007) Differential response of adult and embryonic Mesenchymal progenitor cells to mechanical compression in hydrogels. Stem Cells 25:2730–2738

Tjabringa GS, Zandieh-Doulabi B, Helder MN, Knippenberg M, Wuisman PIJM, Klein-Nulend J (2008) The polymine spermine regulates osteogenic differentiation in adipose stem cells. J Cell Mol Med 12(5A):1710–1717

Toh WS, Lim TC, Kurisawa M, Spector M (2012) Modulation of mesenchymal stem cell chondrogenesis in a tunable hyaluronic acid hydrogel microenvironment. Biomaterials 33:3835e3845

Toh WS, Foldager CB, Olsen BR, Spector M (2013) Basement membrane molecule expression attendant to Chondrogenesis by nucleus Pulposus cells and Mesenchymal stem cells. J Orthop Res 31:1136–1143

Topol L, Chen W, Song H, Day TF, Yang Y (2009) Sox9 inhibits Wnt signaling by promoting beta-catenin phosphorylation in the nucleus. J Biol Chem 284(5):3323–3333

Trindade AB, Therrien J, Garcia JM, Smith LC (2017) Mesenchymal-like stem cells in canine ovary show high differentiation potential. Cell Prolif 50(6):e12391

Varghese S, Hwang NS, Canver AC, Theprungsrikul P, Lin DW, Elisseeff J (2008) Chondroitin sulphate based niches for chondrogenic differentiation of mesenchymal stem cells. Matrix Biol 27:12–21

Versari S, Villa A, Helder MN, Doulabi BZ, van Loon J, Bradamante S (2007) Effects of gravity on proliferation and differentiation of adipose tissue-derived stem cells. J Gravit Physiol 14(1):P127–P128

Vickers SM, Gotterbarm T, Spector M (2010) Cross-linking affects cellular condensation and chondrogenesis in type II collagen-GAG scaffolds seeded with bone marrow-derived mesenchymal stem cells. J Orthop Res 28(9):1184–1192

Volk SW, Wang Y, Hankenson KD (2012) Effects of donor characteristics and ex vivo expansion on canine mesenchymal stem cell properties: implications for MSC-based therapies. Cell Transplant 21(10):2189–2200

Wang L, Gordon RA, Huynh L, Su X, Parkmin KY, Han J, Arthur JS, Kalliolias GD (2010) Ivashkiv L.B. indirect inhibition of toll-like receptor and type I interferon responses by ITAM-coupled receptors and integrins. Immunity 32:518–530

Wang M, Yuan Z, Ma N, Hao C, Guo W, Zou G, Zhang Y, Chen M, Gao S, Peng J et al (2017) Advances and prospects in stem cells for cartilage regeneration. Stem Cells Int 2017:4130607

Wang Q, Wang X, Lai D, Deng J, Hou Z, Liang H, Liu D (2018) BIX-01294 promotes the differentiation of adipose mesenchymal stem cells into adipocytes and neural cells in Arbas cashmere goats. Res Vet Sci 119:9–18

Wei Y, Fang J, Cai S, Lv C, Zhang S, Hua J (2016) Primordial germ cell-like cells derived from canine adipose mesenchymal stem cells. Cell Prolif 49(4):503–511

Wei X, Liu B, Liu G, Yang F, Cao F, Dou X (2019) Mesenchymal stem cell-loaded porous tantalum integrated with biomimetic 3D collagen-based scaffold to repair large osteochondral defects in goats. Stem Cell Res Ther 10:72

Wu KL, Gannon M, Peshavaria M, Offield MF, Henderson E, Ray M et al (1997) Hepatocyte nuclear factor 3beta is involved in pancreatic beta-cell-specific transcription of the pdx-1 gene. Mol Cell Biol 17(10):6002e13

Xiong H, Bai C, Wu S, Gao Y, Lu T, Hu Q, Guan W, Ma Y (2014) Biological characterization of Mesenchymal stem cells from bovine umbilical cord. Anim Cells Syst 1(1):59–67

Yan G, Fan Y, Li P, Zhang Y, Wang F (2015) Ectopic expression of DAZL gene in goat bone marrow-derived mesenchymal stem cells enhances the trans-differentiation to putative germ cells compared to the exogenous treatment of retinoic acid or bone morphogenetic protein 4 signalling molecules. Cell Biol Int 39(1):74–83

Yin H, Wang Y, Sun Z, Sun X, Xu Y, Li P, Meng H, Yu X, Xiao B, Fan T, Wang Y, Xu W, Wang A, Guo Q, Peng J, Lu S (2016) Induction of mesenchymal stem cell chondrogenic differentiation and functional cartilage microtissue formation for in vivo cartilage regeneration by cartilage extracellular matrix-derived particles. Acta Biomater 33:96–109

Zeiter S, Lezuo P, Ito K (2009) Effect of TGF-β1, BMP-2 and hydraulic pressure on chondrogenic differentiation of bovine bone marrow mesenchymal stromal cells. Biorheology 46:45–55

Zhang XL, Jiang XQ (2006) Osteoblastic differentiation of goat bone marrow stromal cells after AdBMP-2 mediated transduction in vitro. Shanghai Kou Qiang Yi Xue 15(6):610–613

Zhang W, Huo Y, Wang X, Jia Y, Su L, Wang C, Li Y, Yang Y, Liu Y (2016) Decellularized ovine arteries as biomatrix scaffold support endothelial of mesenchymal stem cells. Heart Vessel 31(11):1874–1881

Zhang YL, Li PZ, Pang J, Wan YJ, Zhang GM, Fan YX, Wang ZY, Tao NH, Wang F (2019) Induction of goat bone marrow mesenchymal stem cells into putative male germ cells using mRNA for STRA8, BOULE and DAZL. Cytotechnology 71(2):563–572

Zhao Y, Hu J, Buckingham B, Pittenger MF, Wu ZJ, Griffith BP (2014) Pim-1 kinase cooperates with serum signals supporting Mesenchymal stem cell propagation. Cells Tissues Organs 199:140–149

Zhou G, Zheng Q, Engin F, Munivez E, Chen Y, Sebald E, Krakow D, Lee B (2006) Dominance of SOX9 function over RUNX2 during skeletogenesis. Proc Natl Acad Sci U S A 103(50):19004–19009

Zhu HB, Guo DZ, Yang SJ, Zhang YH, Wang H, Guo HT, Zhang Y, Cheng DC (2008) Osteogenic actions of the osteogenic growth peptide on bovine marrow mesenchymal stromal cells in culture. Vet Med 53(9):501–509

Zscharnack M, Poesel C, Galle J, Bader A (2009) Low oxygen expansion improves subsequent Chondrogenesis of ovine bone-marrow-derived Mesenchymal stem cells in collagen type I hydrogel. Cells Tissues Organs 190:81–93

# Mesenchymal Stem Cell Genetic Engineering and Regenerative Medicine

6

M. B. Gugjoo, E. Rasool, and Amar Pal

**Abstract**

Characteristic properties of mesenchymal stem cells (MSCs) make them a suitable candidate in regenerative medicine. There are various associated limitations in MSCs in vivo applications like their short survival and limited expression of particular genes. Genetic engineering of MSCs may be employed to address such issues. Various genetic engineering methods have been incorporated although without any established standard technique. Genetically engineered MSCs are being studied in various kinds of ailments of different tissues with variably promising results. The current chapter focusses on the MSC genetic engineering in relation to the regenerative medicine.

## 6.1 Introduction

Regenerative medicine aims to etch away the concerns related to diseased/non-functional and/or mal-functional tissues/organs. Mesenchymal stem cells (MSCs) are frequently studied for their therapeutic applications attributed to their characteristic properties and lack of any ethical and teratogenic concerns (Gugjoo et al.

M. B. Gugjoo (✉)
Division of Veterinary Clinical Complex, FVSc & AH, SKUAST-Kashmir, Jammu and Kashmir, India

E. Rasool
Division of Surgery & Radiology, FVSc & AH, SKUAST-K, Srinagar, Jammu and Kashmir, India

A. Pal
Division of Surgery, Indian Veterinary Research Institute, Izatnagar, UP, India



M. B. Gugjoo, A. Pal (eds.), *Mesenchymal Stem Cell in Veterinary Sciences*,
https://doi.org/10.1007/978-981-15-6037-8_6

2018). MSCs supposedly work via trophic effects through plethora of compounds, immune-modulatory/anti-inflammatory and/or by cell replacement activities. Their trophic action and differentiation is, however, determined by the available microenvironment upon which a limited control can be exerted (Gugjoo et al. 2019, 2020). MSCs show poor in vivo survival and weak host tissue engraftment especially after local implantation (Wagner and Ho 2007). To realize MSCs full therapeutic potential and tailor them to the specific diseases, the cells are being genetically engineered (Bersenev and Levine 2012, Gage 2012; Roís et al. 2012; Lu et al. 2013; Yan et al. 2015). This avenue has generated much interest among researchers, and the initial reports demonstrated are also promising (Kode et al. 2009; Griffin et al. 2010; Hang et al. 2012).

Genetic engineering/genetic modification/genetic manipulation involves a set of technologies to alter genetic makeup of cells. It is aimed to deliver proteins to neighbouring cells, kill cancer cells, reduce immune rejection, secrete regulatory and pro-healing factors and increase survival of engrafted cells. In genetic engineering, a specific gene cassette is formed to be loaded into a carrier/vehicle (vector). Vectors are the carriers to deliver gene cassette (containing transgene) to the cells (Phillips and Tang 2008; Li et al. 2017; Zhang et al. 2019). These include viruses like retroviruses (Boura et al. 2014), DNA plasmids (Marquez-Curtis et al. 2013; Watkins et al. 2014), mRNA (Rejman et al. 2010), miRNA (Meng et al. 2015) and siRNA (Roís et al. 2012), constructs of minimalistic, immunologically defined gene expression (MIDGE) (Mok et al. 2012; Mählmann et al. 2015) and transposons construct like the Sleeping Beauty System (Martin et al. 2014). Viral vector transduction directly transduces cells while non-viral transfection requires specific methods like lipofection or electroporation to cross the cell membrane (Papanikolaou et al. 2011).

Viral vectors may be genome integrating like lentivirus and adeno-associated virus or non-integrating like adeno and Sendai virus vectors. The integrating viral vectors tend to wield transgene for longer period, while non-integrating vectors express the gene of interest transiently. The genome-integrating viral vectors though may have long-lasting effects but tend to compromise the safety concerns, especially in relation to the insertional mutagenesis and associated potential malignancy (Hacein-Bey-Abina et al. 2003a, b). Inside the cells, the transgene may overexpress unphysiologically (constitutive synthesis) or can express within the physiological limits (controlled by a gene switch). The constitutive transgene expression leads to constant unphysiological oversynthesis of specific proteins. This down-regulates receptors and renders gene expression ineffective. Gene switch transgene expression can be well controlled. The cell responds to the physiological stimulus like oxygen level, concentration of hormones, drugs or chemical agents (Phillips and Tang 2008; Nowakowski et al. 2013). These techniques have their pros and cons and need to be fully weighed before any in vivo application.

MSCs are also being virally transduced but due to the above-mentioned concerns, a safer approach is desired. MSCs characteristic properties including the one of homing make them a suitable non-viral vector candidate. Add-on genetic engineering if well controlled and understood can make MSCs application a game changer in regenerative medicine. The genetically engineered MSCs currently remain to be approved for clinical trials (Nowakowski et al. 2015). The main aim currently remains to develop genetic engineering technique that does not change basic stem cell properties, induce cytotoxicity and should be rescuable after in vivo transplantation (Papanikolaou et al. 2011). The current chapter focuses on the genetically engineered MSCs of veterinary species and their utility in regenerative medicine.

## 6.2 Genetic Engineering of MSCs (GE-MSCs)

Non-viral transfected MSCs in comparison to the viral transduced ones may offer a plausibly safe option in regenerative medicine. The current gene transfection efficiency of MSCs is not at par to that of the MSCs viral vector transduction. One of the reasons that might affect their efficiency is transfection technique. Transfection of cattle bone marrow MSCs (cBM-MSCs) by the electroporation had yielded better (58%) gene transfer efficiency (Colleoni et al. 2005) in comparison to the lipofection (15%) (Díaz et al. 2015) demanding further incites.

GE-MSCs should be ensured to maintain their basic properties (Pinel and Pluhar 2012). MSCs in comparison to other stem cell types have limited culture lifespan. The cell proliferation may cause telomere shortening, and as such the extended culture passaging might lead to genetic changes. In order to improve MSCs survival, genetic engineering has been tried. Sheep bone marrow MSCs (sBM-MSCs) transfected with telomerase reverse transcriptase (TERT) gene had been cultured up to 40 passages. These cells had normal karyotype rate of 88.24% with unaffected differentiation properties (Zhu et al. 2017). Thus, further MSC studies may be conducted to determine the reasons behind their limited self-renewal properties and finally the possible addressal.

Furthermore, it is imperative to confirm MSCs properties remain unaffected with genetic engineering. In a mongrel dog study, VEGF165-transfected BM-MSCs have been demonstrated to express the surface markers as per International Society of Cellular Therapy (ISCT) (Hang et al. 2012). A comparison had been made through the characterization of dog AD-MSCs (dAD-MSCs) and SV40-T retrovirus-transfected dAD-MSCs (tdAD-MSCs). Both these cell lines had characteristic properties as laid down by ISCT. tdAD-MSCs though had higher colony forming potential; pluripotency markers (Sox2 and Nanog) were expressed in both cell lines with stable karyotype. In conclusion, tdAD-MSCs and dAD-MSCs had comparable characteristics (Ayala-Cuellar et al. 2019).

## 6.3 GE-MSCs as Therapeutics

In veterinary science, the genetically engineered MSC studies are very limited. However, the available literature favours their utilization under in vivo conditions. Below are described some of the in vivo GE-MSCs studies conducted in animals.

### 6.3.1 GE-MSCs in Germ Cell Differentiation

There is a growing need to produce elite germplasm in productive animals. Sheep and cattle MSCs have been demonstrated to transdifferentiate into germ cell-like cells (Ghasemzadeh-Hasankolaei et al. 2012, 2014, 2015; Cortez et al. 2018). To boost that potential, genetically engineered MSCs may prove useful. In two goat studies, only a small fraction of Stra8, Boule- and Dazl-transfected BM-MSCs had transdifferentiated to putative germ cells. These differentiated cells could express primordial germ cell-specific genes and premeiotic genes. The cells that co-overexpressed all the three genes (Stra8, Boule and Dazl) were more close to germ cell-like cells (Li et al. 2017; Zhang et al. 2019). This demands further in vitro studies followed by in vivo experimental studies to determine their clinical utility.

### 6.3.2 GE-MSCs and Mastitis Control

Stem cell therapy has a significant potential to regenerate the udder tissues (Sharma et al. 2017; Gugjoo et al. 2019). Mammary stem cells are currently difficult to harvest and as replacement MSCs can be utilized. One of the studies had created bovine lactoferricin (LFcinB) transgene to confirm feasibility of creating genetically engineered cattle or buffalo MSCs. These modified cells had variable expression of the LFcinB, an antibacterial peptide (Sharma et al. 2017). Such LFcinB-cloned MSCs may be injected into the mammary gland for their higher antibacterial properties especially against *Staphylococcus aureus* and *Escherichia coli* (Sharma et al. 2017).

### 6.3.3 GE-MSCs and Bone Healing

Genetically engineered MSCs may be employed to enhance the bone healing. Among various factors, MSCs transfection with BMP-2 has been comparably common. Human BMP-2-transfected goat BM-MSCs had been demonstrated to express and secrete higher concentration of rhBMP-2 as compared to simple BM-MSCs. Goat tibial bone defects treated with calcined bone and hBMP-2-transduced BM-MSCs had complete healing in 5 out of 8 animals. Contrarily, tibial defects treated with scaffold and simple BM-MSCs had slight to almost nil healing. The callus formed in engineered cell-implanted tibial bones had significantly better biomechanical strength as compared to the simple cell-treated bone defects (Dai et al. 2005). In goat femoral head osteonecrosis model, the BMP-2- or β-gal-modified BM-MSCs along with β-tri-calcium phosphate (β-TCP) have been evaluated. Comparing these two cell engineering options, femoral bone density and surface was much improved with BMP-2-transfected cells at 3-month period. The regenerated bone mechanical testing had demonstrated significantly higher compression strength and Young's modulus in the BMP-2 group approaching to that of normal bone (Tang et al. 2007). In dog left radial oligotrophic non-union fracture, BMP-7-transfected dog AD-MSC sheets had led to the early bone healing (9 weeks) (Song et al. 2017). Thus, the combined use of genetic and tissue engineering may provide an innovative way to treat bone defects/fractures. Although these studies have demonstrated improved bone healing results, however, the dose titration of these factors is required. An excess bone formation leading to tumour formation may arise in case higher dose of BMPs is dissipated by these transfected cells (Pinel and Pluhar 2012). Similarly, transplantation of VEGF-transduced autologous BM-MSCs in experimentally induced dog osteonecrotic femoral head had led to better healing in comparison to the simple BM-MSCs (Hang et al. 2012).

### 6.3.4 GM-MSCs in Cartilage and Tendon Repair

Transforming growth factor-beta (TGF-β) plays a crucial role in MSCs proliferation and differentiation, while Nell-1 specifically target cells that are committed to osteochondral lineage. TGF-β up-regulates chondrogenic gene expression through enhanced expression of the Sox9. Under in vitro study, TGF-β1-transduced human synovial-derived MSCs (hSD-MSCs) had undergone effective chondrogenesis as revealed by increased chondrogenic marker gene expression and glycosaminoglycan secretion (Kim et al. 2014). In vitro studies have demonstrated that TGF-β1 transduction of MSCs together with scaffolds like calcium alginate gel creates the chondrogenic micro-environment. This had been confirmed in an in vitro study wherein BM-MSCs had enhanced type II collagen synthesis ability (Zhu et al. 2013). Nell-1-transfected BM-MSCs in the presence of poly-lactic-polyglycolic acid (PLGA) scaffold too had hastened repair of goat mandibular condyle defect. At sixth week fibrocartilage formation while at 24th week complete repair of articular cartilage along with the subchondral bone had been demonstrated. In contrast, simple BM-MSCs/PLGA implants had repaired defects with fibrocartilage at 24 weeks. Overall both the cell lines had improved bone healing as compared to PLGA or empty defects (Zhu et al. 2011).

Insulin-like growth factor 1 (IGF-1) is considered to have stimulating effect on collagen synthesis (Hansen et al. 2013). As such, IGF-1-transfected MSCs may prove useful. In an equine study, IGF-1-transfected BM-MSCs had shown to have significantly higher IGF-1 secretion in comparison to the normal BM-MSCs. Implantation of BM-MSCs and IGF-1-induced BM-MSCs in an equine superficial flexor tendinitis had improved healing of the tendon at 8 weeks. The differences in healing however were statistically non-significant between the cell-treated animals (Schnabel et al. 2009).

### 6.3.5 GM-MSCs in Neural Differentiation and Repair

Brain-derived neurotrophic factor (BDNF) serves as a neurotransmitter modulator and has a crucial role in survival and growth of neurons. It also takes part in neuronal plasticity essentially required in learning and memory (Bathina and Das 2015). HO-1-MSCs reduce inflammation, which favours BDNF-induced neuro-regeneration

(Araujo et al. 2012). In peripheral nerve regeneration experimental study, the BDNF-transfected AD-MSCs were engaged in nerve regeneration. The cells had differentiated into various cell types including Schwann cells and had encouraged functional recovery of the rat sciatic nerve (Tseng and Hsu 2014). Adipose-derived mesenchymal stem cells (AD-MSCs) transfected with BDNF and heme oxygenase-1 (HO-1) were utilized had led to a significant improvement in hind limb functions of spinal cord-injured dogs. Co-transfection of AD-MSCs had been more effective. An increase in clinical evaluation scores and a robust increase in neuronal regeneration were demonstrated. Thus, MSCs transfection with multiple factors may be more effective in promoting the healing of spinal cord injury (SCI) (Khan et al. 2018).

### 6.3.6 GM-MSCs in Cardiac Ailments

Stem cell therapy in myocardial infarction offers an exciting research area. The current issues like stem cell retention and their efficacy in the heart remains to be solved. MSCs genetic modification has been achieved with various factors like glucagon-like peptide-1 (GLP-1), and VEGF and hypoxia-inducible factor 1-α (HIF1-α) are considered to favour better cardiac repair and function. Glucagon-like peptide-1 (GLP-1) is an incretin hormone with cardio-protective properties (Wright et al. 2012), VEGF improves angiogenesis (Locatelli et al. 2015) and hypoxia-inducible factor 1-α is considered to mediate overexpression of erythropoietin, vascular endothelial growth factor, angiopoietin-1 and inducible nitrous oxide synthase (Hnatiuk et al. 2016).

Transfection of human BM-MSC with GLP-1 has been demonstrated as beneficial adjunct in pig myocardial infarction model. GLP-1-transfected human MSCs encapsulated in alginate (bead-GLP-1 MSC) after coronary artery injection had led to the better-left ventricular function and a decrease in infarct area as compared to control. The GLP-1-transfected cell-treated infarcts had reduced inflammation and apoptotic cells together with an increased collagen deposition and fewer myofibroblasts (Wright et al. 2012). In sheep acute myocardial infarction (AMI) model, vascular endothelial growth factor (VEGF)-transfected MSCs had been demonstrated as useful. An increased angiogenesis together with decreased infarct size and improved left ventricular (LV) function was demonstrated with intracardiac implantation of these cells (Locatelli et al. 2015). Similarly, HIF1-α-transfected allogeneic BM-MSCs as compared to the simple BM-MSCs had led to significant reduction of infarct size in a sheep acute myocardial infarction (AMI) model. BM-MSCs too had decreased infarct size although lower than transfected cells but better than the control (Hnatiuk et al. 2016).

### 6.3.7 GE-MSCs in Skin Wounds

In skin wounds the main concern remains to improve angiogenesis along with the collagen deposition. Although skin healing is good, but in some cases it takes longer time especially in diabetes or at the joint areas. VEGF-modified human umbilical cord-derived MSCs (VEGF-UC-MSCs) along with acellular dermal matrix scaffolds evaluated in pig skin wounds had effectively promoted vascularization of the graft and led to an improved quality of wound healing (Han et al. 2014).

Apart from these studies, numerous other animal studies like corneal regeneration, specific diseases among others may be conducted to evaluate beneficial effects of engineered cells. The preliminary studies involving angiopoietin-1 (Piao et al. 2010) and ACE2 (He et al. 2014) among others also have been conducted in human. These studies are preliminary, case uncontrolled and without any uniformity to be able to draw any conclusion.

## 6.4 Conclusion

MSCs are being either transduced virally or transfected by non-viral methods. Either of the methods have their prospects and constraints. Currently, the genetic engineering of MSCs is under experimental phase, and the standardization of these techniques needs further understanding. The genetically engineered MSCs have been studied under in vitro conditions with reasonably better results. Their in vivo applications also have appear promising although variably. Further studies that are aimed at their safe and effective genetic engineering need to be conducted. The engineered cells should then be evaluated under in vivo conditions for their actual safety, utility and efficacy.

## References

Araujo JA, Zhang M, Yin F (2012) Heme oxygenase-1, oxidation, inflammation, and atherosclerosis. Front Pharmacol 3:119

Ayala-Cuellar AP, Kim CW, Hwang KA, Kang JH, Lee G, Cho J, Choi KC (2019) Characterization of canine adipose tissue-derived mesenchymal stem cells immortalized by SV40-T retrovirus for therapeutic use. J Cell Physiol 234(9):16630–16642

Bathina S, Das UN (2015) Brain-derived neurotrophic factor and its clinical implications. Arch Med Sci 11(6):1164–1178

Bersenev A, Levine BL (2012) Convergence of gene and cell therapy. Regen Med 7:50–56

Boura JS, Vance M, Yin W, Madeira C, Lobato da Silva C, Porada CD, Almeida Porada G (2014) Evaluation of gene delivery strategies to efficiently overexpress functional HLA-G on human bone marrow stromal cells. Mol Ther Methods Clin Dev 2014(1):14041

Colleoni S, Donofrio G, Lagutina I, Duchi R, Galli C, Lazzari G (2005) Establishment, differentiation, electroporation, viral transduction, and nuclear transfer of bovine and porcine mesenchymal stem cells. Cloning Stem Cells 7(3):154–166. https://doi.org/10.1089/clo.2005.7.154

Cortez J, Bahamonde J, De los Reyes M, Palomino J, Torres CG, Peralta OA (2018) In vitro differentiation of bovine bone marrow-derived mesenchymal stem cells into male germ cells by exposure to exogenous bioactive factors. Reprod Domest Anim 53(3):700–709

Dai KR, Xu XL, Tang TT, Zhu ZA, Yu CF, Lou JR, Zhang XL (2005) Repairing of goat Tibial bone defects with BMP-2 gene-modified tissue-engineered bone. Calcif Tissue Int 77:55–61

Díaz P, Cuevas F, Peralta OA (2015) GFP labelling and epigenetic enzyme expression of bone marrow-derived mesenchymal stem cells from bovine foetuses. Res Vet Sci 99:120–128

Gage FH (2012) Transplantation in the future. Prog Brain Res 200:7–13
Ghasemzadeh-Hasankolaei M, Sedighi-Gilani MA, Eslaminejad MB (2014) Induction of ram bone marrow mesenchymal stem cells into germ cell lineage using transforming growth factor-b superfamily growth factors. Reprod Dom Anim 49:588–598
Ghasemzadeh-Hasankolaei M, Eslaminejad BM, Sedighi-Gilani M (2015) Derivation of male germ cells from ram bone marrow mesenchymal stem cells by three different methods and evaluation of their fate after transplantation into the testis. In Vitro Cell DevBiol Anim 52(1):49–61
Ghasemzadeh-Hasankolai M, Batavani R, Eslaminejad BM, Sedighi-Gilani M (2012) Effect of zinc ions on differentiation of bone marrow-derived mesenchymal stem cells to male germ cells and some germ cell-specific gene expression in rams. Biol Trace Elem Res 150:137–146
Griffin M, Greiser U, Barry F, O'Brien T, Ritter T (2010) Genetically modified mesenchymal stem cells and their clinical potential in acute cardiovascular disease. Discov Med 9:219–223
Gugjoo MB, Amarpal, Chandra V, Wani MY, Dhama K, Sharma GT (2018) Mesenchymal stem cell research in veterinary medicine. Current Stem Cell Research and Therapy 13(8):645–657
Gugjoo MB, Amarpal, Fazili MR, Shah RA, Sharma GT (2019) Mesenchymal stem cell: basic research and potential applications in cattle and buffalo. J Cell Physiol 234(6):8618–8635
Gugjoo MB, Amarpal, Fazili MR, Shah RA, Mir MS, Sharma GT (2020) Goat mesenchymal stem cell basic research and potential applications. Small Rum Res 183:106045
Hacein-Bey-Abina S, von Kalle C, Schmidt M, Le Deist F, Wulffraat N, McIntyre E (2003a) A serious adverse event after successful gene therapy for X-linked severe combined immunodeficiency. N Engl J Med 348:255–256
Hacein-Bey-Abina S, Von Kalle C, Schmidt M, McCormack MP, Wulffraat N, Leboulch P (2003b) LMO2-associated clonal T cell proliferation in two patients after gene therapy for SCID-X1. Science 302:415–419
Han Y, Tao R, Han Y, Sun T, Chai J, Xu G, Liu J (2014) Microencapsulated *VEGF* gene–modified umbilical cord mesenchymal stromal cells promote the vascularization of tissue-engineered dermis: an experimental study. Cytotherapy 16(2):160–169
Hang D, Wang Q, Guo C, Chen Z, Yan Z (2012) Treatment of osteonecrosis of the femoral head with VEGF165 transgenic bone marrow mesenchymal stem cells in mongrel dogs. Cells Tissues Organs 195:495–506
Hansen M, Boesen A, Holm L, Flyvbjerg A, Langberg H, Kjaer M (2013) Local administration of insulin-like growth factor-I (IGF-I) stimulates tendon collagen synthesis in humans. Scand J Med Sci Sports 23(5):614–619
He HL, Liu L, Chen QH, Cai SX, Han JB, Hu SL, Chun P, Yang Y, Guo FM, Huang YZ, Qiu HB (2014) MSCs modified with ACE2 restore endothelial function following LPS challenge by inhibiting the activation of RAS. J Cell Physiol 230(3):691–701
Hnatiuk AP, Ong S-G, Olea FD, Locatelli P, Riegler J, Lee WH, Jen CH, De Lorenzi A, Gimenez CS, Laguens R, Wu JC, Crottogini A (2016) Allogeneic Mesenchymal stromal cells overexpressing mutant human hypoxia-inducible factor 1-a (HIF1-a) in an sheep model of acute myocardial infarction. J Am Heart Assoc 5:e003714
Khan IU, Yoon Y, Kim A, Jo KR, Choi KU, Jung T, Kim N, Son Y, Kim WH, Kweon O (2018) Improved healing after the co-transplantation of HO-1 and BDNF overexpressed mesenchymal stem cells in the subacute spinal cord injury of dogs. Cell Transplant 27:1140–1153
Kim YI, Ryu JS, Yeo JE, Choi YJ, Kim YS, Ko K, Koh YG (2014) Overexpression of TGF-β1 enhances chondrogenic differentiation and proliferation of human synovium-derived stem cells. Biochem Biophys Res Commun 450(4):1593–1599
Kode JA, Mukherjee S, Joglekar MV, Hardikar AA (2009) Mesenchymal stem cells: immunobiology and role in immunomodulation and tissue regeneration. Cytotherapy 11:377–391
Li PZ, Yan GY, Han L, Pang J, Zhong BS, Zhang GM, Wang F, Zhang YL (2017) Overexpression of STRA8, BOULE, and DAZL genes promotes goat bone marrow-derived mesenchymal stem cells in vitro trans-differentiation toward putative male germ cells. Reprod Sci 24(2):300–312
Locatelli P, Olea FD, Hnatiuk A, De Lorenzi A, Cerdá M, Giménez CS, Sepúlveda D, Laguens R, Crottogini A (2015) Mesenchymal stromal cells overexpressing vascular endothelial growth factor in sheep myocardial infarction. Gene Ther 22(6):449–457

Lu W, Yaoming N, Boli R, Jun C, Changhai Z, Yang Z, Zhiyuan S (2013) mHCN4 genetically modified canine mesenchymal stem cells provide biological pacemaking function in complete dogs with atrioventricular block. Pacing Clin Electrophysiol 36(9):1138–1149

Mählmann K, Feige K, Juhls C, Endmann A, Schuberth H-J, Oswald D, Hellige M, Doherr M, Cavalleri J-MV (2015) Local and systemic effect of transfection-reagent formulated DNA vectors on equine melanoma. BMC Vet Res 11:107

Marquez-Curtis LA, Gul-Uludag H, Xu P, Chen J, Janowska-Wieczorek A (2013) CXCR4 transfection of cord blood mesenchymal stromal cells with the use of cationic liposome enhances their migration toward stromal cell-derived factor-1. Cytotherapy 15(7):840–849

Martin PK, Stilhano RS, Samoto VY, Takiya CM, Peres GB, da Silva Michelacci YM, da Silva FH, Pereira VG, D'Almeida V, Marques FL, Otake AH, Chammas R, Han SW (2014) Mesenchymal stem cells do not prevent antibody responses against human α-L-iduronidase when used to treat mucopolysaccharidosis type I. PLoS One 9(3):e92420

Meng YB, Li X, Li ZY, Zhao J, Yuan XB, Ren Y, Cui ZD, Liu YD, Yang XJ (2015) microRNA-21 promotes osteogenic differentiation of mesenchymal stem cells by the PI3K/β catenin pathway. J Orthop Res 33(7):957–964

Mok PL, Cheong SK, Leong CF, Chua KH, Ainoon O (2012) Extended and stable gene expression via nucleofection of MIDGE construct into adult human marrow mesenchymal stromal cells. Cytotechnology 64(2):203–216

Nowakowski A, Andrzejewska A, Janowski M, Walczak P, Lukomska B (2013) Genetic engineering of stem cells for enhanced therapy. Acta Neurobiol Exp 73:1–18

Nowakowski A, Walczak P, Janowski M, Lukomska B (2015) Genetic engineering of mesenchymal stem cells for regenerative medicine. Stem Cells Dev 24(19):2219–2242

Papanikolaou E, Pappa KI, Anagnou NP (2011) Genetic manipulation of stem cells. Gynecol Obstetric S6:001

Phillips MI, Tang YL (2008) Genetic modification of stem cells for transplantation. Adv Drug Deliv Rev 60(2):160–172

Piao W, Wang H, Inoue M, Hasegawa M, Hamada H, Huang J (2010) Transplantation of Sendai viral angiopoietin-1-modified mesenchymal stem cells for ischemic limb disease. Angiogenesis 13(3):203–210

Pinel CB, Pluhar GE (2012) Clinical application of recombinant human bone morphogenetic protein in cats and dogs: a review of 13 cases. Can Vet J 53:767–774

Rejman J, Tavernier G, Bavarsad N, Demeester J, de Smedt SC (2010) mRNA transfection of cervical carcinoma and mesenchymal stem cells mediated by cationic carriers. J Control Release 147(3):385–391

Roís CN, Skoracki RJ, Mathur AB (2012) GNAS1 and PHD2 short-interfering RNA support bone regeneration in vitro and in an in vivo sheep model. Clin Orthop Relat Res 470(9):2541–2553

Schnabel L, Lynch M, Van der Meulen M, Yeager A, Kornatowski M, Nixon A (2009) Mesenchymal stem cells and insulin-like growth factor-I gene-enhanced mesenchymal stem cells improve structural aspects of healing in equine flexor digitorum superficialis tendons. J Orthop Res 27:1392–1398

Sharma N, SWK DLH, Ghosh M, Sodhi SS, Singh AK, Kim NE et al (2017) A PiggyBac mediated approach for lactoferricin gene transfer in bovine mammary epithelial stem cells for management of bovine mastitis. Oncotarget 8(61):104272

Song J, Kim Y, Kweon O-K, Kang B-J (2017) Use of stem-cell sheets expressing bone morphogenetic protein-7 in the management of a nonunion radial fracture in a toy poodle. J Vet Sci 18(4):555–558

Tang TT, Lu B, Yue B, Xie XH, Xie YZ, Dai KR, Lu JX, Lou JR (2007) Treatment of osteonecrosis of the femoral head with hBMP-2-gene-modified tissue-engineered bone in goats. J Bone Joint Surg Br 89(1):127–129

Tseng TC, Hsu SH (2014) Substrate-mediated nanoparticle/gene delivery to MSC spheroids and their applications in peripheral nerve regeneration. Biomaterials 35(9):2630–2641

Wagner W, Ho AD (2007) Mesenchymal stem cell preparations—comparing apples and oranges. Stem Cell Rev 3:239–248

Watkins DJ, Zhou Y, Matthews MA, Chen L, Besner GE (2014) HB-EGF augments the ability of mesenchymal stem cells to attenuate intestinal injury. J Pediatr Surg 49(6):938–944

Wright EJ, Farrell KA, Malik N, Kassem M, Lewis AL, Wallrapp C, Holt CM (2012) Encapsulated glucagon-like peptide-1-producing mesenchymal stem cells have a beneficial effect on failing pig hearts. Stem Cells Transl Med 1(10):759–769

Yan G, Fan Y, Li P, Zhang Y, Wang F (2015) Ectopic expression of DAZL gene in goat bone marrow-derived mesenchymal stem cells enhances the trans-differentiation to putative germ cells compared to the exogenous treatment of retinoic acid or bone morphogenetic protein 4 signalling molecules. Cell Biol Int 39(1):74–83

Zhang YL, Li PZ, Pang J, Wan YJ, Zhang GM, Fan YX, Wang ZY, Tao NH (2019) Induction of goat bone marrow mesenchymal stem cells into putative male germ cells using mRNA for STRA8. BOULE and DAZLCytotechnol 71(2):563–572

Zhu S, Zhang B, Man C, Ma Y, Hu J (2011) NELL like molecule-1 modified bone marrow mesenchymal stem cells/poly-lactic-co-glycolic acid composite improves repair of large osteochondral defects in mandibular condyle. Osteoarthr Cart 19:743–750

Zhu S, Zhang T, Sun C, Yu A, Qi B, Cheng H (2013) Bone marrow mesenchymal stem cells combined with calcium alginate gel modified by hTGF-β1 for the construction of tissue-engineered cartilage in three-dimensional conditions. Exp Ther Med 5(1):95–101

Zhu X, Liu Z, Deng W, Zhang Z, Liu Y, Wei L, Zhang Y, Zhou L, Wang Y (2017) Derivation and characterization of sheep bone marrow-derived mesenchymal stem cells induced with telomerase reverse transcriptase. Saudi J Biolog Sci 24:519–525

# Sheep Mesenchymal Stem Cell Basic Research and Potential Applications

# 7

M. B. Gugjoo and Amar Pal

**Abstract**

Mesenchymal stem cells (MSCs) act as reservoir of repair system in an adult tissue. These cells are available in almost all the tissues. The cells have been harvested and isolated from numerous tissue sources. Due to their miniscule number in tissues culture expansion followed by characterization becomes imperative. Sheep MSCs (sMSCs) too have been obtained from numerous sources including foetal membranes. These cells have been evaluated in various human translational models and have variably proven to be useful. Such studies additionally form basis of the MSCs therapeutic applications in veterinary practice. The present chapter throws light on various aspects of sheep mesenchymal stem cells including their potential therapeutic applications.

## 7.1 Introduction

Veterinary and human medicine is intricately dependant and linked to each other, forming important constituents of "One Health" initiative (Gugjoo et al. 2018). Various ailments among the animals and human have similar pathophysiology and established therapeutics in one species can form basis of its applications in other (Gugjoo et al. 2018). Sheep acts as a suitable model animal for various ailments of epithelium, cardiovascular, skeletal and neurological systems for human (DiVincenti

M. B. Gugjoo
Division of Veterinary Clinical Complex, FVSc & AH, SKUAST-Kashmir, Jammu and Kashmir, India

A. Pal (✉)
Division of Surgery, Indian Veterinary Research Institute, Izatnagar, UP, India



M. B. Gugjoo, A. Pal (eds.), *Mesenchymal Stem Cell in Veterinary Sciences*, https://doi.org/10.1007/978-981-15-6037-8_7

et al. 2014; Pelagalli et al. 2018). Sheep being neurologically less developed than carnivores and equines has an edge in the particular area. Sheep body features like long bone size and its macrostructure structure are approximate to that of human (Anderson et al. 1999; Van Der Donk et al. 2001). Due to their comparable bone biomechanics, biochemistry and histology sheep bone fracture healing turn over markers may be adopted for human (Sousa et al. 2015). Sheep have sufficient superficial space on their backs that may be utilized to study MSCs regeneration potential in experimental lesions (Chevrier et al. 2009; Music et al. 2018; Yr et al. 2018). The current chapter details sheep mesenchymal stem cell (sMSCs) characteristics and their potential therapeutic applications.

## 7.2 MSCs Characterization

sMSCs are characterized as per the recommendations of International Society for Cellular Therapy (ISCT) (Dominici et al. 2006). Plastic adherent sMSCs show surface marker(s) expression, although variably and differentiate into adipogenic, chondrogenic and osteogenic lineages (Table 7.1) (Colosimo et al. 2013; Liu et al. 2014 Tian et al. 2016).

Sheep bone marrow derived mesenchymal stem cells (sBM-MSCs) grow as morphologically large adherent cells and exhibit a fibroblastic morphology. sBM-MSCs are bigger (diameter: 18.1 ± 3.3 μm volume: 3.4 ± 1.7pl) than the mononuclear cells (diameter: 8.6 ± 1.0 μm in volume: 0.34 ± 0.12pl) (Caminal et al. 2017). Cell morphology under transmission electron microscopy (TEM) had appeared as cuboidal electron-lucent and spindle shaped electron-dense cells similar to human MSCs. The surface marker expression especially CD90 was comparable in these cell types. The characterization of glycoprotein secretory system in these cells demonstrates their potential secretory properties (Desantis et al. 2015).

## 7.3 sMSCs Number and Proliferation

Different tissue sources have variable sMSCs population and proliferation potential. These cells become senescent after certain passages. Among various sources, sheep peripheral blood harboured lowest number of MSCs (sPB-MSCs) and had shown lower ability to proliferate (Lyahyai et al. 2012). MSCs from various sources may have different proliferation properties. Comparative studies have demonstrated that sBM-MSCs had higher proliferation potential as compared to cotyledon derived MSCs (Cot-MSCs) (Ribitsch et al. 2017) and sheep adipose tissue MSCs (sAD-MSCs) (Heidari et al. 2013). Apart from source based differences, culture expansion may also affect their proliferation. In initial passages, sheep dermis derived MSCs (sD-MSCs) in comparison to sBM-MSCs may show lower proliferation potential but that becomes higher with extended passaging (Jahroomishirazi et al. 2014). Likewise sheep liver-MSCs had shorter population doubling time as compared to sAD-MSCs (Heidari et al. 2013). Further incites in this regard are required to confirm the reason behind such variabilities.

**Table 7.1** Sheep mesenchymal stem cell (MSC) sources and their characteristics (Gugjoo and Amarpal 2018)

| Positive markers | Negative markers | Differentiation characteristics | References |
|---|---|---|---|
| *Adipose tissue MSCs (AD-MSCs)* | | | |
| CD29, CD44 | – | – | Grzesiak et al. (2011) |
| CD90, CD105, CD73, CD29 | HLA-DR, CD45,CD34, STRO-1 | Adipogenic, chondrogenic, osteogenic | Zorzi et al. (2015) |
| CD44, CD90 | CD31, CD45 | | Vahedi et al. (2016) |
| CD44 | CD31, CD45 | Adipogenic, chondrogenic, osteogenic | Feng et al. (2018) |
| *Amniotic fluid MSCs (am-MSCs)* | | | |
| CD13, CD29, CD44, CD90, CD106, OCT4 | CD45 | Adipogenic, chondrogenic, osteogenic | Tian et al. (2016) |
| CD44, CD58, CD166 | CD14, CD31, CD45 | – | Shaw et al. (2011) |
| *Amnion MSCs (A-MSCs)* | | | |
| CD29, CD44, CD105, CD90 | CD31 | Adipogenic, osteogenic | Klein et al. (2011) |
| CD166 antigen at low levels and CD29 and CD58 intermediate levels | CD14, CD31, CD45, CD49f | – | Colosimo et al. (2013) |
| *Bone marrow MSCs (BM-MSCs)* | | | |
| CD44, CD105, Vimentin | CD45, CD34 | Adipogenic, chondrogenic, osteogenic | Mrugala et al. (2008) |
| CD29, CD44, CD166 | CD14, CD31, CD45 | Adipogenic, chondrogenic, osteogenic | Mccarty et al. (2009) |
| CD9, CD44, CD54,CD73, CD90, CD105, CD166 | CD45 | – | Rentsch et al. (2010) |
| CD44 | CD34 | Adipogenic, chondrogenic, osteogenic | Czernik et al. (2013) |
| CD44, CD90 (low expression) | CD45, HLA-DR | Adipogenic, chondrogenic, osteogenic | Caminal et al. (2014) |
| CD44, CD58, HLAI, CD90 | CD31, CD34, CD45, HLA-DR | – | Desantis et al. (2015) |
| CD29, CD36, CD73, CD90, CD166, CD29 | CD45 and CD34 (variable expression) | Adipogenic, chondrogenic, osteogenic, neuronal like cells (downregulation) | Mediano et al. (2015a, b) |
| CD44, CD90, CD140a, CD105 and CD166 | CD45,CD73 | Adipogenic, chondrogenic, osteogenic | Caminal et al. (2017) |
| CD146 | CD34, CD45 | Adipogenic, chondrogenic, osteogenic | Hindle et al. (2016) |

(continued)

**Table 7.1** (continued)

| Positive markers | Negative markers | Differentiation characteristics | References |
|---|---|---|---|
| CD73 (96.9 ± 5.9), CD90 (99.6% ± 0.3), CD105 (99.16 ± 1.5), CD271 (97.762.0) and MHCI (94.0% 67.2) | Negative (CD117 (0.1%60.1) or low (MHCII (10.5% ± 16.0); CD31 (14.6% ± 4.2) and CD45 (39.4% ± 31.8) | Adipogenic, chondrogenic, osteogenic | Khan et al. (2016) |
| CD90 | – | Adipogenic, chondrogenic, osteogenic | Landa-Solis et al. (2016) |
| CD44, CD90, CD140a, CD105 and CD166 | – | Adipogenic, chondrogenic, osteogenic | Vivas et al. (2018) |
| *Cotyledonary MSCs* | | | |
| CD29, CD33 and CD166 | CD31, CD45 | Adipogenic (weak), chondrogenic, osteogenic | Ribitsch et al. (2017) |
| *Dermal MSCs (D-MSCs)* | | | |
| β-integrin, CD71, CD44 and CD73 | – | Adipogenic, chondrogenic, osteogenic, neurogenic | Cui et al. (2014) |
| CD271 | CD45 | Adipogenic, chondrogenic, osteogenic | Jahroomishirazi et al. (2014) |
| *Endometrium MSCs (end-MSCs)* | | | |
| CD271 | CD49f | Adipogenic, chondrogenic, osteogenic, smooth muscle | Letouzey et al. (2015) |
| *Kidney MSCs (K-MSCs)* | | | |
| Oct-4, VIM, CD44, FN1, CD73, CD44 and PAX2 | CD34 | Adipogenic, chondrogenic, hepatocellular | Ji et al. (2016) |
| *Lung MSCs (L-MSCs)* | | | |
| Oct4, Nanog, CD29, CD44, CD71, CD73 and CD90 | CD34, CD45 | Adipogenic, chondrogenic, osteogenic, hepatocyte | Ma et al. (2017) |
| *Periodontal ligament MSCs (Pl-MSCs)* | | | |
| CD106, CD44, CD166, CBFA-1, collagen-I, bone sialoprotein | – | – | Gronthos et al. (2006) |
| *Peripheral blood MSCs (PB-MSCs)* | | | |
| CD29, CD73, CD90 and CD105 (low expression) | CD45, and CD34 (variable expression) | Adipogenic, chondrogenic, osteogenic, neurogenic | Lyahyai et al. (2012) |
| CD29, CD36, CD73, CD90, CD166, CD105 | CD45 and CD34 (variable expression) | Adipogenic, chondrogenic, osteogenic, neuronal like cells (downregulation) | Mediano et al. (2015a, b) |
| Stro-1, CD90, CD106, CD105, CD146, CD166, CD44 | CD34, CD45 | Adipogenic, Chondrogenic, Osteogenic, | Hopper et al. (2015) |

(continued)

**Table 7.1** (continued)

<table>
<tr><th>Positive markers</th><th>Negative markers</th><th colspan="2">Differentiation characteristics</th><th>References</th></tr>
<tr><td colspan="5">Synovial MSCs (Sy-MSCs)</td></tr>
<tr><td>CD73, CD90, CD105</td><td>–</td><td colspan="2">Adipogenic, chondrogenic, osteogenic</td><td>Burk et al. (2017)</td></tr>
<tr><td colspan="5">Umbilical cord MSCs (UC-MSCs)</td></tr>
<tr><td>CD44</td><td>CD38, CD45, CD41/61.</td><td colspan="2">–</td><td>Fadel et al. (2011)</td></tr>
<tr><td>CD29, CD44, CD73, CD90</td><td>CD34</td><td colspan="2">Adipogenic, chondrogenic, osteogenic</td><td>Chen et al. (2018)</td></tr>
<tr><td colspan="5">Placental MSCs (Pl-MSCs)</td></tr>
<tr><td>CD29, CD44, CD73, CD90, CD105, Sox2</td><td>CD31, CD45</td><td>Early stage gestation</td><td rowspan="2">Adipogenic, chondrogenic, osteogenic</td><td rowspan="2">Brown et al. (2016)</td></tr>
<tr><td>CD29, CD44, CD73, CD90, CD105</td><td>CD31, CD45, Sox2</td><td>Late stage gestation</td></tr>
</table>

*AD-MSCs* adipose tissue MSCs, *Am-MSCs* amniotic fluid MSCs, *A-MSCs* amnion MSCs, *BM-MSCs* bone marrow MSCs, *D-MSCs* dermal MSCs, *End-MSCs* endometrium MSCs, *K-MSCs* Kidney MSCs, *L-MSCs* lung MSCs, *Pl-MSCs* periodontal ligament MSCs, *PB-MSCs* peripheral blood MSCs, *Sy-MSCs* synovial MSCs, *UC-MSCs* umbilical cord MSCs, *Pl-MSCs* placental MSCs

sMSCs proliferation potential decreases with extended passaging (Colosimo et al. 2013; Ji et al. 2016; Tian et al. 2016). sAD-MSCs and umbilical cord derived MSCs (sUC-MSCs) had normally proliferated up to passage 6 (Vahedi et al. 2016) and passage 10 (Chen et al. 2018), respectively, and became senescent afterwards. This depicts variability in proliferation abilities of the sMSCs at different passages, though various other factors like culture conditions and donor tissue type may also affect these properties.

sMSCs karyotypic features may remain unaffected in >90% of the cells up to as high as 20 passages (Ji et al. 2016; Tian et al. 2016). The cells (sD-MSCs) harvested from young ones (1 month old) may show even more extended passaging (48 passages) with normal karyotype (Cui et al. 2014). Among various reasons of limited self-renewal property, telomere length shortening with cell proliferation appears to be important. This has been confirmed in telomerase reverse transcriptase (TERT) transfected sBM-MSCs. These transfected cells could be cultured up to 40 passages with normal karyotype rate of 88.24% (Zhu et al. 2017).

## 7.4 Tissue Source and Marker Expression

sMSCs express various surface markers and lack haematopoietic markers. Variability in surface markers of the sMSCs with respect to the tissue source has been demonstrated (Table 7.1). The variability in marker expression is attributed to the lack of species specific antibodies, besides differences in the type of tissue sources and harvesting methods (Gugjoo et al. 2020a, b). Additionally, sMSCs have also been demonstrated to express various pluripotency markers like Oct4, Nanog and Stro-1 (Hopper et al. 2015; Brown et al. 2016; Ji et al. 2016; Ma et al. 2017).

## 7.5 Source and Differentiation Potential

sMSCs exhibit variability in their differentiation potential on the basis of their tissue source. The musculoskeletal differentiation potential of sCot-MSCs and sBM-MSCs had been akin but sBM-MSCs adipogenic potential was more pronounced (Ribitsch et al. 2017). Foetal MSCs usually show low PPARγ pathway and as such may be less demanding for adipogenic activity (Ragni et al. 2013). Among various musculoskeletal tissues, sheep synovial fluid derived MSCs (sSy-MSCs) had shown higher chondrogenic expression in comparison to sBM-MSCs (Burk et al. 2017). However, MSCs from adipose tissue, bone marrow and liver-MSCs had comparable differentiation activity (Heidari et al. 2013). Such variabilities in differentiation potential of MSCs with respect to source need further investigation especially on molecular basis. This may help to better utilize MSCs for therapeutic applications.

## 7.6 Miscellaneous Factors Affecting MSCs Characteristics

Scrapie infected sheep carry MSCs with normal tri-lineage differentiation ability. However, their proliferation and neurocyte like cell differentiation may be reduced (Mediano et al. 2015a, b) and as such may offer limited therapeutic benefits in neural ailments. The cellular characteristics of MSCs may also be influenced by the factors like feeding standard, age of the animal and disease status of the animal. Low feeding status of ewe may affect their foetal MSCs characteristics (Pillai et al. 2016). Ageing may or may not decrease MSCs proliferation and clonogenic potential (Rhodes et al. 2004; Han et al. 2010). In sheep unlike dog, breed variation had non-significant effect on BM-MSCs proliferation (Rhodes et al. 2004).

## 7.7 Potential Therapeutic Applications

MSCs in sheep have been implanted through various routes including parenteral considering their homing property (Gugjoo and Amarpal 2019). MSCs secretory properties and differentiation in particular tissue is being determined by the available local milieu. The cells (including of xenogenic and/or allogeneic sources) have been well tolerated under *in vivo* conditions. In a pre-immune sheep foetus, xenogenic MSCs (human source) had differentiated into hepatic cells upon intrahepatic/intraperitoneal implantation (Chamberlain et al. 2007) or secreted insulin upon pancreatic transplantation (Ersek et al. 2011). Xenogeneic MSCs too had been demonstrated to have tissue specific trans-differentiation even in immuno-potent sheep (Liechty et al. 2000; Mackenzie and Flake 2001; Shaw et al. 2011). Likewise, allogeneic applications of these sMSCs had been demonstrated to be safe (Berner et al. 2013; Dooley et al. 2015; McDonald et al. 2015; Zorzi et al. 2015). Apart from being immuno-compromised, the cells have been demonstrated to have immune-modulatory properties (Dooley et al. 2015). Considering such immuno-compromised and immuno-modulatory properties, allogeneic MSCs are increasingly being studied to make ready-to-use therapy possible.

sMSCs have been mainly studied in experimental human translational models without any direct sheep clinical trial (Gugjoo and Amarpal 2018). This poses an impediment in their definitive therapeutic applications in veterinary practice, in general and in sheep clinical practice, in particular. For *in vivo* identification, sMSCs may be labelled with superparamagnetic iron oxide (SIPO), though its concentration must be standardized (Scharf et al. 2015).

### 7.7.1 Wound Healing

In foetal tissues wound healing occurs through regeneration and in adults by repair (Klein et al. 2011). Sheep has been chosen as wound model as it has sufficient space on back and limited neurological development in comparison to the canine and equine (Martinello et al. 2018). In utero implanted xenogeneic or autologous MSCs have been demonstrated to home to the site of foetal wounds. The implanted cells have been able to expedite wound healing by increasing the secretion of extracellular matrix (Table 7.2) (Mackenzie and Flake 2001; Klein et al. 2011). Apart from foetal wounds, adult sheep wounds too tend to heal faster with allogeneic PB-MSCs. The wound had closed within 15 days and might have been due to the cell induced inflammation inhibition and enhanced re-epithelialization, neo-vascularization and contraction. These treated wounds had significantly higher expression of hair-keratine (hKER) and collagen type I (Col1$\alpha$1) being observed from day 15 up to day 42 (Martinello et al. 2018). It may be concluded that MSCs application are quite promising for wound management. Further studies are desired to confirm the posology of the MSCs applications.

### 7.7.2 Osteochondral Defects

An extensive literature on MSCs osteochondral studies exist. The results have been mostly favourable and without any serious adverse effect as detailed in Table 7.3. A single study however had failed to demonstrate any obvious improvement with MSCs as compared to the control (Delling et al. 2015a).

MSCs intra-articular implantation remains safe (Caminal et al. 2014; Abdulmula et al. 2017). Intra-articularly implanted cells remain viable and attached to the joint structures (Delling et al. 2015b; Feng et al. 2018; Markides et al. 2019). The available inflammatory environment however, may reduce their chondrogenic potential, although without affecting their phenotype (Ando et al. 2012). MSCs possible pathway of neo-cartilage formation may be ensued via endochondral ossification process (Lydon et al. 2019). MSCs tend to undergo hypertrophic chondrocyte formation and may lead to osteogenesis (Gugjoo et al. 2018).

Intravenous implantation of BM-MSCs had been able to reduce lameness, joint pain and swelling in a sheep mono-arthritis model. The experimental joints had less cartilage erosions, limited inflammatory cells and were supplemented with activated stromal cells and angiogenesis (Abdalmula et al. 2017). In chronic induced osteoarthritis models, improved parameters (histological and/micro-CT scores) of healed

**Table 7.2** In vivo preclinical experimental mesenchymal stem cell studies on cutaneous wounds

| Model type | Number of animals | Model defect size/study period | Biomaterial used | Cell dose/ assembly | Evaluation criteria | Overall result | References |
|---|---|---|---|---|---|---|---|
| In utero tracking of cells to wounds or developing organs | 29 animals | 13 months | Xenogeneic (human) MSCs | 5–20 × $10^6$ early gestation (65 days gestation); 1 × $10^8$ cells/ kg (mid-gestation, 85 days gestation) | Polymerase chain reaction (PCR), immuno-histochemistry and in situ hybridization | PCR confirmed cellular engraftment to one or more tissues. Relatively large engraftment was seen in wounded areas. Cells persisted in foetal tissue up to a year | Mackenzie and Flake (2001) |
| Full thickness square wound (4 × 4 cm, six wounds) | Six wounds (three wounds treated with conventional topical cream or gel, fourth with cold ionized plasma, fifth with PBS and sixth with MSCs) | 15 and 42 days | Allogeneic PB-MSCs, hyaluronic acid | 1 × $10^6$ (injected in wound margins), 1 × $10^6$ (loaded with hyaluronic acid and implanted at wound centre) | Clinical, histopathological, and molecular analysis | At 15 days post-lesion, the wounds treated with MSCs showed a higher degree of wound closure, a higher percentage of re-epithelialization, proliferation, neo-vascularization and increased contraction in comparison to a control group. Later, the wounds treated with MSCs had more mature and denser cutaneous adnexa as compared to the control group with absence of inflammation and expression of CD3+ and CD20+. Moreover, the mRNA expression of hair-keratine (hKER) and collagen I was observed in the MSCs treated group 15 days | Martinello et al. (2018) |

**Table 7.3** In vivo chondrogenic preclinical experimental mesenchymal stem cell studies in sheep

| Model type | Number of animals | Model defect size/ study period | Biomaterial used | Cell dose | Evaluation criteria | Overall result | References |
|---|---|---|---|---|---|---|---|
| Medial femoral condyle defect | 28 ($n = 16$, cell + β-TCP; $n = 8$ in β-TCP only and $n = 4$ in control) | 8 mm (diameter) and 4 mm (depth)/24 weeks | Autologous BM-MSCs + beta-tricalcium phosphate (β-TCP) | $3 \times 10^7$ | Macroscopic observation, histological, immuno-histochemical, biochemical analysis | Experimental animal defects were resurfaced with hyaline-like tissue. An ideal interface formed between the engineered cartilage, adjacent normal cartilage and the underlying bone | Guo et al. (2004) |
| Osteonecrosis of femoral head | 8 animals ($n = 4$ in control group; $n = 2$ each in sheep MSC group and human MSC group treated after 8 weeks of induced necrosis) | 10 ml of absolute ethanol induced | Sheep BM-MSCs (transfected) and human dental stem cells | $1 \times 10^6$ (each cell type) | Light microscopy | Better bone regeneration in cell treated group animals | Feitosa et al. (2010) |
| Chronic model of medial femoral condyles osteochondral lesions | 10 animals ($n = 40$ defects; group I: chondrogenically differentiated MSC/ hydrogel constructs; group II: Undifferentiated ovine MSC/hydrogel constructs; group III: Cell-free hydrogel; group IV: control) | 7 mm/6 months | Autologous BM-MSCs /collagen I hydrogel constructs | $4 \times 10^5$ MSCs mixed with collagen I | Histopathology | Group I had significantly better histologic scores with morphologic characteristics of hyaline cartilage such as columnarization and presence of collagen type II as compared to others. However, each group showed variability in results | Zscharnack et al. (2010) |

(continued)

**Table 7.3** (continued)

| Model type | Number of animals | Model defect size/ study period | Biomaterial used | Cell dose | Evaluation criteria | Overall result | References |
|---|---|---|---|---|---|---|---|
| Chronic model of anterior cruciate ligament excision | 16 animals ($n = 6$ animals in treatment group; group I: pre-differentiated MSCs and group II: undifferentiated MSCs) and $n = 4$ (control group) | 6 weeks | Chondrogenically differentiated MSCs or undifferentiated MSCs | $10 \times 10^6$ per joint | Gross, histological and clinical observation | Retardation of osteoarthritis in cell treated groups. Non-significant difference in group I and II except for macroscopic observations of meniscus repair. Severe osteoarthritis in control | Al Faqeh et al. (2012) |
| Chronic model full thickness medial femoro-tibial condyles and meniscal tear | 10 ($n = 20$ defects; 10 studied at 6 months period while other 10 at 12 months period) one of the limbs remained control | 60 mm defect size/6 month or 12 months | BM-MSCs | $1{,}1 \times 10^7$ (6 month period animals) or $1.2 \times 10^7$ (12 month period animals) | Radiography, MRI, ultrasound, macroscopic and histological analyses | Regeneration of articular cartilage and meniscus was case-dependent but statistically significant improvement was found in specific macroscopic and histological parameters | Caminal et al. (2014) |

| | | | | | | | |
|---|---|---|---|---|---|---|---|
| Medial femoro-tibial condyle defect | 9 ($n$ = 18 defects) | 7 mm defect size/4 months and 12 months | BM-MSCs alone or seeded on co-polymeric poly-lactide:Polyglycolic acid scaffolds either | $3.3 \times 10^6 \pm 0.4 \times 10^6$ cells | Biomechanical testing, macroscopic and histological analyses | Better macroscopic scores at 4 months in cell treated as compared to 12 months evaluation period. Non-significant histopathological scores at 12 months between cell treated and cell-free groups | Caminal et al. (2014a) |
| Chronic anterior cruciate ligament transection and medial meniscectomy | 18 ($n$ = 6 animals in each group) group I: BM-MSCs; group II: Bone marrow mononuclear cells; group III: Control | 8 weeks | Autologous BM-MSCs | $10 \times 10^6$ after 12 weeks of model creation | Macroscopically and histologically, and glycosaminoglycan (GAG) contents, gene expression levels (collagen II, Aggrecan and matrix metalloproteinase-13), tumour necrosis factor-α (TNF-α) and transforming growth factor beta | Significantly higher cartilage regeneration and lower proteoglycan loss in group I than group II. Comparable inhibition of PGE2, TNF-α and TGF-β levels in synovial fluid and promotion of higher levels of aggrecan and col II in two cell treated groups. Downregulation of MMP-13 also comparable. Both the cell treated groups had significantly better cartilage than control | Song et al. (2014) |

(continued)

**Table 7.3** (continued)

| Model type | Number of animals | Model defect size/ study period | Biomaterial used | Cell dose | Evaluation criteria | Overall result | References |
|---|---|---|---|---|---|---|---|
| Full thickness lateral femoral condyle defect | 12 animals ($n = 3$ in each group): group I (amniotic membrane); group II (cryopreserved amniotic membrane previously cultivated with BM-MSCs); Group III (cryopreserved amniotic membrane alone); group IV (control) | 7 × 5 mm/8 weeks | BM-MSCs and amniotic membrane | $2 \times 10^6$ cells and amniotic membrane | Gross and histopathology | Significant difference in healing between treatment and control groups. Non-significant differences between treatment groups. The samples of groups II, III and IV reported a ICRS grade II (similar to healthy cartilage). The mean value of O'Driscoll scale in the control group (3.3) was significantly lower as compared to the treatment groups (group II: 10.7; G III: 8; G IV: 11.3). Non-significant differences were found between the experimental groups | Garcia et al. (2015) |
| Partial thickness medial femoral condyle defect | 15 animals ($n = 30$ knees (group I: Scaffold plus cell; group II; scaffold only; group III control) | 10 mm/6 months | Xenogenic AD-MSCs and collagen/chitosan scaffold | $1 \times 10^6$ along with scaffold | Microscopic and macroscopic analysis | Significantly higher histological scores in cell treated group as compared to others | Zorzi et al. (2015) |

| | | | | | | | |
|---|---|---|---|---|---|---|---|
| Unilateral medial meniscectomy | 20 (group I: $n = 6$ animals, BM-MSCs + scaffold; group II: $n = 6$ animals BM concentrate + scaffold; group III: $n = 4$ animals scaffold treated group IV: $n = 4$ animals, control) | 12 weeks | BM-MSCs + scaffold (Hyaff®–11) and BM concentrate + scaffold (Hyaff®–11) | $6 \times 10^6$ seeded on scaffold | Macroscopy, histology, immuno-histochemistry and micro-computed tomography | BM concentrate better inhibited inflammation in cartilage, meniscus and synovium. It also improved cartilage healing. Subchondral bone thickness decreased in both the cell treated groups | Desando et al. (2016) |
| Meniscal cartilage tear model | 30 animals (three groups with $n = 10$ animals in each group). Group I: Scaffold laden MSCs; group II: Scaffold only and group III: Suturing only | 5 mm × 3 mm/13 weeks and 6 months | BM-MSCs collagen I scaffold | $1x10^6/cm^2$ | Macroscopy and histopathology | Statistically significant improvement in cell treated as compared to control at 13 weeks. But no difference at 6 months period | Whitehouse et al. (2017) |
| Anterior cruciate ligament resection and medial meniscectomy | 24 ($n = 6$ in each group) group I: Control; group II: HA; group III: Low dose AD-MSCs + HA; group IV: High dose MSCs + hyaluronic acid | 14 weeks after treatment | Allogeneic AD-MSCs and hyaluronic acid | $5 \times 10^7$ (high dose) at 3 weeks and ($1 \times 10^7$ cells (low dose) at 6 weeks | Magnetic resonance imaging (MRI), macroscopy, micro-computed tomography and cartilage-specific staining | AD-MSCs + hyaluronic acid could efficiently block osteoarthritis progression and promoted cartilage regeneration | Feng et al. (2018) |

tissue had been observed with the implantation of BM-MSCs (Caminal et al. 2014) and AD-MSCs (Feng et al. 2018). The variable concentrations of AD-MSCs implanted along with hyaluronic acid at 3 and 6 weeks post injury were viable for more than 2 months and had restricted osteoarthritis progression (Feng et al. 2018). Even xenogenic AD-MSCs (human) along with scaffold (chitosan/collagen) had significantly improved ICRS-I score in a femoral condyle partial thickness defect as compared to the scaffold treated and control (Zorzi et al. 2015).

The hyaline tissue formation has not been reported in all the studies and individual cases with some demonstrating chondroid metaplasia. The integration of healed tissue to native cartilage too had been the problem (Caminal et al. 2014). It may thus, be concluded that MSCs may repair the osteochondral injuries but does not guarantee the same (Gugjoo et al. 2020a, b).

One of the sheep osteoarthritis model studies had failed to yield improved results at 12 weeks post autologous MSCs implantation (Delling et al. 2015a). This may have been due to the weak joint injury failing to induce obvious osteoarthritic changes or else due to the less concentration of implanted cells (Feng et al. 2018).

#### 7.7.2.1 Comparative osteochondral Studies

The *ex vivo* studies have demonstrated higher chondrogenic gene expression of BM-MSCs cultured under hypoxia (chondrogenically primed) as compared to the normoxia cultured BM-MSCs (Bornes et al. 2017). These chondrogenically primed MSCs however, have failed to yield any significant improvement in tissue healing as compared to non-primed cells (Al Faqeh et al. 2012; Bornes et al. 2017). The cryopreservation does not seem to have any adverse effect on MSCs. Amniotic-MSCs as fresh and cryopreserved cells had shown comparable healing in a sheep full thickness lateral femoral condyle defect model (Garcia et al. 2015).

Cultured BM-MSCs and AD-MSCs tend to have an edge over bone marrow mononuclear cells (MNCs) or stromal vascular fraction (SVF), respectively. Implantation of both the former cultured cell types had shown improved cartilage regeneration together with little loss of proteoglycan in a sheep osteoarthritis model (Song et al. 2014; Lv et al. 2018) in comparison to the non-cultured cell mix. The mononuclear cells though still had some positive impact on healing as compared to control. The inflammatory mediators were inhibited in synovial fluid (Song et al. 2014). Since MNCs and SVF contain little concentration of MSCs, comparably little beneficial effects with these cell types may be expected. Contrarily, bone marrow concentrate (BMC) along with hyaluronan had shown greater repair ability in comparison to the BM-MSCs/hyaluronan. However, both cells types had reduced inflammation and thus, may be effective to prevent fibrosis and hypertrophy (Desando et al. 2016). Scaffolds as such may play an important role in stem cell effectiveness.

There is need to conduct studies that incorporate similar source, passage cells at uniform concentrations. The studies have to be case controlled with an extended study period. Furthermore, role of incorporated scaffolds need to be considered.

Comparative studies against the established surgical techniques like microfracture must also be made to evaluate feasibility and efficacy of MSC therapy.

### 7.7.3 Bone Defects

Bone tissue engineering is increasing by each passing day to ensure early and better healing (Chandran et al. 2017). MSCs due to their characteristic properties are being made subject. Most of the sheep preclinical experimental bone model studies have shown better and early bone healing upon MSCs implantation as enlisted in table (Table 7.4). One of the sheep study had shown early osteogenic synthesis with scaffold laden sMSCs, implanted subcutaneously (Boos et al. 2014). Apart from culture expanded MSCs, bone marrow mononuclear cells (contains MSCs) along with sheep bone mineral too have favoured improved sinus augmentation at 8 and 16 weeks in comparison to the autogenous graft (Gutwald et al. 2010).

MSCs bone healing potential has been evaluated in different sheep bone models. Among various bones studied are metatarsal-, tibial- mandible- and cranial-critical bone defects. Mostly autologous or allogeneic BM-MSCs have been evaluated in these studies. BM-MSCs along with coral granules (Viateau et al. 2007; Manassero et al. 2013), cortical allograft (Fernandes et al. 2014), polycaprolactone and hydroxyapatite (Berner et al. 2015), serum scaffold (Gallego et al. 2015) and calcium alginate (Shang et al. 2001) have been implanted to evaluate healing potential in critical size bone defects. The implantation of scaffold laden cells had led to an improved early bone healing with better structural and morphological features (Fernandes et al. 2014; Gallego et al. 2015). The healed tissue had better quality and approximated to normal bone tissue (Gallego et al. 2015). The healed tissue strength could commensurate with clinical standards (Berner et al. 2015). In all these cell and scaffold treated defects healing was much improved as compared to those of scaffold treated. Incorporation of growth factor like rhBMP-2 to BM-MSCs and/autologous serum loaded on to the scaffold and implanted subcutaneously had led to subcutaneous ossification (Boos et al. 2014).

Cell immunogenicity may or may not affect healing potential of MSCs. Autologous and allogeneic MSCs had shown comparable tibial bone healing (Berner et al. 2013) while xenogeneic BM-MSCs had failed to improve sheep tibial bone healing (Niemeyer et al. 2010). Mononuclear cells (MNCs) may show improved healing over raw bone marrow concentrate (Ardjomandi et al. 2015). MSCs may also be effective to repair osteoporotic bone (Chandran et al. 2017). MSCs along with the tricalcium phosphate scaffold or graft may also be used to promote interbody spinal fusion and spinal arthrodesis (Table 7.4).

Combination of MSCs and other implants/scaffolds like TiAl6V4 constructs with a calcium–phosphate coating (Garcıa-Gareta et al. 2015) and polyetherketoneketone (Adamzyk et al. 2016) had failed to enhance healing in a sheep trabecular bone and calvarial defects, respectively. The scaffolds although had favoured *in vitro* cell adhesion, growth and osteogenic differentiation of MSCs (Adamzyk et al.

**Table 7.4** In vivo preclinical experimental osteogenic mesenchymal stem cell studies in sheep

| Model type | Number of animals | Model defect size/study period | Biomaterial used | Cell dose/assembly | Evaluation criteria | Overall result | References |
|---|---|---|---|---|---|---|---|
| Tibial defects | 4 animals (cell-loaded implants, $n = 2$; cell-free HAC cylinders, $n = 2$) | 2-month | Autologous BM-MSCs, hydroxyapatite ceramic (HAC) | 0.5–1.0 × $10^8$ in fibrinogen | Indentation assay | In cell-loaded implants, bone formation occurred both within the internal macropore space and around the HAC cylinder while in control cell-free implants, bone formation was limited mostly to the outer surface and was not observed in most of the inner pores. As tested in an indentation assay, the stiffness of the complex HAC-bone material was found to be higher in cell-loaded implants as compared to controls | Kon et al. (2000) |
| Cranial bone defect (parietal bone, 20 mm in diameter, bilateral) | 8 animals with 16 defects (MSCs and calcium alginate, $n = 8$; calcium alginate, $n = 4$; control unrepaired, $n = 4$) | 6 weeks and 18 weeks post repair | Autologous BM-MSCs and calcium alginate | – | Gross, histological and computerized tomography (CT) scan | New bone tissues were observed with MSCs/scaffold as early as 6 weeks post-repairing but not in control groups. The engineered bone tissue became more mature at 18 weeks post-repairing. CT scan revealed an almost complete repair of the defect of MSCs/scaffold at 18 weeks | Shang et al. (2001) |

| | | | | | | | |
|---|---|---|---|---|---|---|---|
| Left metatarsal bone defect (25 mm) stabilized with plate and PMMA space being replaced after 6 weeks | 21 animals (group I: Empty, $n = 5$; group II: Coral scaffold, $n = 4$; group III: Coral scaffold/MSCs, $n = 6$; autogenous cortico-cancellous graft, $n = 6$) | 6 months | Autologous BM-MSCs | 138 ± 22 (coral pieces), $8.28 \pm 1.32 \times 10^6$ (BM-MScs) | Radiology, histology and CT | Non-significant differences in newly formed bone in defects were seen in coral/MSCs and autograft groups. Significant differences in radiological scores in MSCs/coral and autograft groups were seen (21% and 100%, respectively) | Viateau et al. (2007) |
| Postero-lateral lumbar fusion | 24 (autograft (AG), $n = 6$; cell enriched tricalcium phosphate (TCP), $n = 6$; TCP alone, $n = 6$; TCP/10 ml bone marrow, $n = 6$) | 6 months | Bone marrow (selective cell retention technology), tricalcium phosphate | – | Micro-CT, manual palpation and histology | 33% and 25% of the MSCs and AG groups had fused vertebrae while bone marrow/TCP had 8% and TCP alone had 0% fusion | Gupta et al. (2007) |
| Critical sized tibial defect (stabilized by internal locking compression plate and external skeletal fixator) | 45 animals ($n = 15$ in each group) autologous bone chips of iliac crest graft; ceramic scaffold only; ceramic scaffold + cells | 20–24 weeks | Autologous BM-MSCs, silicon stabilized tricalcium phosphate | $70–100 \times 10^6$ | Micro-radiography, histology and vascular density | Radiography revealed best healing in cell/scaffold treated group. Histology showed better vascular density in cell/scaffold group. Partial bone deposition occurred at periphery of bone stumps (scaffold group) | Giannoni et al. (2008) |

(continued)

**Table 7.4** (continued)

| Model type | Number of animals | Model defect size/study period | Biomaterial used | Cell dose/ assembly | Evaluation criteria | Overall result | References |
|---|---|---|---|---|---|---|---|
| Anterior cervical discectomy (C3-C4 and C4-C5) | 18 animals (gelfoam sponge, $n = 6$; MSCs + gelfoam, $n = 6$; gelfoam sponge + MSCs + pentosan polysulphate, $n = 6$) | 3 months | Allogeneic MSCs, Gelfoam sponge, pentosan polysulphate (PPS) | $1 \times 10^6$ | Radiology and histology | New bone formation in gelfoam sponge only and MSCs/gelfoam group was observed in 9/12, 11/12 animals, respectively. In group additionally implanted with PPS had significantly higher cartilaginous tissue within the interbody cages as compared to other groups. Only 1/12 in the PPS group had new bone formation | Goldschlager et al. (2010) |
| Critical sized tibial bone defect (3 cm, stabilized by plate) | 21 animals (treated with mineralized collagen, $n = 7$; human MSCs, $n = 7$ and ovine MSCs, $n = 7$) | 3 months ($n = 2$ each group) and 6 months ($n = 5$) | Human BM-MSCs and oBM-MSCs and mineralized collagen scaffold (matrices) | $2 \times 10^7$ (oBM-MSCs or human BM-MSCs) | Radiology and histology | Radiology and histology revealed significantly better bone formation upon oBM-MSCs seeded on matrices implantation compared to matrices and human BM-MSCs seeded matrices implantation. However, human BM-MSCs did not evoke any local or systemic reaction | Niemeyer et al. (2010) |

| | | | | | | | |
|---|---|---|---|---|---|---|---|
| Critical size tibial segmental defect (3cmm length and 8 mm diameter) (stabilized by interlocking nail) | 24 animals (control scaffold only, $n = 12$; cell + scaffold) | 9 months | BM-MSCs | $225 \times 10^6$ | Radiology, CT, biomechanical, histology | The cell treated group had significantly higher callus formation and rate of bone apposition in the defect | Field et al. (2011) |
| Cervical interbody fusion (C3-C4) | 30 animals (autograft (AG) implantation, $n = 6$; hydroxyapatite/ tricalcium phosphate (HA/TCP), $n = 6$; HA/ TCP + 5million cells, $n = 6$; HA/TCP + ten million cells, $n = 6$; control unoperated sham, $n = 6$) | 3 months | Allogeneic BM-MSCs + hydroxyapatite/ tricalcium phosphate (HA/TCP) | $5 \times 10^6$ and $10 \times 10^6$ | CT, functional radiography, histomorphological, biomechanical analysis | 9/12 of the cell/scaffold treated animals had continuous bony bridging while AG had 1/6 and HA/TCP had 2/6. New bone density was 121% higher in cell groups compared to HA/TCP and 128% as compared to AG groups. Functional radiography revealed significantly reduced macromotion between C3/C4 | Goldschlager et al. (2011) |
| Critical sized tibial segmental bone defect | 24 animals (6 in allogeneic MSCs treated + scaffold; 6 in autologous cell treated + scaffold; 6 scaffold only; 6 autologous graft | 12 weeks | BM-MSCs + scaffold (polycaprolactone + β-tricalcium phosphate) | $35 \times 10^6$ | Radiology, micro-computed tomography (micro-CT), biomechanical testing, histology | No significant differences in bone formation in allogeneic and autologous groups but better bone formation than scaffold treated. However, biomechanical testing failed to show any differences in bone formation in cell loaded and scaffold loaded | Berner et al. (2013) |

(continued)

**Table 7.4** (continued)

| Model type | Number of animals | Model defect size/study period | Biomaterial used | Cell dose/ assembly | Evaluation criteria | Overall result | References |
|---|---|---|---|---|---|---|---|
| Metatarsal bone defect (25 mm) supported by 3.5 mm dynamic compression plate | 10 animals (scaffold, $n = 4$; MSCs on scaffold, $n = 5$) | 6 months | BM-MSCs + acropora coral scaffold | $7.5 \pm 1.2 \times 10^6$ | Micro-CT and histology | Two-fold increase in bone formation with cell + scaffold transplantation as compared to scaffold only at 6 months | Manassero et al. (2013) |
| Radial and tibial nerve section-resection (1 cm) | 7 animals ($n = 2$ tibial nerve and $n = 2$ radial nerves: Treated with scaffold + PRP; $n = 4$ tibial nerve and $n = 2$ radial nerves; control $n = 1$ tibial nerve and $n = 3$ radial nerves: Scaffold + pooled human serum) | 3 and 6 months | BM-MSCs, PRP and scaffold (neurolac$^{RX}$ biodegradable scaffold) | $30–50 \times 10^6$ (repair of nerve after 3 weeks) | Clinical evaluation, neuro-physiological study, morphology study | At 6 months significant myelinated nerve fibres and increased conduction in MSCs treated with or without PRP | Casanas et al. (2014) |
| Postero-lateral lumbar fusion (L4–L5) | 34 animals (lumbar fusions: Treated with autograft $n = 17$ on right half and allograft on left half $n = 17$; hydroxyapatite $n = 17$ on left half; hydroxyapatite + cells (hybrid scaffold) on right half $n = 17$) | 6 months | Osteo-induced BM-MSCs, scaffolds, mineral scaffold and hybrid scaffold | $50 \times 10^6$ | Clinical observation, radiography, CT, histology, histomorphometry | Lumbar fusion rates were higher for auto and allograft (70%) than mineral scaffold (22%) and hybrid scaffold (35%) based on histology and CT. Quantity of bone formation is also higher in autograft and allograft. Hybrid scaffold had higher but statistically insignificant differences from hydroxyapatite scaffold based on histology | Cuenca-Lopez et al. (2014) |

| | | | | | | | |
|---|---|---|---|---|---|---|---|
| Midshaft critical sized tibial bone defect | 8 animals (*n* = 4 control; *n* = 4 treatment group) | 18 weeks | Autologous BM-MSCs on allogeneic graft | $6 \times 10^7$ | Radiography, CT and histology | More conspicuous healing and union between host bone and allograft in cell treated group. Complete remodelling at 18 weeks in cell treated groups. Histologically marrow cavity re-established, intertrabecular spaces filled with adipose marrow | Fernandes et al. (2014) |
| Femoral defect (10 mm diameter and 3 cm deep) | 12 animals (right femur treated with scaffold plus cells; left femur with scaffold) | 8 and 16 week | BM-MSCs derived osteoblasts + β-TCP | $2 \times 10^6$ loaded on β-TCP | Radiography, histology | Bone defect was successfully repaired at 16 week. Sufficient volume of bone tissue volume formed in cell groups compared to scaffold only | Li et al. (2014) |
| Postero-lateral spinal fusion (three-level bilateral postero-lateral inter-transverse fusion, L1–L6) | 6 animals (group I: Cell-seeded hydroxyapatite; at L-L2; group II: Cell-seeded HA/tricalcium phosphate at L2–L3; autograft at L5–L6) | 12 weeks | BM-MSCs, hydroxyapatite (HA), tricalcium phosphate (TCP) | $5–6 \times 10^7$ | Radiography, manual palpation, histology, scanning electron microscopy | TCP/HA achieved superior lumbar inter-transverse fusion compared to HA but autograft had best fusion | Shamsul et al. (2014) |

(continued)

**Table 7.4** (continued)

| Model type | Number of animals | Model defect size/study period | Biomaterial used | Cell dose/ assembly | Evaluation criteria | Overall result | References |
|---|---|---|---|---|---|---|---|
| Sinus augmentation | 16 animals (ficoll density isolated MSCs, $n = 8$; bone marrow aspirate, $n = 8$) | 8 and 16 weeks | BM-MSCs, bone marrow aspirate, bovine bone mineral | – | Histological and histomorphometrical, CT | The approach of using BBM and MSCs in combination with fibrin adhesive was sufficient for new bone formation as the FICOLL experiment indicated. However, due to significantly lower cell numbers isolated using the BMAC in sheep, less new bone was formed in the test arm | Ardjomandi et al. (2015) |
| Critical sized tibial defect (3 cm, plate stabilized) | 24 animals (polycaprolactone-hydroxyapatite scaffold (PCL/HA), $n = 8$; PCL/HA + $100x10^6$, $n = 8$; autologous bone defect, $n = 6$) | 12 months | Allogeneic BM-MSCs, PCL/HA (4 weeks after defect surgery) | $100 \times 10^6$ cells | Biomechanical testing and micro-CT, histology and immune-histochemistry, scanning electron microscopy | Improved bone regeneration in cell/scaffold group comparable to autograft group | Berner et al. (2015) |
| Segmental mandibular bone defect | 15 animals ($n = 5$ control, serum scaffold; $n = 10$ treatment: cell + scaffold) | 12 and 32 weeks | Autologous BM-MSCs and autologous serum scaffold | $25–30 \times 10^6$ | Computed and micro-CT | Complete bone healing in treatment group was observed at 12 and 32 weeks but only at 32 week in control | Gallego et al. (2015) |

| | | | | | | | |
|---|---|---|---|---|---|---|---|
| Lumbar interbody spine fusion | 36 animals [$n = 8$ cortico-cancellous autograft; n = 8 in lower concentration ($2.5 \times 10^6$) cell group; $n = 8$ in medium cell concentration ($6.25 \times 10^6$) group; $n = 8$ in higher cell concentration ($12.5 \times 10^6$) group] | 16 weeks | Allogeneic BM-MSCs, scaffold (hydroxyapatite, tricalcium phosphate and type I collagen) | $2.5 \times 10^6$; $6.25 \times 10^6$; $12.5 \times 10^6$ | Plain radiography, CT, histopathology, histomorphometric and biomechanical competency | No adverse reaction to cell implantation. The cell/scaffold implantation achieved similar or better fusion as that of autograft after 16 weeks | Wheeler et al. (2016) |
| Tibial (2 holes each in right and left) and femoral defects (2 holes each in right and left) (8 mm diameter and 4 mm deep) | 10 ($n = 8$ defects per animals, group I: Scaffold without cells; group II: Scaffold seeded with cells in static conditions; group III: Scaffold seeded with cells in dynamic conditions; group IV: Control empty defects) | 2 and 4 months | Autologous osteodifferentiated BM-MSCs + hydroxyapatite | $2 \times 10^6$ seeded on scaffold | Macroscopy, radiography, micro-CT and histology | Tibial bone is characterized by limited repair capability as compared to femur in which cell implantation is not crucial. Dynamic cell-loaded implants performed better compared to groups I and II | Lovati et al. (2016) |

(continued)

**Table 7.4** (continued)

| Model type | Number of animals | Model defect size/study period | Biomaterial used | Cell dose/ assembly | Evaluation criteria | Overall result | References |
|---|---|---|---|---|---|---|---|
| Osteoporotic bone model (ovariectomy induced, 12 mm × 4 mm defect size) | 8 animals | 10 months | AD-MSCs, strontium hydroxyapatite (SrHA) | – | Histological, histomorphometry, micro-CT and elemental analysis | cSrHA along with MSCs enhanced osteogenic ability with mature lamellar bone formation as compared to bare HA, SrHA and tissue engineered HA implanted groups. Histomorphometry data substantiated improved osteogenesis on par with material resorption, as cSrHA implanted group exhibited highest regeneration ratio. Density histograms from micro-CT further signified the enhanced osteointegrative ability of cSrHA implants. The therapeutic potential of cSrHA in osteoporotic bone healing and proposes the use of allogeneic AD-MSCs for fabricating "off the shelf tissue engineered products" | Chandran et al. (2017) |

2016). Thus, *ex vivo* cell compatible tissue engineered constructs may not necessarily prove useful under *in vivo* environment.

### 7.7.4 Tendon Defects

The tendons and ligaments have limited ability to self-heal due to low cell concentration and associated low matrix turnover. For proper functioning, the healed tissue should be free of adhesions (Gugjoo et al. 2018a). Healing potential of stem cells has been studied in various types of sheep tendon models. An improved healing but variably has been demonstrated (Table 7.5).

MSCs from bone marrow (Crovace et al. 2008; Lacitignola et al. 2014), amniotic fluid (Colosimo et al. 2013), and peripheral blood (Martinello et al. 2013) has been demonstrated to improve tendon healing in sheep collagenase-induced Achilles and digital flexor tendinitis, respectively. Tissue remodelling and improved structural organization was observed in healed tendons (Martinello et al. 2013). Even mononuclear cells (MNCs) too had improved tendon healing (Crovace et al. 2008). BM-MSCs implanted locally were demonstrated to survive up to 6 weeks and engraft efficiently (Colosimo et al. 2013; Lacitignola et al. 2014). MSCs and MNCs had shown comparable fibre and extracellular matrix restoration in treated tendons (Table 7.5). However, the CD34 positive cells were highly appreciable in the BM-MNCs treated cases (Crovace et al. 2008). Similarly, PB-MSCs implantation in sheep digital flexor tendon (DFT) model had improved the healing of tendon.

MSCs from amnion (Am-MSCs) together with scaffold had improved diaphragmatic tendon repair. The repaired tendon tissues had better functional and mechanical parameters in comparison to the tendons treated with acellular grafts (Fuchs et al. 2004; Turner et al. 2011). Implantation of MSCs along with demineralised bone matrix in a sheep tendon-bone defect model also had promoted healing. Tissue repaired with MSCs laden on allogeneic graft had better mechanical and histological properties as compared to that of the xenogeneic graft assembly (Thangarajah et al. 2016).

### 7.7.5 Cardiac Ailments

Sheep outperform other species as cardiovascular model animal (DiVincenti et al. 2014). Among various implantation routes, intracoronary implantation of allogeneic BM-MSCs had been demonstrated to be safe with marked efficacy in sheep model of acute myocardial infarction (AMI) (Table 7.6) (Zhao et al. 2012; Houtgraaf et al. 2013). 3-D echo guided intramyocardially delivered BM-MSCs also offer safety with better results (Cheng et al. 2013). Intra-myocardial BM-MSCs delivery in MI however, may attain therapeutic threshold. The higher cell concentration may not always be useful rather might further damage the area (Dixon et al. 2009; Hamamoto et al. 2009).

**Table 7.5** In vivo preclinical experimental tenogenic mesenchymal stem cell studies in sheep

| Model type | Number of animals | Model defect size/ study period | Biomaterial used | Cell dose/ assembly | Evaluation criteria | Overall result | References |
|---|---|---|---|---|---|---|---|
| Tendinitis of the Achilles tendon (collagenase-induced) | 30 ($n = 6$ each group: BM-MSCs; BM-MNCs; fibrin; saline; sham control) | 8 weeks after treatment | BM-MSCs, BM-MNCs and fibrin glue after 2 weeks of induced tendinitis | $1.1 \times 10^6 \pm 1.4 \times 10^5$ (MSCs); $101 \times 10^6 \pm 26 \times 10^6$ (MNCs) | Histomorphology, immuno-histochemistry of collagen type I, III, cartilage Oligomeric matrix protein (COMP) and CD34 positive cells expression | BM-MSCs and BM-MNCs showed similar restoration of the architecture of fibres and extra cellular matrix (ECM), with a high expression of collagen type I and COMP and a very low expression of collagen type III in treated tendons. The presence of CD34 positive cells was appreciable in the BMMNC group while few cBMSC showed this cluster of differentiation, not expressed in tendons treated with fibrin or saline | Crovace et al. (2008) |

| | | | | | | | |
|---|---|---|---|---|---|---|---|
| Deep digital flexor tendons lesions (bilateral) | 18 animals (groups 1 and 4 received injections of 1 ml of PRP into the tendon; groups 2 and 5 1 ml of PRP + autologous MSCs; groups 3 and 6 received autologous PB-MSCs in 1 ml of hyaluronic acid) | 30 and 120 days | Autologous PB-MSCs, PRP, hyaluronic acid (only left limb lesions treated while contralateral limb served as control) | PRP ($882 \pm 199 \times 10^3$/ml); MSCs ($10 \times 10^6$) | Clinical, ultrasonographic, histology, immuno-histochemical expression | Significant differences were found between treated groups and their corresponding controls (placebo) regarding tendon morphology and extracellular matrix (ECM) composition. However, our results indicate that the combined use of PRP and MSCs did not produce an additive or synergistic regenerative response and highlighted the predominant effect of MSCs on tendon healing, enhanced tissue remodelling and improved structural organization | Martinello et al. (2013) |

(continued)

**Table 7.5** (continued)

| Model type | Number of animals | Model defect size/ study period | Biomaterial used | Cell dose/ assembly | Evaluation criteria | Overall result | References |
|---|---|---|---|---|---|---|---|
| Tendinitis of the Achilles tendon (collagenase-induced, bilateral) | 9 animals (BM-MSCs/fibrin glue, $n = 9$; fibrin glue, $n = 9$) | 3, 4 and 6 weeks | Allogeneic BM-MSC (transfected with red fluorescent protein), fibrin glue | $6 \times 10^7$ after 2 weeks | Evaluation for morphology, collagen I deposition and presence of CD34 + cells | BM-MSCs into tendon lesions had positive effects on the injured tendons. The BM-MSCs survived at three, four and 6 weeks after treatment. These cells enhanced quality healing of tendons as compared to the controls, where no labelled cells were detected. Interestingly, we demonstrated high expression of CD34+ cells in tendons that had been treated with BM-MSCs | Lacitignola et al. (2014) |

| | | | | | | | |
|---|---|---|---|---|---|---|---|
| Tendon defect model of retraction | 10 animals (cells + Xenogenic demineralised bone matrix, $n = 5$; cells + allogeneic demineralised bone matrix, $n = 5$) | 6, 9 and 12 weeks | BM-MNCs | – | Force plate analysis, CT, histology | Allograft was associated with significantly higher weight bearing as compared to xenograft. Greater remodelling of demineralised bone matrix into tendon like tissue at defect area and more direct type of enthesis characterized by significantly more fibrocartilage. No failure of tendon-bone healing was noted in either group | Thangarajah et al. (2016) |

(continued)

**Table 7.5** (continued)

| Model type | Number of animals | Model defect size/ study period | Biomaterial used | Cell dose/ assembly | Evaluation criteria | Overall result | References |
|---|---|---|---|---|---|---|---|
| Lesion of lateral branch of the ovine deep digital flexor tendon (lateral border) | 50 (group 1: Sacrificed at 1 day ($n = 2$), treated with MION-labelled MSCs. Group 2: Sacrificed 1–2 weeks ($n = 4$), three sheep treated with MION-labelled MSCs and one sheep with PBS. Group 3: 4 weeks ($n = 12$), six sheep treated with non-labelled MSCs and six sheep with PBS. Group 4: 12 weeks ($n = 16$), eight sheep treated with non-labelled MSCs and eight sheep with PBS. Group 5: 24 weeks ($n = 16$), eight sheep treated with non-labelled MSCs and eight sheep with PBS) | 1 day, and 1–2, 4 12 and 24 weeks | Magnetic iron-oxide nanoparticles labelled autologous BM-MSCs (implanted 2 weeks after lesion) | $5 \times 10^6$ | MRI, histology and flow cytometry | Cell treated limbs indicated cellular distribution throughout the tendon synovial sheath but restricted to the synovial tissues, with no MSCs detected in the tendon or surgical lesion. The lesion was associated with negligible morbidity with minimal inflammation post-surgery. Evaluation of both cell treated and control lesions showed no evidence of healing of the lesion at 4, 12 and 24 weeks on gross and histological examination | Khan et al. (2018) |

**Table 7.6** In vivo preclinical mesenchymal stem cell studies on cardiac/arterial models in sheep

| Model type | Number of animals | Model defect size/study period | Biomaterial used | Cell dose/ assembly | Evaluation criteria | Overall result | References |
|---|---|---|---|---|---|---|---|
| Foetal cardiac development (intraperitoneal implantation of MSCs into sheep foetus at 55–62 days of gestation) | 21 animals | Analysis done at 115–131 days of gestation (appx 50–55 days later) | GFP transducted MSCs from human BM, foetal brain and liver | – | Immuno-histochemistry | Cells were detected in 20/21 foetal hearts. At least 10 areas of cell engraftment were reported. Human areas represented 43.2% areas of the total Purkinje fibre areas. No difference in engraftment could be observed based on cell type. ~0.01% cardiomyocytes were from human origin in sheep foetal heart. In conclusion, stem cells take part in development of heart | Airey et al. (2004) |
| Myocardial infarction (coronary ligation) | 47 animals (group I: 25 million cells, $n = 7$; group II: 75 Million cells, $n = 7$; group III: 225 Million cells, $n = 10$; group IV: 450 Million cells, $n = 8$; group V: Free media group) | 8 weeks | Stro-3 positive cells derived from BM | $25 \times 10^6$; $75 \times 10^6$; $225 \times 10^6$; $450 \times 10^6$ (injected at border zone after 1 h) | Echocardiography, histology, collagen content and MMPs level of myocardium | End diastolic volume increased in all groups but in group I and II, it was attenuated. ICTP, an index of collagen degradation was more in group I. Matrix metalloproteinases (MMPs) were higher in group I as compared to other groups including group V | Dixon et al. (2009) |
| Transmural myocardial infarction (coronary ligation) | 46 animals (group I: 25 million cells, $n = 7$; group II: 75 Million cells, $n = 7$; group III: 225 million cells, $n = 10$; group IV: 450 Million cells, $n = 8$; group V: Free media group) | 4 and 8 weeks | Stro-3 positive cells derived from BM | $25 \times 10^6$; $75 \times 10^6$; $225 \times 10^6$; $450 \times 10^6$ (injected at border zone after 1 h) | Echocardiography, CD31 and α smooth muscle actin immune-histochemistry | No evidence of myocardial regeneration. Low dose cell groups had better ejection fraction and CD31 and α smooth muscle actin immune-histochemistry showed increased vascular density at border zone | Hamamoto et al. (2009) |

(continued)

**Table 7.6** (continued)

| Model type | Number of animals | Model defect size/study period | Biomaterial used | Cell dose/assembly | Evaluation criteria | Overall result | References |
|---|---|---|---|---|---|---|---|
| Myocardial infarction (MI) | 10 animals (MSCs treated, $n = 5$; MI only, $n = 5$) | 12 weeks | Allogeneic BM-MSCs | $2 \times 10^8$ (six injections each containing $3.33 \times 10^7$) | Histology, determination of apoptosis, Western blot, determination of SERCA2a activity and calcium uptake | Remodelling and contractile strain alteration was reduced in cell treated cases. Cardiomyocyte hypertrophy was significantly attenuated with normalization of hypertrophy of related signalling proteins like phosphatidylinositol 3-kinase-α (PI3K- α), PI3K-γ, extracellular signal-related kinase (ERK), phosphorylated ERK as compared to control. Imbalance of calcium handling proteins was prevented by MSCs. Upregulation of pro-apoptotic protein BCL-2 was normalized, apoptosis of cardiomyocyte was reduced and there was less fibrosis | Zhao et al. (2012) |
| Chronic myocardial injury (anterior wall myocardial infarction) | 39 animals ($n = 10$: $25 \times 10^6$; $n = 9$: $75 \times 10^6$; $n = 10$: $225 \times 10^6$; $n = 10$ non-conditioned control media) | 8 weeks after study design | Allogeneic MSCs | $25 \times 10^6$; $75 \times 10^6$; $225 \times 10^6$ | Histological, echocardiography | Highest cell dose lead to incremental increase in ejection fraction and wall thickening in infarct and border zones as compared to control. Highest density of arteriole density in infarct and border zone in highest cell treated as compared to control | Cheng et al. (2013) |

| | | | | | | | |
|---|---|---|---|---|---|---|---|
| Acute myocardial infarction | 68 [(34 died and 4 died at 8 weeks); placebo (10 animals); 12.5 (7 animals); 25 (7 animals); 37.5 (6 animals) million cells] | 8 weeks | Mesenchymal precursor cells | $12.5 \times 10^6$; $25 \times 10^6$; $37.5 \times 10^6$ | Invasive haemodynamics, echocardiography | Global left ventricular ejection fraction as measured by pressure–volume loop analysis deteriorated in controls to 40.7 ± 2.6% after 8 weeks. In contrast, MPC treatment improved cardiac function to 52.8 ± 0.7%. Echocardiography revealed significant improvement of both global and regional cardiac functions. Infarct size decreased by 40% in treated sheep, whereas infarct and border zone thickness were enhanced. Left ventricular adverse remodelling was abrogated by MPC therapy, resulting in a marked reduction of left ventricular volumes. Blood vessel density increased by >50% in the infarct and border areas. Compensatory cardiomyocyte hypertrophy was reduced in border and remote segments, accompanied by reduced collagen deposition and apoptosis. No microinfarctions in remote myocardial segments or histological abnormalities in unrelated organs were found | Houtgraaf et al. (2013) |

(continued)

**Table 7.6** (continued)

| Model type | Number of animals | Model defect size/study period | Biomaterial used | Cell dose/ assembly | Evaluation criteria | Overall result | References |
|---|---|---|---|---|---|---|---|
| Acute myocardial infarction (anteroapical) | 32 animals ($n = 8$ each in transfected MSCs treated; MSCs treated and PBS injected; VEGF treated groups) | 1 month | VEGF transfected BM-MSCs (1 week after defect) | $2 \times 10^7$ | Magnetic resonance, echocardiography | 31% decrease in infarct size post treatment. Myocardial salvage occurred mostly at subendocardium level. Ejection fraction recovered to baseline value. All these effects were predominantly observed in transfected MSCs treated animals | Locatelli et al. (2015) |
| Acute myocardial infarction (diagonal branch) | 10 animals ($n = 5$: MSCs; $n = 5$: PBS control) | 1 month | Human MSCs | $1 \times 10^6$ cells per kg body weight | Echocardiography and perfusion scintigraphy | Cell treated animals had significant improvement in myocardial perfusion compared with baseline measurements. No change in ventricular dimensions in either group | Dayan et al. (2016) |
| Acute myocardial infarction (anterior apical infarct) | 18 animals (control, PBS $n = 6$; transfected BM-MSCs, $n = 6$; BM-MSCs, $n = 6$) | 2 months | Allogeneic human hypoxia-inducible factor 1-α transfected BM-MSCs, simple BM-MSCs | $2 \times 10^7$ (BM-MSCs); $2x10^7$ (transfected BM-MSCs) | Cardiac magnetic resonance, echocardiography, histology | Infarct volume decreased and increased ejection fraction with transfected BM-MSCs compared to BM-MSCs was seen on cardiac magnetic resonance and echocardiography. Histology showed improved angio/ arteriogenesis in transfected cell group. In control neither parameter improved | Hnatiuk et al. (2016) |
| Acute myocardial infarction | 18 animals (MSCs treated group, $n = 6$; endothelial cell (ECs) treated, $n = 6$; control, $n = 6$) | 2 months | BM-MSCs and endothelial cells | $27 \times 10^6$ (MSCs); $27 \times 10^6$ (ECs) | Echocardiography, electron microscopy and immune-histochemistry | Cell treated groups showed significant improvement in ejection fraction as compared to control. Vascular density also improved with cardiomyocytes in variable stages of development | Rabbani et al. (2017) |

| | | | | | | | |
|---|---|---|---|---|---|---|---|
| Pulmonary artery patching | 7 animals (1–2 weeks, $n = 4$; 4–6 weeks, $n = 3$) | 2 and 6 weeks | Autologous MSCs and endothelial progenitor cells (EPCs) seeded onto poly-4-hydroxybutyrate (P4HB)-coated polyglycolic acid (PGA) nonwoven biodegradable mesh scaffolds (10 × 20 mm) | $2 \times 10^5$ MSCs and $2 \times 10^6$ EPCs per square centimetre. | Histology and immune-histochemistry | One week after implantation, seeded patches had surface thrombus formation and macrophage infiltration. Seeded patches implanted for 2 weeks showed granulation tissue, early pannus formation, macrophages, foreign body giant cells around disintegrating polymer and early angiogenesis (microvessel formation). After 4 weeks in vivo, seeded patches contained glycosaminoglycans, collagen and coverage of the luminal surface by host artery-derived pannus containing α-SMA-positive cells and laminated elastin; polymer scaffold degradation was almost complete with replacement by fibrous tissue containing viable cells | Mendelson et al. (2007) |
| Carotid artery resection (40 mm) | 7 animals ($n = 4$, treated with cells/scaffold and each 2 animals evaluated at 2 month and 5 month; control, $n = 3$) | 2 and 5 months | BM-MSCs differentiated smooth muscle like cells (SMCs) and endothelial cells (ECs) | SMCs ($2 \times 10^7$) in collagen gel and ECs ($1 \times 10^7$) implanted on decellularized scaffold | Histology, immuno-histochemistry, electron microscopy | Tissue engineered blood vessels (TEBVs) were patent, anti-thrombotic and stable up to 5 months. TEBVs had smooth muscle, endothelium and collagen and elastin at both the time points (2 and 5 months). Non-seeded grafts occluded within 2 weeks | Zhao et al. (2010) |

(continued)

**Table 7.6** (continued)

| Model type | Number of animals | Model defect size/study period | Biomaterial used | Cell dose/assembly | Evaluation criteria | Overall result | References |
|---|---|---|---|---|---|---|---|
| Surgical resection of carotid artery (5 cm) | 24 animals (n=6 animals, groups implanted either with PVA/dx along with MSCs or with cells and anticoagulant or PTFE grafts[Rx]) | 24 weeks (each animal sacrificed at 4, 8, 12, 16, 20 and 24 weeks) | Human WJ-MSCs along with polyvinyl alcohol and dextran (PVA/dx; 90/10 v/v) | $10^4$ | Graft explanation, Doppler studies, histopathology, immune-histochemistry, scanning electron microscopy, digital image analysis | PVA/dx grafts could maintain similar or higher patency rate, systolic/diastolic laminar blood flow velocities comparable to PTFE grafts. CD12 and α-actin staining presented similar results in PVA/dx/MSCs and PTFE groups. Fibrosis layer was lower and endothelial cells were only detected at graft artery transitions near the MSCs implantation sites | Alexandre et al. (2014) |

The infusion rate may also affect the results. In normal heart, 1.25 million cells per minute had appeared safer as compared to the 2.5 million cells that had led to the myocardial necrosis. In an infarcted myocardium, MSCs at the rate of 01 million cells/min had led to sluggish coronary flow after 25 million cell implantation while the cell implantation at 0.5 million cells had led to sluggish flow after 40 million dose (Houtgraaf et al. 2013).

Intraperitoneal implantation of xenogeneic MSCs from BM, foetal liver at 55–62 days of gestation had engrafted into heart and exhibited properties of various cardiac cells (Airey et al. 2004). These xenogenic MSCs too had enhanced myocardial perfusion although without any significant effect on ejection fraction (Dayan et al. 2016). To evaluate long term effects of MSCs extended studies are desired. Bone marrow implantation may not be feasible option to treat myocardial infarction (Bel et al. 2003).

Even transfection of MSCs with some specific factors may improve their healing potential. Among notable options are hypoxia-inducible factor 1-α (HIF1-α) and vascular endothelial growth factor (VEGF). BMSCs transfected with these factors had significantly reduced infarct size in a sheep acute myocardial infarction (AMI) model. Improved results might have been due to enhanced angio−/arteriogenesis through cell paracrine secretion (Locatelli et al. 2015; Hnatiuk et al. 2016).

Apart from myocardial repair, MSCs have been evaluated in skeletal and blood vessel repair. MSCs had significantly improved rotator muscle cuff repair irrespective of their source. Experimental muscles had increased contraction force, vascularity, and increased myocyte concentration over adipocytes (Coleman et al. 2011). For blood vessel repair, MSCs have been employed along with tissue engineered scaffold patches (Mendelson et al. 2007; Zhao et al. 2010). The cell-seeded scaffold patches implanted in sheep pulmonary artery had remodelled to a layered and viable tissue. The tissue had been well integrated into the native arterial wall. The key remodelling processes included are intimal overgrowth at the luminal surface (pannus formation; neointima) and granulation tissue formation, and fibrosis with foreign body reaction (Mendelson et al. 2007). MSCs laden tissue engineered blood vessels (TEBVs) had also been implanted in case of carotid artery resection model. The constructs were patent, anti-thrombotic and stable up to 5 months. TEBVs had smooth muscle, endothelium and collagen and elastin up to 5 months. Non-seeded grafts contrarily were occluded within 2 weeks (Zhao et al. 2010).

### 7.7.6 Nervous System

MSCs may trans-differentiate into neuron-like and/glial-like cells, in addition to PAX3 expressing progenitor cells (Chen et al. 2001; Ceccarelli et al. 2015). The differentiated cell functionality remains controversial (Przyborski et al. 2008). Due to these properties MSCs have been evaluated in various nerve ailments as detailed in Table 7.7. Additionally MSCs carry potential to be evaluated for scrapie and post hypoxia sequelae of brain (Mediano et al. 2015a, b; Ophelders et al. 2016).

**Table 7.7** In vivo preclinical experimental studies of mesenchymal stem cell studies on neurological ailments in sheep

| Model type | Number of animals | Model defect size/ study period | Biomaterial used | Cell dose/ assembly | Evaluation criteria | Overall result | References |
|---|---|---|---|---|---|---|---|
| Radial and tibial nerve resection (1 cm) | 7 animals (MSCs + scaffolds) | 3 and 6 months | BM-MSCs, repairmen with Neurolac™ biocamera scaffold | 30–50 × $10^6$ | Histological, immuno-histochemistry and Morphometry | At 3 months period results were not good. At 6 months results showed significantly myelined nerve fibres with increased conduction as compared to control in both radial and tibial nerves | Casañas et al. (2014) |
| In utero repair of Myelomeningocele (at 75 days of pregnancy) | 4 animals (1 case treated with term derived placental MSCs + foetal membrane (FM) patch; other with early gestation placental MSCs + FM patch; 1 case with FM only; 1 case with skin closure only) | 145 days of pregnancy at delivery time | Placenta derived MSCs in collagen gel | 5 × $10^5$ (early gestation MSCs); 1 × $10^6$ (term gestation MSCs) | Histology and motor assessment | Increased preservation of spinal cord architecture and large neurons in early gestation MSCs treated compared to others. The animal treated with early gestation MSCs was capable of normal ambulation. Limited distal motor function in others | Brown et al. (2016) |
| Myelomeningocele (created at 75 day of pregnancy) | 12 animals (cell treated, $n = 6$; control, $n = 6$) | At term (145 days) | Human placental MSCs | 5 × $10^5$ | Motor functional, histological and immune-histochemical analysis | Significant and clinically relevant improvement of motor function with increased preservation of large neurons with spinal cord | Wang et al. (2015) |

| | | | | | | | |
|---|---|---|---|---|---|---|---|
| Foetal Myelomeningocele (created midgestational-76–81 days) | 21 animals | 25 days | Placental MSCs + porcine small intestinal derived extracellular matrix | | Sheep Locomotor rating (SLR) scale and histology of spinal cord | Foetal viability of 71% (5/8) in group treated with cells while only 17% (1/6) in matrix treated group. SLR and large neuron density was higher in cell treated animals. The cross sectional areas of spinal cord and grey matter were equally preserved. Cell treated animals improve hind limb motor function in lambs | Kabagambe et al. (2017) |
| Global hypoxia-ischemic injury in preterm foetus (umbilical cord occlusion, UCO) | 24 animals (sham UCO, $n = 6$; sham UCO and cell derived vesicle treatment, $n = 6$; UCO treated with saline, $n = 6$; UCO with cell derived vesicle treatment, $n = 6$) | 7 days | Human BM-MSCs derived extracellular vesicles | $4 \times 10^7$ ($2 \times 10^7$ intrauterine injection twice; one 1 h after UCO and other 4 days later) | Electrophysiological parameters and histology of brain | Improved brain function by reducing number and duration of seizures and by preserving baroreceptor reflex sensitivity. Cerebral inflammation remained unaffected | Ophelders et al. (2016) |

(continued)

**Table 7.7** (continued)

| Model type | Number of animals | Model defect size/ study period | Biomaterial used | Cell dose/ assembly | Evaluation criteria | Overall result | References |
|---|---|---|---|---|---|---|---|
| Intervertebral disc degeneration (IVD) (Postero-lateral annulotomy at 3 levels: L3/L4, L4/L5, L5/L6) | 8 animals (MSCs into nucleosus pulposus; MSCs into annulus fibrosus and phosphate saline control) | 6, 9 and 12 months | Allogeneic BM-MSCs 6 months after defect creation | $1 \times 10^6$ | Radiography, MRI, biochemical and histology | Significant improvement in disc height index and disc height in both cell treatment groups at 6 months. Mean Pfirrmann and histopathological grade improved in cell treated animals. Spontaneous repair of antero-lateral annular lesion in cell treated groups | Freeman et al. (2016) |
| Intervertebral disc degeneration (antero-lateral annulus fibrosus lesion; 20 mm wide 6 mm deep: Lesion at L1–L2, L3–L4 and L5–L6) | 45 animals (control $n = 10$; early treatment group at 4 weeks after lesion creation, $n = 24$; late treatment group at 12 weeks, $n = 12$) | 3 and 6 months | Allogeneic BM-MSCs | $1 \times 10^7$ | Histology and gene expression | MSC treatment prevented increased adipose and connective tissue formation, normally forms after damage. MSCs treatment did not prevent slow-to-fast muscle fibre formation. Alteration in gene expression of pro-inflammatory cytokines within the muscle. Increased interleukin-1 expression was reduced by early MSCs treatment. TNF and TGF-β1 expression was increased at 6 months | Dowdell et al. (2017) |

| | | | | | | | |
|---|---|---|---|---|---|---|---|
| Lumbar disc herniation (repair by microdiscectomy) | 18 animals (pMSCs, $n = 6$; MSCs, $n = 6$; control, $n = 6$) | 6 months | Pentosan polysulphate-primed mesenchymal progenitor cells (pMSCs) and MSCs | $5 \times 10^5$ | 3 T MRI, standard radiography, morphologically, histologically, biochemically for their proteoglycans (PGs), collagen and DNA content | Both the MSCs- and pMSCs-groups exhibited significantly less reduction in disc height and lower Pfirrmann grades as compared to the control, but morphologic scores for the pMPC-injected discs were significantly lower. The PG content of the annulus Fibrosus injury site region 1 of pMSCs discs was significantly higher than MSCs and injury control AF1. At the AF1 and contralateral AF2 regions, the DNA content of pMSCs discs was significantly lower than injured control discs and MSCs-injected discs. Histologic and birefringent microscopy revealed increased structural organization and reduced degeneration in pMSC discs as compared to the MSCs and controls | Daly et al. (2018) |

sBM-MSCs may steadily improve radial and tibial nerve repair. Initially up to 3 months poor nerve regeneration and conduction upon MSCs implantation had been demonstrated. However, later at 6 months period myelinated nerve fibres had generated both proximally and distally. Further, nerve action potential had also improved (Casañas et al. 2014).

Amniotic fluid MSCs (AF-MSCs) may improve motor functions and increase large neuron preservation in spinal cord (Wang et al. 2015). The early gestation AF-MSCs in comparison to late stage gestation had better ability to preserve spinal cord architecture and large neurons in myelomeningocele (MMC). Lambs that received early gestation cells had normal ambulation while only limited distal motor function was demonstrated in animals that received late gestation MSCs (Brown et al. 2016). This questions if MSCs are able to maintain their neural specific properties with developmental advancement.

MSCs may offer therapeutics in intervertebral disc degeneration (Shamsul et al. 2014). Sheep intervertebral disc disease model treated with BM-MSCs had quantifiable improvement at 6 months period. Annulus fibrosis implantation may be better than nucleus pulposus transplantation (Freeman et al. 2016). Their disc regeneration potential may be promoted through pentosan polysulphate priming (Daly et al. 2018).

MSCs application in neurological ailments is in infancy and extensive research studies are desired for any conclusion.

### 7.7.7 Respiratory and Urinary Systems

Sheep has been used as an animal model for human respiratory ailments. Acute respiratory distress syndrome (ARDS) in these animals has been created through endotoxaemia. BM-MSCs or lung derived MSCs in combination with other support system had better managed the ARDS. The inflammation along with oedema had subsided post cell implantation (Ingenito et al. 2012; Asmussen et al. 2014; Kocyildirim et al. 2017) (Table 7.8).

MSCs have also been studied in acute renal injuries. The cells had ability to engraft although without any observable beneficial effect (Behr et al. 2007, 2009) (Table 7.8).

### 7.7.8 Reproductive System

As MSCs are able to trans-differentiate into germinal cells, these may be utilized for *in vivo* applications. One available study had demonstrated that transdifferentiated cells carry ability to home to seminiferous tubules of ram testis. Even these cells had formed colonies at the area (Ghasemzadeh-Hasankolaei et al. 2015) (Table 7.9). Initially, it is mandatory to develop differentiation media cocktail that ensures the typical germ cells properties of transdifferentiated cells. This may be followed by *in vivo* studies of these differentiated cells.

**Table 7.8** In vivo preclinical experimental mesenchymal stem cell studies on respiratory and renal affections

| Model type | Number of animals | Model defect size/study period | Biomaterial used | Cell dose/assembly | Evaluation criteria | Overall result | References |
|---|---|---|---|---|---|---|---|
| Emphysema | 10 animals (control scaffold treated $n = 5$; cell treated $n = 5$) | 4 weeks | Autologous lung MSCs + scaffold (1% fibrinogen + 20 µg/ml fibronectin + 20 µg/ml poly-L-lysine) | 5–10 × $10^6$ | Clinical toxicity, tissue mass (CT), lung physiology, scintigraphy, histology | Cells were well tolerated, increased perfusion and tissue mass as compared to control. Histology confirmed cell retention, increased cellularity and extracellular matrix in cell treated animals | Ingenito et al. (2012) |
| Acute respiratory distress syndrome (ARDS, bacterial pneumonia) | 19 animals (control, PlasmaLyte A, $n = 8$; low dose MSCs, $n = 7$; high dose MSCs, $n = 4$) | 24 h | Human BM-MSCs | 5 × $10^6$/kg (low dose); 10 × $10^6$/kg (high dose) | Post-mortem lung wet-to-dry weight ratio, broncho-alveolar lavage, serum biochemical parameters | Cells were well tolerated. $PaO_2/FiO_2$ ratio was significantly improved in cell treated groups compared to control. Median lung water content was significantly lower in high cell treated group than control | Asmussen et al. (2014) |
| Acute respiratory distress syndrome (Escherichia coli exdotoxin induced) | 11 animals (control, $n = 5$; extracorporeal membrane oxygenation (ECMO), $n = 3$; MSCs + ECMO, $n = 3$) | – | BM-MSCs | 40 × $10^6$ | Blood gas results, radiography, histopathology | Better histopathologic changes with less inflammation in combined cell and ECMO treated group | Kocyildirim et al. (2017) |

(continued)

**Table 7.8** (continued)

| Model type | Number of animals | Model defect size/ study period | Biomaterial used | Cell dose/ assembly | Evaluation criteria | Overall result | References |
|---|---|---|---|---|---|---|---|
| Renal ischaemic reperfusion injury (percutaneous transluminal placement of balloon catheter in renal artery) | 24 animals [sham group (balloon catheterization performed but no ischaemic reperfusion injury (IRI), $n = 3$; control group (IRI induced without any cell treatment, $n = 3$); cell treated but no IRI induced, $n = 3$; IRI induced and cells injected 1 h. later, $n = 5$; IRI infused and cells infused 2 weeks afterwards. This group divided into two, one evaluated at 3 week, $n = 5$ and other at 5 week, $n = 5$] | 3–5 weeks | Autologous MSCs | $80 \times 10^6$ | Identification of engrafted cell, immune-phenotype analysis | Renal (tubules and glomeruli) engraftment of MSCs. MSCs expressed tubular epithelial cell markers and podocyte phenotype. Significant increase in engraftment of MSCs was seen when injected immediately than at later period | Behr et al. (2007) |
| Renal ischaemic reperfusion injury (bilateral) (balloon catheter) | Sham group; control injected culture medium; cell treated group | 7 days | BM-MSCs | $50 \times 10^6$ | Immuno-phenotypic analysis, renal function, histology and urine analysis | Transplanted cells were found in glomeruli but not tubules. Cell did not express glomerular cell markers (podocin and von Willebrand factor). Functional evaluated revealed no beneficial effects of implanted cells. MSCs failed to repair renal parenchyma and failed to modulate cell death and proliferation or cytokine release | Behr et al. (2009) |

**Table 7.9** In vivo preclinical experimental models of different types in sheep

| Model type | Number of animals | Model defect size/study period | Biomaterial used | Cell dose/assembly | Evaluation criteria | Overall result | References |
|---|---|---|---|---|---|---|---|
| Intraperitoneal implantation of MSCs (76 days) | 7 animals | 2 weeks | Autologous amniotic fluid MSCs | $1 \times 10^5$ to $1.8 \times 10^7$ | PCR, western blotting, cytofluorometry, immunofluorescence | Liver, heart, placenta, umbilical cord, adrenal gland and muscle | Shaw et al. (2011) |
| Transverse subtotal ear resection at 123.4 ± 2.3 days (intact cartilage left to heal by second intention) | 10 animals (2 wounds in each animal covered and uncovered) | 7–14 days prior to term | GFP transfected autologous amniotic-MSCs | 3.22–$4.32 \times 10^9$ (implanted in amniotic cavity) | Histomorphologic, collagen and hyaluronan content | Covered wounds carry significantly lower healing rate compared to non-covered. Significantly lower elastin levels in non-covered wounds. Insignificant difference in hyaluronic acid levels in wound types | Klien et al. (2011) |
| In vivo testicular homing and colony formation of in vitro MSCs derived germ-like cells | 4 animals ($n = 2$ animals in each treatment group; 3 of the 4 testis were treated with treated cells and one non-injected control; 2 testis of lamb also as control. One testis treated with undifferentiated MSCs and other with only medium | 2 months | In vitro MSCs derived germ-like cell through 14 day treatment of retinoic acid; 21 day treatment of retinoic acid; transforming growth factor-β | 15–$20 \times 10^6$ | Histological study | TGF-β transformed group exhibited highest efficiency of homing and colony formation followed by 21 day treatment of retinoic acid. 14 day treatment of retinoic acid exhibited lowest homing and colony formation | Ghasamzadeh-Hasankolaei et al. (2015) |

## 7.8 Conclusion(s)

sMSCs isolation and characterization process is similar to that of other species. sMSCs characteristics are source dependant and are also affected by donor age and health status. A good amount of literature on sMSCs human translational studies on bone, skin, osteochondral and cardiovascular system could be observed. These cells had offered promising therapeutic effects although further extensive studies are desired for their actual utility. Additionally, inadequate MSCs studies on reproductive, nervous, respiratory and urinary system models have failed to provide any conclusion.

## References

Abdalmula A, Dooley LM, Kaufman C, Washington EA, House JV, Blacklaws BA, Ghosh P, Itescu S, Bailey SR, Kimpton WG (2017) Immunoselected STRO-3+ mesenchymal precursor cells reduce inflammation and improve clinical outcomes in a large animal model of monoarthritis. Stem Cell Res Therapy 8:22

Adamzyk C, Kachel P, Hoss M, Gremse F, Modabber A, Hölzle F, Tolba R, Neuss S, Lethaus B (2016) Bone tissue engineering using polyetherketoneketone scaffolds combined with autologous mesenchymal stem cells in a sheep calvarial defect model. J Craniomaxillofac Surg 44(8):985–994

Airey JA, Almeida-Porada G, Colletti EJ, Porada CD, Chamberlain J, Movsesian M, Sutko JL, Zanjani ED (2004) Human mesenchymal stem cells form Purkinje fibers in fetal sheep heart. Circulation 109:1401–1407

Al Faqeh H, Nor Hamdan BM, Chen HC, Aminuddin BS, Ruszymah BH (2012) The potential of intra-articular injection of chondrogenic-induced bone marrow stem cells to retard the progression of osteoarthritis in a sheep model. Exp Gerontol 47:458–464

Alexandre N, Ribeiro J, Gärtner A, Pereira T, Amorim I, Fragoso J, Lopes A, Fernandes J, Costa E, Santos-Silva A, Rodrigues M, Santos JD, Maurício AC, Luís AL (2014) Biocompatibility and hemocompatibility of polyvinyl alcohol hydrogel used for vascular grafting—in vitro and in vivo studies. Biomed Mater Res 102(12):4262–4275

Anderson ML, Dhert WJ, De Bruijn JD, Dalmeijer RA, Leenders H, Van Blitterswijk CA, Verbout AJ (1999) Critical size defect in the goat's os ilium. A model to evaluate bone grafts and substitutes. Clin Orthop Relat Res 364:231–239

Ando W, Heard BJ, Chung M, Nakamura N, Frank CB, Hart DA (2012) Sheep synovial membrane-derived mesenchymal progenitor cells retain the phenotype of the original tissue that was exposed to in-vivo inflammation: evidence for a suppressed chondrogenic differentiation potential of the cells. Inflamm Res 61:599–608

Ardjomandi N, Duttenhoefer F, Xavier S, Oshima T, Kuenz A, Sauerbier S (2015) In vivo comparison of hard tissue regeneration with sheep mesenchymal stem cells processed with either the FICOLL method or the BMAC method. J Craniomaxillofac Surg 43(7):1177–1183

Asmussen S, Ito H, Traber DL, Lee JW, Cox RA, Hawkins HK, McAuley DF, McKenna DH, Traber LD, Zhuo H, Wilson J, Herndon DN, Prough DS, Liu KD, Matthay MA, Enkhbaatar P (2014) Human mesenchymal stem cells reduce the severity of acute lung injury in a sheep model of bacterial pneumonia. Thorax 69(9):819–825

Behr L, Hekmati M, Fromont G, Borenstein N, Noel LH, Lelievre-Pegorier M, Laborde K (2007) Intra renal arterial injection of autologous mesenchymal stem cells in an sheep model in the postischemic kidney. Nephron Physiol 107:65–76

Behr L, Hekmati M, Houcinet K, Faussat A-M, Borenstein N, Noel L-H, Lelievre-Pegorier M, Laborde K (2009) Evaluation of the effect of autologous mesenchymal stem cell injection in a large-animal model of bilateral kidney ischaemia reperfusion injury. Cell Prolif 42:284–297

Bel A, Messas E, Agbulut O, Richard P, Samuel JL, Bruneval P, Hagège AA, Menasché P (2003) Transplantation of autologous fresh bone marrow into infarcted myocardium: a word of caution. Circulation 108(Suppl 1):247–252

Berner A, Reichert JC, Woodruff MA, Saifzadeh S, Morris AJ, Epari DR, Nerlich M, Schuetz MA, Hutmacher DW (2013) Autologous vs. allogenic mesenchymal progenitor cells for the reconstruction of critical sized segmental tibial bone defects in aged sheep. Acta Biomater 9(8):7874–7884

Berner A, Henkel J, Woodruff MA, Steck R, Nerlich M, Schuetz MA, Hutmacher DW (2015) Delayed minimally invasive injection of allogenic bone marrow stromal cell sheets regenerates large bone defects in an sheep preclinical animal model. Stem Cells Transl Med 4:503–512

Boos AM, Weigand A, Deschler G, Gerber T, Arkudas A, Kneser U, EHorch R, Beier JP (2014) Autologous serum improves bone formation in a primary stable silica-embedded nanohydroxyapatite bone substitute in combination with mesenchymal stem cells and rhBMP-2 in the sheep model. Int J Nanomedicine 9:5317–5339

Bornes TD, Adesida AB, Jomha NM (2017) Articular cartilage repair with mesenchymal stem cells after chondrogenic priming: a pilot study. Tissue Eng 24:9–10. https://doi.org/10.1089/ten.TEA.2017.0235

Brown EG, Keller BA, Lankford L, Pivetti CD, Hirose S, Farmer DL, Wang A (2016) Age does matter: a pilot comparison of placenta-derived stromal cells for in utero repair of myelomeningocele using a lamb model. Fetal Diagn Ther 39(3):179–185

Burk J, Glauche SM, Brehm W, Crovace A, Francioso E, Hillmann A, Schubert S, Lacitignola L (2017) Characterisation and intracellular labelling of mesenchymal stromal cells derived from synovial fluid of horses and sheep. Vet J 222:1–8

Caminal M, Fonseca C, Peris D, Moll X, Rabanal RM, Barrachina J, Codina D, García F, Cairó JJ, Gódia F, Pla A, Vives J (2014) Use of a chronic model of articular cartilage and meniscal injury for the assessment of long-term effects after autologous mesenchymal stromal cell treatment in sheep. N Biotechnol 31(5):492–498

Caminal M, Moll X, Codina D, Rabanal RM, Morist A, Barrachina J, Garcia F, Pla A, Vives J (2014a) Transitory improvement of articular cartilage characteristics after implantation of polylactide:polyglycolic acid (PLGA) scaffolds seeded with autologous mesenchymal stromal cells in a sheep model of critical-sized chondral defect. Biotechnol Lett 36:2143–2153

Caminal M, Vélez R, Rabanal RM, Vivas D, Batlle-Morera L, Aguirre M, Barquinero J, García J, Vives J (2017) A reproducible method for the isolation and expansion of sheep mesenchymal stromal cells from bone marrow for use in regenerative medicine preclinical studies. J Tissue Eng Regen Med 11(12):3408–3416

Casañas J, de la Torre J, Solerc F, Garcíab F, Rodellard C, Pumarolac M, Climente J, Solerf R, Orozco L (2014) Peripheral nerve regeneration after experimental section in sheep radial and tibial nerves using synthetic nerve grafts, including expanded bone marrow mesenchymal cells: morphological and neurophysiological results. Injury 45(S4):S2–S6

Ceccarelli G, Pozzo E, Scorletti F, Benedetti L, Cusella G, Ronzoni FL, Sahakyan V, Zambaiti E, Mimmi MC, Calcaterra V, Deprest J, Sampaolesi M, Pelizzo G (2015) Molecular signature of amniotic fluid derived stem cells in the fetal sheep model of myelomeningocele. J Pediatr Surg 50(9):1521–1527

Chamberlain J, Yamagami T, Colletti E, Theise ND, Desai J, Frias A, Pixley J, Zanjani ED, Porada CD, Almeida-Porada G (2007) Efficient generation of human hepatocytes by the intrahepatic delivery of clonal human mesenchymal stem cells in fetal sheep. Hepatology 46(6):1935–1945

Chandran S, Shenoy SJ, Babu SS, Nairm RP, Varma HK, John A (2017) Strontium Hydroxyapatite scaffolds engineered with stem cells aid osteointegration and osteogenesis in osteoporotic sheep model. Colloids Surf B Biointerfaces 163:346–354

Chen J, Li Y, Wang L, Lu M, Zhang X, Chopp M (2001) Therapeutic benefit of intracerebral transplantation of bone marrow stromal cells after cerebral ischemia in rats. J Neurol Sci 189:49–57

Chen F, Zhao C, Zhao Y, Li L, Liu S, Zhu Z, Guan W (2018) The biological characteristics of sheep umbilical cord mesenchymal stem cells. Can J Vet Res 82:216–224
Cheng Y, Yi G, Conditt GB, Sheehy A, Kolodgie FD, Tellez A, Polyakov I, Gu A, Aboodi MS, Wallace-Bradley D, Schuster M, Martens T, Itescu S, Kaluza GL, Basu S, Virmani R, Granada JF, Sherman W (2013) Catheter-based endomyocardial delivery of mesenchymal precursor cells using 3D echo guidance improves cardiac function in a chronic myocardial injury sheep model. Cell Transplant 22(12):2299–2309
Chevrier A, Nelea M, Hurtig MB, Hoemann CD, Buschmann MD (2009) Meniscus structure in human, sheep, and rabbit for animal models of meniscus repair. J Orthop Res 27(9):1197–1203
Coleman SH, Ehteshami J, Kisiday JD, Warren RF, Turner AS (2011) The effects of mesenchymal stem cells on rotator cuff muscle in a chronic injury model in sheep. American Academy of Orthopaedic Surgeons, San Diego
Colosimo A, Russo V, Mauro A, Curini V, Marchisio M, Bernabò N, Alfonsi M, Mattioli M, Barboni B (2013) Prolonged in vitro expansion partially affects phenotypic features and osteogenic potential of sheep amniotic fluid-derived mesenchymal stromal cells. Cytotherapy 15:930–950
Crovace A, Lacitignola L, Francioso E, Rossi G (2008) Histology and immunohistochemistry study of sheep tendon grafted with cBMSCs and BMMNCs after collagenase-induced tendinitis. Vet Comp Orthop Traumatol 21(4):329–336
Cuenca-López MD, Andrades JA, Gómez S, Zamora-Navas P, Guerado E, Rubio N, Blanco J, Becerra J (2014) Evaluation of posterolateral lumbar fusion in sheep using mineral scaffolds seeded with cultured bone marrow cells. Int J Mol Sci 15(12):23359–23376
Cui P, He X, Pu Y, Zhang W, Zhang P, Li C, Guan W, Li X, Ma Y (2014) Biological characterization and pluripotent identification of sheep dermis-derived mesenchymal stem/progenitor cells. Biomed Res Int 2014:786234
Czernik M, Fidanza A, Sardi M, Galli C, Brunetti D, Malatesta D, Della Salda L, Matsukawa K, Ptak GE, Loi P (2013) Differentiation potential and GFP labeling of sheep bone marrow-derived mesenchymal stem cells. J Cell Biochem 114:134–143
Daly CD, Ghosh P, Zannettino ACW, Badal T, Shimmon R, Jenkin G, Oehme D, Jain K, Sher I, Vais A, Cohen C, Chandra RV, Goldschlager T (2018) Mesenchymal progenitor cells primed with pentosan polysulfate promote lumbar intervertebral disc regeneration in an sheep model of microdiscectomy. Spine J 18(3):491–506
Dayan V, Sotelo V, Delfina V, Delgado N, Rodriguez C, Suanes C, Langhain M, Ferrando R, Armand K, Benech A, Touriño C (2016) Human mesenchymal stromal cells improve cardiac perfusion in an sheep immunocompetent animal model. J Invest Surg 29(4):1–8
Delling U, Brehm W, Ludewig E, Winter K, Jülke H (2015a) Longitudinal evaluation of effects of intraarticular mesenchymal stromal cell administration for the treatment of osteoarthritis in a sheep model. Cell Transplant 24:2391–2407
Delling U, Brehm W, Metzger M, Ludewig E, Winter K, Jülke H (2015b) In vivo tracking and fate of intra-articularly injected superparamagnetic iron oxide particle-labeled multipotent stromal cells in a sheep model of osteoarthritis. Cell Transplant 24:2379–2390
Desando G, Giavaresi G, Cavallo C, Bartolotti I, Sartoni F, Aldini NN, Martini L, Parrilli A, Mariani E, Fini M, Grigolo B (2016) Autologous bone marrow concentrate in a sheep model of osteoarthritis: new perspectives for cartilage and meniscus repair. Tissue Eng 22(6):608–619
Desantis S, Accogli G, Zizza S, Mastrodonato M, Blasi A, Francioso E, Rossi R, Crovace A, Resta L (2015) Ultrastructural study of cultured sheep bone marrow-derived mesenchymal stromal cells. Ann Anat 201:43–49
DiVincenti LJ, Westcott R, Lee C (2014) Sheep (*Ovis aries*) as a Model for Cardiovascular Surgery and Management before, during, and after cardiopulmonary bypass. J Am Assoc Lab Anim Sci 53(5):439–448
Dixon JA, Gorman RC, Stroud RE, Bouges S, Hirotsugu H, Gorman JH, Martens TP, Itescu S, Schuster MD, Plappert T, St John-Sutton MG, Spinale FG (2009) Mesenchymal cell transplantation and myocardial remodeling after myocardial infarction. Circulation 120(11):S220–S229
Dominici M, Le Blanc K, Mueller I, Slaper-Cortenbach I, Marini FC, Krause D, Deans R, Keating A, Prockop DJ, Horwitz E (2006) Minimal criteria for defining multipotent mesenchymal

stromal cells. The International Society for Cellular Therapy position statement. Cytotherapy 8(4):315–317

Dooley LM, Abdalmula A, Washington EA, Kaufman C, Tudor EM, Ghosh P, Itescu S, Kimpton WG, Bailey SR (2015) Effect of mesenchymal precursor cells on the systemic inflammatory response and endothelial dysfunction in an sheep model of collagen-induced arthritis. PLoS One 10(5):e0124144

Dowdell J, Erwin M, Choma T, Vaccaro A, Iatridis J, Cho SK (2017) Intervertebral disk degeneration and repair. Neurosurgery 80(3S):S46–S54

Ersek A, Pixley JS, Goodrich AD, Porada CD, Almeida-Porada G, Thain DS, Zanjani ED, Fadel L, Viana BR, Feitosa MLT, Ercolin ACM, Roballo KCS, Casals JB, Pieri NC, Meirelles FV, Martins DS, Miglino MA, Ambrósio CE (2011) Protocols for obtainment and isolation of two mesenchymal stem cell sources in sheep. Acta Cir Bras 26:267–273

Fadel L, VianaII BR, Feitosa MLT, ErcolinII ACM, RoballoII KCS, Casals JB, Pieri NCG, Meirelles FV, dos Santos MD, Miglino MA, Ambrósio CE (2011) Protocols for obtainment and isolation of two mesenchymal stem cell sources in sheep. Acta Cir Bras 26(4):267–273

Feitosa MLT, Fadel L, Beltrão-Braga PCB, Wenceslau CV, Kerkis I, Kerkis A, Júnior EHB, Martins JFP, Martins DS, Miglino MA, Ambrósio CE (2010) Successful transplant of mesenchymal stem cells in induced osteonecrosis of the sheep femoral head. Preliminary results. Acta Cir Bras 25(5):416–422

Feng C, Luo X, He N, Xia H, Lv X, Zhang X, Li D, Wang F, He J, Zhang L, Lin X, Lin L, Yin H, He J, Wang J, Cao W, Wang R, Zhou G, Wang W (2018) Efficacy and persistence of allogeneic adipose derived mesenchymal stem cells combined with hyaluronic acid in osteoarthritis after intra-articular injection in a sheep model. Tissue Eng 24(3–4):219–233

Fernandes MBC, Guimarães JAM, Casadom PL, Cavalcantim AS, Gonçalves NN, Ambrósio CE, Rodrigues F, Pinto ACF, Miglino MA, Duarte MEL (2014) The effect of bone allografts combined with bone marrow stromal cells on the healing of segmental bone defects in a sheep model. BMC Vet Res 10:36

Field JR, McGee M, Stanley R, Ruthenbeck G, Papadimitrakis T, Zannettino A, Gronthos S, Itescu S (2011) The efficacy of allogeneic mesenchymal precursor cells for the repair of an sheep tibial segmental defect. Vet Comp Orthop Traumatol 24:113–121

Freeman BJ, Kuliwaba JS, Jones CF, Shu CC, Colloca CJ, Zarrinkalam MR, Mulaibrahimovic A, Gronthos S, Zannettino AC, Howell S (2016) Allogeneic mesenchymal precursor cells promote healing in postero-lateral annular lesions and improve indices of lumbar intervertebral disc degeneration in an sheep model. Spine 41(17):1331–1339

Fuchs JR, Kaviani A, Oh JT, LaVan D, Udagawa T, Jennings RW, Wilson JM, Fauza DO (2004) Diaphragmatic reconstruction with autologous tendon engineered from mesenchymal amniocytes. J Pediatr Surg 39(6):834–838

Gallego L, Perez-Basterrechea M, Garcıa-Consuegra L, Alvarez-Viejo M, Megas J, Novoa A, Costilla S, Meana A, Junquera L (2015) Repair of segmental mandibular bone defects in sheep using bone marrow stromal cells and autologous serum scaffold: a pilot study. J Clin Periodontol 42:1143–1151

Garcia D, Longo UG, Vaquero J, Forriol F, Loppini M, Khan WS, Denaro V (2015) Amniotic membrane transplant for articular cartilage repair: an experimental study in sheep. Curr Stem Cell Res Therapy 10:77–83

Garcıa-Gareta E, Hua J, Blunn GW (2015) Osseointegration of acellular and cellularized osteoconductive scaffolds: Is tissue engineering using mesenchymal stem cells necessary for implant fixation? J Biomed Mater Res 103A:1067–1076

Ghasemzadeh-Hasankolaei M, Eslaminejad BM, Sedighi-Gilani M (2015) Derivation of male germ cells from ram bone marrow mesenchymal stem cells by three different methods and evaluation of their fate after transplantation into the testis. In Vitro Cell Dev Biol Anim 52(1):49–61

Giannoni P, Mastrogiacomo M, Alini M, Pearce SG, Corsi A, Santolini F, Muraglia A, Bianco P, Cancedda R (2008) Regeneration of large bone defects in sheep using bone marrow stromal cells. J Tissue Eng Regen Med 2:253–262

Goldschlager T, Ghosh P, Zannettino A, Gronthos S, Rosenfeld JV, Itescu S, Jenkin G (2010) Cervical motion preservation using mesenchymal progenitor cells and pentosan polysulfate, a novel chondrogenic agent: preliminary study in an sheep model. Neurosurg Focus 28(6):E4

Goldschlager T, Jeffrey V, Rosenfeld JV, Ghosh P, Itescu S, Blecher C, McLean C, Jenkin G (2011) Cervical interbody fusion is enhanced by allogeneic mesenchymal precursor cells in an sheep model. Spine 36(8):615–623

Gronthos S, Mrozik K, Shi S, Bartold PM (2006) Sheep periodontal ligament stem cells: isolation, characterization, and differentiation potential. Calcif Tissue Int 79(5):310–317

Grzesiak J, Marycz K, Wrzeszcz K, Czogała J (2011) Isolation and morphological characterisation of sheep adipose-derived mesenchymal stem cells in culture. Int J Stem Cells 4(2):99–104

Gugjoo MB, Amarpal (2018) Mesenchymal stem cell research in sheep: current status and future prospects. Small Rum Res 169:46–56. https://doi.org/10.1016/j.smallrumres.2018.08.002

Gugjoo MB, Amarpal, Makhdoomi DM, Sharma GT (2019) Equine mesenchyaml stem cells: properties, sources, characterization and potential therapeutic applications. J Equine Vet Sci 72:16–27. https://doi.org/10.1016/j.jevs.2018.10.007

Gugjoo MB, Amarpal, Chandra V, Wani MY, Dhama K, Sharma GT (2018) Mesenchymal stem cell research in Veterinary Medicine. Curr Stem Cell Res Therapy 13(8):645–657. https://doi.org/10.2174/1574888X13666180517074444

Gugjoo MB, Gayas MA, Ahmad RA, Handoo N (2018a) Mesenchymal stem cell studies in selected musculoskeletal tissues. In: Jalal H, Vinus, Gugjoo MB, Kawareti PK (eds) Advances in veterinary sciences. Akinik Publications, New Delhi, pp 51–77

Gugjoo MB, Amarpal, Fazili MR, Shah RA, Mir MS, Sharma GT (2020a) Goat mesenchymal stem cell basic research and potential applications. Small Rum Res 183:106045

Gugjoo MB, Amarpal, Ismail AA, Aithal HP, Kinjavdekar P, Kumar GS, Sharma GT (2020b) Allogeneic mesenchymal stem cells and growth factors in gel scaffold repair osteochondral defect in rabbit. Regen Med 15:2. https://doi.org/10.2217/rme-2018-0138

Guo X, Wang C, Zhang Y, Xia R, Hu M, Duan C, Zhao Q, Dong L, Lu J, Qing S, Y. (2004) Repair of large articular cartilage defects with implants of autologous mesenchymal stem cells seeded into beta-tricalcium phosphate in a sheep model. Tissue Eng 10(11–12):1818–1829

Gupta MC, Theerajunyaporn T, Maitra S, Schmidt MB, Holy CE, Kadiyala S, Bruder SP (2007) Efficacy of mesenchymal stem cell enriched grafts in an sheep posterolateral lumbar spine model. Spine 32(7):720–726

Gutwald R, Haberstroh J, Kuschnierz J, Kister C, Lysek DA, Maglione M, Xavier SP, Oshima T, Schmelzeisen R, Sauerbier S (2010) Mesenchymal stem cells and inorganic bovine bone mineral in sinus augmentation: comparison with augmentation by autologous bone in adult sheep. Br J Oral Maxillofac Surg 48(4):285–290

Hamamoto H, Gorman JH, Ryan LP, Hinmon R, Martens TP, Schuster MD, Plappert T, Kiupel M, St John-Sutton MG, Itescu S, Gorman RC (2009) Allogeneic mesenchymal precursor cell therapy to limit remodeling after myocardial infarction: the effect of cell dosage. Ann Thorac Surg 87(3):794–801

Han J, Liu JY, Swartz DD, Andreadis ST (2010) Molecular and functional effects of organismal ageing on smooth muscle cells derived from bone marrow mesenchymal stem cells. Cardiovasc Res 87:147–155

Heidari B, Shirazi A, Akhondi MM, Hassanpour H, Behzadi B, Naderi MM, Sarvari A, Borjian S (2013) Comparison of proliferative and multilineage differentiation potential of sheep mesenchymal stem cells derived from bone marrow, liver, and adipose tissue. Avicenna J Med Biotechnol 5(2):104–117

Hindle P, Baily J, Khan N, Biant LC, Simpson A, Hamish R, Péault B (2016) Perivascular mesenchymal stem cells in sheep: characterization and autologous transplantation in a model of articular cartilage repair. Stem Cells Dev 25(21):1659–1669

Hnatiuk AP, Ong S-G, Olea FD, Locatelli P, Riegler J, Lee WH, Jen CH, De Lorenzi A, Gimenez CS, Laguens R, Wu JC, Crottogini A (2016) Allogeneic mesenchymal stromal cells overexpressing mutant human hypoxia-inducible factor 1-a (HIF1-a) in an sheep model of acute myocardial infarction. J Am Heart Assoc 5:e003714

Hopper N, Wardale J, Brooks R, Power J, Rushton N, Henson F (2015) Peripheral blood mononuclear cells enhance cartilage repair in in vivo osteochondral defect model. PLoS One 10(8):e0133937

Houtgraaf JH, de Jong R, Kazemi K, de Groot D, van der Spoel TI, Arslan F, Hoefer I, Pasterkamp G, Itescu S, Zijlstra F, Geleijnse ML, Serruys PW, Duckers HJ (2013) Intracoronary infusion of allogeneic mesenchymal precursor cells directly after experimental acute myocardial infarction reduces infarct size, abrogates adverse remodeling, and improves cardiac function. Circ Res 113(2):153–166

Ingenito EP, Tsai L, Murthy S, Tyagi S, Mazan M, Hoffman A (2012) Autologous lung-derived mesenchymal stem cell transplantation in experimental emphysema. Cell Transplant 21(1):175–189

Jahroomishirazi R, Bader A, Ebert S, Schmidt C, Sedaghati B, Schulz-Siegmund M, Zscharnack M (2014) Isolation and characterization of CD271 + stem cells derived from sheep dermal skin. Cells Tissues Organs 200:141–152

Ji M, Bai C, Li L, Fan Y, Ma C, Li X, Guan W (2016) Biological characterization of sheep kidney-derived mesenchymal stem cells. Exp Therap Med 12:3963–3971

Kabagambe SK, Chen YJ, Farmer DL (2017) Fetal surgery for myelomeningocele: a review of current clinical practice and translational research. Minerva Pediatr 69:59–65

Khan MR, Chandrashekran A, Smith RKW, Dudhia J (2016) Immunophenotypic characterization of sheep mesenchymal stem cells. Cytometry A 89(5):443–450

Khan MR, Dudhia J, David FH, De Godoy R, Mehra V, Hughes G, Dakin SG, Carr AJ, Goodship AE, Smith RKW (2018) Bone marrow mesenchymal stem cells do not enhance intra-synovial tendon healing despite engraftment and homing to niches within the synovium. Stem Cell Res Ther 9:169

Klein JD, Turner CGB, Steigman SA, Ahmed A, Zurakowski D, Eriksson A, Fauza DO (2011) Amniotic mesenchymal stem cells enhance normal fetal wound healing. Stem Cells Dev 20(6):969–976

Kocyildirim E, Cárdenes N, Ting A, Cáceres E, BermUdez C, Rojas M (2017) The use of GMP-produced bone marrow-derived stem cells in combination with extracorporeal membrane oxygenation in ARDS: An animal model. ASAIO J 63(3):324–332

Kon E, Muraglia A, Corsi A, Bianco P, Marcacci M, Martin I, Boyde A, Ruspantini I, Chistolini P, Rocca M, Giardino R, Cancedda R, Quarto R (2000) Autologous bone marrow stromal cells loaded onto porous hydroxyapatite ceramic accelerate bone repair in critical-size defects of sheep long bones. J Biomed Mater Res 49(3):328–337

Lacitignola L, Staffieri F, Rossi G, Francioso E, Crovace A (2014) Survival of bone marrow mesenchymal stem cells labelled with red fluorescent protein in an sheep model of collagenase-induced tendinitis. Vet Comp Orthop Traumatol 27(3):204–209

Landa-Solis C, Olivos-Meza A, Olivos-Díaz B, Ibarra C (2016) Isolation and culture of mobilized mesenchymal stem cells in preclinical models for tissue engineering: basic science. Adult Stem Cells: Recent Adv. www.smgebooks.com

Letouzey V, Tan KS, Deane JA, Ulrich D, Gurung S, Ong YR, Gargett CE (2015) Isolation and characterisation of mesenchymal stem/stromal cells in the sheep endometrium. PLoS One 10(5):e0127531

Li D, Li M, Liu P, Zhang Y, Lu J, Lin J (2014) Tissue-engineered bone constructed in a bioreactor for repairing critical-sized bone defects in sheep. Int Orthop 38(11):2399–2406

Liechty KW, Mackenzie TC, Shaaban AF, Radu A, Moseley AB, Deans R, Marshak DR, Flake AW (2000) Human mesenchymal stem cells engraft and demonstrate site specific differentiation after *in utero* transplantation in sheep. Nat Med 6(11):1282–1286

Liu Z, Wang W, Gao J, Zhou H, Zhang Y (2014) Isolation, culture, and induced multiple differentiation of Mongolian sheep bone marrow-derived mesenchymal stem cells. In Vitro Cell Dev Biol Anim 50(5):464–474

Locatelli P, Olea FD, Hnatiuk A, De Lorenzi A, Cerdá M, Giménez CS, Sepúlveda D, Laguens R, Crottogini A (2015) Mesenchymal stromal cells overexpressing vascular endothelial growth factor in sheep myocardial infarction. Gene Ther 22(6):449–457

Lovati AB, Lopa S, Recordati C, Talò G, Turrisi C, Bottagisio M, Losa M, Scanziani E, Moretti M (2016) In vivo bone formation within engineered hydroxyapatite scaffolds in a sheep model. Calcif Tissue Int 99(2):209–223

Lv X, He J, Zhang X, Luo X, He N, Sun Z, Xia H, Liu V, Zhang L, Lin X, Lin L, Yin H, Jiang D, Cao W, Wang R, Zhou G, Wang W (2018) Comparative efficacy of autologous stromal vascular fraction and autologous adipose-derived mesenchymal stem cells combined with hyaluronic acid for the treatment of sheep osteoarthritis. Cell Transplant 27(7):1111–1125

Lyahyai J, Mediano DR, Ranera B, Sanz A, Remacha AR, Bolea R, Zaragoza P, Rodellar C, Martín-Burriel I (2012) Isolation and characterization of sheep mesenchymal stem cells derived from peripheral blood. BMC Vet Res 8:169

Lydon H, Getgood A, Henson FMD (2019) Healing of osteochondral defects via endochondral ossification in an ovine model. Cartilage 10(1):94–101

Ma C, Liu C, Li X, Lu T, Bai C, Fan Y, Guan W, Guo Y (2017) Cryopreservation and multipotential characteristics evaluation of a novel type of mesenchymal stem cells derived from Small Tailed Han Sheep fetal lung tissue. Cryobiology 75:7–14

Mackenzie TC, Flake AW (2001) Human mesenchymal stem cells persist, demonstrate site-specific multipotential differentiation, and are present in sites of wound healing and tissue regeneration after transplantation into fetal sheep. Blood Cells Mol Dis 27(3):601–604

Manassero M, Viateau V, Deschepper M, Oudina K, Logeart-Avramoglou D, Petite H, Bensidhoum M (2013) Bone regeneration in sheep using acropora coral, a natural resorbable scaffold, and autologous mesenchymal stem cells. Tissue Eng 19(13–14):1554–1563

Markides H, Newell KJ, Rudorf H, Ferreras LB, Dixon JE, Morris RH, Graves M, Kaggie J, Henson F, El Haj AJ (2019) Ex vivo MRI cell tracking of autologous mesenchymal stromal cells in an sheep osteochondral defect model. Stem Cell Res Ther 10:25

Martinello T, Bronzini I, Perazzi A, Testoni S, De Benedictis GM, Negro A, Caporale G, Mascarello F, Iacopetti I, Patruno M (2013) Effects of in vivo applications of peripheral blood-derived mesenchymal stromal cells (PB-MSCs) and platelet-rich plasma (PRP) on experimentally injured deep digital flexor tendons of sheep. J Orthop Res 31(2):306–314

Martinello T, Gomiero C, Perazzi A, Iacopetti I, Gemignani F, DeBenedictis GM, Ferro S, Zuin M, Martines E, Brun P, Maccatrozzo L, Chiers K, Spaas JH, Patruno M (2018) Allogeneic mesenchymal stem cells improve the wound healing process of sheep skin. BMC Vet Res 14:202

Mccarty R, Gronthos S, Zannettino AC, Foster BK, Xian CJ (2009) Characterisation and developmental potential of sheep bone marrow derived mesenchymal stem cells. J Cell Physiol 219:324–333

McDonald CA, Oehme D, Pham Y, Kelly K, Itescu S, Gibbon A, Jenkin R (2015) Evaluation of the safety and tolerability of a high-dose intravenous infusion of allogeneic mesenchymal precursor cells. Cytotherapy 17(9):1178–1187

Mediano DR, Sanz-Rubio D, Bolea R, Marín B, Vázquez FJ, Remacha AR, pez-Pérez O, Fernández-Borges N, Castilla J, Zaragoza P, Badiola JJ, Rodellar C, Martín-Burriel I (2015a) Characterization of mesenchymal stem cells in sheep naturally infected with scrapie. J. General Virol 96:3715–3726

Mediano DR, Sanz-Rubio D, Ranera B, Bolea R, Martın-Burriel I (2015b) The potential of mesenchymal stem cell in prion research. Zoonoses Public Health 62(3):165–178

Mendelson K, Aikawa E, Mettler BA, Sales V, Martin D, Mayer JE, Schoen FJ (2007) Healing and remodeling of bioengineered pulmonary artery patches implanted in sheep. Cardiovasc Pathol 16:277–282

Mrugala D, Bony C, Neves N, Caillot L, Fabre S, Moukoko D, Jorgensen C, Noël D (2008) Phenotypic and functional characterisation of sheep mesenchymal stem cells: application to a cartilage defect model. Ann Rheum Dis 67(3):288–295

Music E, Futrega K, Doran MR (2018) Sheep as a model for evaluating mesenchymal stem/stromal cell (MSC)-based chondral defect repair. Osteoarthr Cartil 26(6):730–740

Niemeyer P, Schönberger TS, Hahn J, Kasten P, Fellenberg J, Suedkamp N, Mehlhorn AT, Milz S, Pearce S (2010) Xenogenic transplantation of human mesenchymal stem cells in a critical size defect of the sheeptibia for bone regeneration. Tissue Eng 16(1):33–43

Ophelders DRMG, Wolfs TGAM, Jellema RK, Zwanenburg A, Andriessen P, Delhaas T, Ludwig A-K, Radtke S, Peters V, Janssen L, Giebel B, Kramer BW (2016) Mesenchymal stromal cell-derived extracellular vesicles protect the fetal brain after hypoxia-ischemia. Stem Cells Transl Med 5:754–763

Pelagalli A, Nardelli A, Lucarelli E, Zannetti A, Brunetti A (2018) Autocrine signals increase sheep mesenchymal stem cells migration through Aquaporin-1 and CXCR4 overexpression. J Cell Physiol 233(8):6241–6249

Pillai SM, Sereda NH, Hoffman ML, Valley EV, Crenshaw TD, Park Y-K, Lee J-Y, Zinn SA, Govoni KE (2016) Effects of poor maternal nutrition during gestation on bone development and mesenchymal stem cell activity in offspring. PLoS One 11(12):e0168382

Przyborski SA, Hardy SA, Maltman DJ (2008) Mesenchymal stem cells as mediators of neural differentiation. Curr Stem Cell Res Ther 3:43–52

Rabbani S, Soleimani M, Sahebjam M et al (2017) Effects of endothelial and mesenchymal stem cells on improving myocardial function in a sheep animal model. J Tehran Heart Cent 12(2):65–71

Ragni E, Vigano M, Parazzi V, Montemurro T, Montelatici E, Lavazza C, Budelli S, Vecchini A, Rebulla P, Giordano R, Lazzari L (2013) Adipogenic potential in human mesenchymal stem cells strictly depends on adult or foetal tissue harvest. Int J Biochem Cell Biol 45:2456

Rentsch C, Hess R, Rentsch B, Hofmann A, Manthey S, Scharnweber D, Biewener A, Zwipp H (2010) Sheep bone marrow mesenchymal stem cells: isolation and characterization of the cells and their osteogenic differentiation potential on embroidered and surface-modified polycaprolactone-co-lactide scaffolds. In Vitro Cell Dev Biol Anim 46:624–634

Rhodes NP, Srivastava JK, Smith RF, Longinotti C (2004) Heterogeneity in proliferative potential of sheep mesenchymal stem cell colonies. J Mater Sci Mater Med 15(4):397–402

Ribitsch I, Chang-Rodriguez S, Egerbacher M, Gabner S, Gueltekin S, Huber J, Schuster T, Jenner F (2017) Sheep placenta cotyledons: a noninvasive source of sheep mesenchymal stem cells. Tissue Eng 23(5):298–310

Scharf A, Holmes S, Thoresen M, Mumaw J, Stumpf A, Peroni J (2015) Superparamagnetic iron oxide nanoparticles as a means to track mesenchymal stem cells in a large animal model of tendon injury. Contrast Media Mol Imag 10(5):388–397

Shamsul BS, Tan KK, Chen HC, Aminuddin BS, Ruszymah BHI (2014) Posterolateral spinal fusion with osteogenesis induced BMSC seeded TCP/HA in a sheep model. Tissue Cell 46:152–158

Shang Q, Wang Z, Liu W, Shi Y, Cui L, Cao Y (2001) Tissue-engineered bone repair of sheep cranial defects with autologous bone marrow stromal cells. J Craniofac Surg 12(6):586–593

Shaw SW, Bollini S, Nader KA, Gastaldello A, Mehta V, Filppi E, Cananzi M, Gaspar HB, Qasim W, De Coppi P, David AL (2011) Autologous transplantation of amniotic fluid-derived mesenchymal stem cells into sheep fetuses. Cell Transplant 20(7):1015–1031

Song F, Tang J, Geng R, Hu H, Zhu C, Cui W, Fan W (2014) Comparison of the efficacy of bone marrow mononuclear cells and bone mesenchymal stem cells in the treatment of osteoarthritis in a sheep model. Int J Clin Exp Pathol 7:1415–1426

Sousa CP, Dias IR, Lopez-Peña M, Camassa JA, Lourenço PJ, Judas FM, Gomes ME, Reis RL (2015) Bone turnover markers for early detection of fracture healing disturbances: a review of the scientific literature. An Acad Bras Cienc 87:1049–1061

Thangarajah T, Shahbazi S, Pendegrass CJ, Lambert S, Alexander S, Blunn GW (2016) Tendon reattachment to bone in an sheep tendon defect model of retraction using allogenic and xenogenic demineralised bone matrix incorporated with mesenchymal stem cells. PLoS One 11(9):e0161473

Tian Y, Tao L, Zhao S, Tai D, Liu D, Liu P (2016) Isolation and morphological characterization of sheep amniotic fluid mesenchymal stem cells. Exp Anim 65(2):125–134

Turner CG, Klein JD, Steigman SA, Ahmed A, Zurakowski D, Eriksson E, Fauza DO (2011) Preclinical regulatory validation of an engineered diaphragmatic tendon made with amniotic mesenchymal stem cells. J Pediatr Surg 46(1):57–61

Vahedi P, Soleimanirad J, Roshangar L, Shafaei H, Jarolmasjed S, Charoudeh HN (2016) Advantages of sheep infrapatellar fat pad adipose tissue derived stem cells in tissue engineering. Adv Pharm Bull 6(1):105–110

Van Der Donk S, Buma P, Aspenberg P, Schreurs BW (2001) Similarity of bone ingrowth in rats and goats: a bone chamber study. Comp Med 51:336–340

Viateau V, Guillemin G, Bousson V, Oudina K, Hannouche D, Sedel L, Logeart-Avramoglou D, Petite H (2007) Long-bone critical-size defects treated with tissue-engineered grafts: a study on sheep. J Orthop Res 25(6):741–749

Vivas D, Caminal M, Oliver-Vila I, Vives J (2018) Derivation of multipotent mesenchymal stromal cells from sheep bone marrow. Curr Protoc Stem Cell Biol 44(2B):9.1–9.22

Wang A, Brown EG, Lankford L, Keller BA, Pivetti CD, Sitkin NA, Beattie MS, Bresnahan JC, Farmer DL (2015) Placental mesenchymal stromal cells rescue ambulation in sheep myelomeningocele. Stem Cells Transl Med 4:659–669

Wheeler DL, Fredericks DC, Dryer RF, Bae HW (2016) Allogeneic mesenchymal precursor cells (MPCs) combined with an osteoconductive scaffold to promote lumbar interbody spine fusion in an sheep model. Spinal J 16:389–399

Whitehouse M, Howells NR, Parry M, Austin E, Kafienah W, Brady K, Goodship AE, Eldridge JD, Blom AYW, Hollander AP (2017) Repair of torn avascular meniscal cartilage using undifferentiated autologous mesenchymal stem cells: from in vitro optimization to a first-in-human study. Stem Cells Transl Med 6(4):1237–1248

Yr KJ, Evans RG, May CN (2018) An sheep model for studying the pathophysiology of septic acute kidney injury. Methods Mol Biol 1717:207–218

Zhao Y, Zhang S, Zhou J, Wang J, Zhen M, Liu Y, Chen J, Qi Z (2010) The development of a tissue-engineered artery using decellularized scaffold and autologous sheep mesenchymal stem cells. Biomaterials 31(2):296–307

Zhao Y, Li T, Wei X, Bianchi G, Hu J, Sanchez PG, Xu K, Zhang P, Pittenger MF, Wu ZJ, Griffith BP (2012) Mesenchymal stem cell transplantation improves regional cardiac remodeling following sheep infarction. Stem Cells Transl Med 1:685–695

Zhu X, Liu Z, Deng W, Zhang Z, Liu Y, Wei L, Zhang Y, Zhou L, Wang Y (2017) Derivation and characterization of sheep bone marrow-derived mesenchymal stem cells induced with telomerase reverse transcriptase. Saudi J Biol Sci 24:519–525

Zorzi AR, Amstalden EMI, Plepis AMG, Martins VCA, Ferretti M, Antonioli E, Duarte ASS, Luzo ACM, Miranda JB (2015) Effect of human adipose tissue mesenchymal stem cells on the regeneration of sheep articular cartilage. Int J Mol Sci 16:26813–26831

Zscharnack M, Hepp P, Richter R, Aigner T, Schulz R, Somerson J, Josten C, Bader A, Marquass B (2010) Repair of chronic osteochondral defects using predifferentiated mesenchymal stem cells in an sheep model. Am J Sports Med 38(9):1857–1869

# Goat Mesenchymal Stem Cell Basic Research and Potential Applications

# 8

M. B. Gugjoo, Amar Pal, M. R. Fazili, R. A. Shah, M. S. Mir, and G. T. Sharma

**Abstract**

Innumerable goat MSC studies focussing on their culture properties in relation to their therapeutic properties have been conducted. The direct translation of ex vivo results into the in vivo application, however, remains distant. Goat MSCs (gMSCs) have been studied from various tissue sources, and their therapeutic evaluation in experimental models has been conducted mostly in relation to humans. Most of these experimental studies have favoured their applications. The current chapter details gMSCs sources, their characterization, and potential therapeutic applications.

M. B. Gugjoo (✉) · M. R. Fazili
Division of Veterinary Clinical Complex, FVSc & AH, SKUAST-Kashmir, Jammu and Kashmir, India

A. Pal
Division of Surgery, Indian Veterinary Research Institute, Izatnagar, UP, India

R. A. Shah
Division of Animal Biotechnology, FVSc & AH, SKUAST-K, Srinagar, Jammu and Kashmir, India

M. S. Mir
Directorate of Extension Education, SKUAST-K, Srinagar, Jammu and Kashmir, India
e-mail: masoodmir1@skuastkashmir.ac.in

G. T. Sharma
Division of Physiology & Climatology, Indian Veterinary Research Institute, Izatnagar, UP, India


M. B. Gugjoo, A. Pal (eds.), *Mesenchymal Stem Cell in Veterinary Sciences*, https://doi.org/10.1007/978-981-15-6037-8_8

## 8.1 Introduction

There exists a considerable body of literature that has focussed on goat mesenchymal stem cells (gMSCs) properties to be translated to their therapeutic applications. An understanding of the basic cellular physiological properties, although currently very limited, may pave the way for their applications in veterinary medicine. Utilization of goat as translational animal model and application of gMSCs in these models has been aimed to provide further incites on stem cell applications in human medicine. The current chapter focuses on the gMSCs culture characteristics and their in vivo applications.

## 8.2 In Vitro Studies

### 8.2.1 Sources of gMSCs

In comparison to other species, gMSCs have been harvested from a limited number of tissue/organs. gMSCs from adult tissue organs and foetal membranes are enlisted in Table 8.1.

### 8.2.2 Characterization of gMSCs

gMSCs are characterized as per the standard criteria given for human and as followed by other species. gMSCs attach to the plastic surfaces and appear spindle-shaped fibroblasts. Short or long spindle size at various passages may be seen as has been reported for goat umbilical cord MSCs (gUC-MSCs). These cells may self-renew and grow as colonies for certain passages (Martins et al. 2017). The surface marker expression in general follows pattern as per standard recommendations (Table 8.1). However, weak expression of markers like CD90, CD166, and CD105 has also been reported in few studies (Knippenberg et al. 2005; Ghaneialvar et al. 2018). In addition to the surface markers, gMSCs express pluripotency markers (Table 8.1) and have been demonstrated to show plasticity and undergo extended differentiation into putative germ cells and myocyte-like cells (Tripathi et al. 2010;Li et al. 2017; Zhang et al. 2019).

### 8.2.3 Donor Tissue Source and gMSCs

Donor tissue determines the gMSCs initial concentration and the ability to proliferate and differentiate. Additionally, available culture environment too affects gMSCs characteristics. Bone tissue (ilium bone) in comparison to bone marrow may provide a higher yield (Akram et al. 2017). The presence of the periosteal layer (inner cambium) in bone chip may have led to the higher cell yield. gMSCs tend to proliferate for certain passages (p 20) beyond which these become

**Table 8.1** Mesenchymal stem cell sources and characterization in goats

| Positive markers | Negative markers | Differentiation potential | References |
|---|---|---|---|
| *Adipose tissue MSCs (AD-MSCs)* | | | |
| CD166 (30%), CD105 (35%) | – | Osteoblasts | Knippenberg et al. (2005) |
| CD166 (30%), CD105 (35%) | – | Osteoblasts | Knippenberg et al. (2006) |
| CD44, CD106, Vimentin | CD34, CD13 | Adipogenic | Wang et al. (2018) |
| *Amniotic fluid MSCs (Am-MSCs)* | | | |
| CD-73, CD-90 and CD-105, Oct-4, Nanog, Sox-2, SSEA-1, SSEA-4 | CD34 | Adipogenic, chondrogenic, and osteogenic | Pratheesh et al. (2013) |
| CD73, CD90 and CD105 Pluripotency markers (Oct4, Sox2, Nanog, KLF, cMyc, FoxD3) | CD34 | Adipogenic, chondrogenic, and osteogenic | Somal et al. (2016, 2017) |
| *Amniotic sac MSCs (AS-MSCs)* | | | |
| CD73, CD90 and CD105 Pluripotency markers (Oct4, Sox2, Nanog, KLF, cMyc, FoxD3) | CD34 | Adipogenic, chondrogenic, and osteogenic | Somal et al. (2016, 2017) |
| *Bone marrow MSCs (BM-MSCs)* | | | |
| SH4, CD44 | CD34, CD14 | Adipogenic, chondrogenic, and osteogenic | Murphy et al. (2003) |
| CD44, CD29, CD90, CD166, OCT4, Nanog | CD45 | Adipogenic, osteogenic | Zhang et al. (2006) |
| Oct-3/4, PCNA, Ck-Pan, vimentin and Nanog | – | – | da Silva Filho et al. (2014) |
| CD44 | CD45 | Adipogenic, chondrogenic, and osteogenic | Kwong et al. (2014) |
| CD29, CD90 | CD34, CD45 | Adipogenic, chondrogenic, and osteogenic | Elkhenany et al. (2015) |
| CD90, CD105 and CD73, OCT4 | CD45 | Adipogenic, chondrogenic, and osteogenic | Mohamad-Fauzi et al. (2015) |
| CD105, CD90 | CD34, CD45 | Adipogenic, chondrogenic, and osteogenic | Murphy et al. (2003) |
| – | – | Adipogenic, chondrogenic, and osteogenic | Burdzinska et al. (2017) |
| CD73, CD105, Stro-1 | CD34 | Adipogenic, chondrogenic, and osteogenic | Pratheesh et al. (2017) |

(continued)

**Table 8.1** (continued)

| Positive markers | Negative markers | Differentiation potential | References |
|---|---|---|---|
| CD29, CD44 | CD34 | – | Xu et al. (2017) |
| CD44 and CD166, CD105, CD90 (weak expression) | CD34, CD45 | – | Ghaneialvar et al. (2018) |
| *Cord blood MSCs (CB-MSCs)* | | | |
| CD73, CD90 and CD105 Pluripotency markers (Oct4, Sox2, Nanog, KLF, cMyc, FoxD3) | CD34 | Adipogenic, chondrogenic, and osteogenic | Somal et al. (2016, 2017) |
| *Deciduous teeth MSCs (Dt-MSCs)* | | | |
| CD29, CD146 | – | Osteogenic | Zhao et al. (2017) |
| *Endometrium MSCs (End-MSCs)* | | | |
| – | – | Adipogenic, chondrogenic, and osteogenic | Tamadon et al. (2017) |
| *Skeletal muscle MSCs (S-MSCs)* | | | |
| – | – | Osteocytes, adipocytes, and myogenic | Tripathi et al. (2010) |
| *Umbilical cord MSCs (UC-MSCs)* | | | |
| CD73, Thy-1, CD105 | CD34 | Adipogenic, osteogenic | Kumar et al. (2016) |
| CD90, CD105, CD44 | CD34 | Adipogenic, chondrogenic, and osteogenic | Martins et al. (2017) |
| CD90, CD105 | CD34, CD44 | Adipogenic, chondrogenic, and osteogenic | |
| *Wharton's jelly MSCs (WJ-MSCs)* | | | |
| CD-73, STRO-1, CD-105 | CD34 | Adipogenic, chondrogenic, and osteogenic | Pratheesh et al. (2014) |
| CD73, CD90 and CD105 Pluripotency markers (Oct4, Sox2, Nanog, KLF, cMyc, FoxD3) | CD34 | Adipogenic, chondrogenic, and osteogenic | Somal et al. (2016, 2017) |

*AD-MSCs* adipose-derived MSCs, *AF-MSCs* amniotic fluid-derived MSCs, *AS-MSCs* amniotic sac-derived MSCs, *BM-MSCs* bone marrow-derived MSCs, *CB-MSCs* cord blood-derived MSCs, *Dt-MSCs* deciduous teeth-derived MSCs, *End-MSCs* endometrium-derived MSCs, *Skeletal M-MSCs* skeletal muscle-derived MSCs, *UC-MSCs* umbilical cord-derived MSCs, *WJ-MSCs* Wharton's jelly-derived MSCs

senescent. With each passage, the population doubling time (PDT) tends to increase (Mohamad-Fauziet al. 2015). Average PDT of gUC-MSCs had increased from 33.49 h at passage 5 to 34.91 h at passage 12 (Qiuet al. 2012). gMSCs proliferation differences with tissue source also occur. The mean PDT of goat bone marrow (gBM)-, amniotic fluid (gAF)-, and Wharton's jelly (gWJ)-derived MSCs had been 24.94 ± 2.67 h (Eslaminejad et al. 2009); 33.1 h (Pratheesh et al. 2013); 36.06 ± 1.2 h (Pratheesh et al. 2014), respectively, at specific passage. MSCs have also been harvested from goat foetal adnexa. The cells from particular adnexal tissues (gWJ-MSCs and goat cord blood-derived MSCs) tend to show faster proliferation as compared to others (gAF-MSCs and amniotic sac-derived MSCs) (Somal et al. 2016). Additionally, gWJ-MSCs had significantly higher expression of some of the surface and pluripotency markers as compared to MSCs from other sources. The expression of few pluripotency markers (Klf and cMyc), however, had remained comparable between gWJ-MSCs and goat amniotic fluid MSCs (gAF-MSCs) (Somal et al. 2017).

Self-renewal and differentiation properties of MSCs may be regulated by epigenetic modifications. MSCs derived from adipose tissues, bone marrow, and muscle may not carry identical promoter methylation profiles (Sorensen et al. 2010). The culture condition sensitivity of gMSCs might change their methylation pattern. Goat adipose tissue MSCs (gAD-MSCs) tend to undergo higher epigenetic changes as compared to goat bone marrow (BM)-MSCs (Wang et al. 2017). Likewise gAD-MSCs may have superior adipogenic but weaker osteogenic differentiation potential as compared to musculoskeletal sources like gBM-MSCs (Mohamad-Fauzi et al. 2015). The upregulation of osteogenic pathways in these cells had been variable; wap38 MAPK is upregulated in gBM-MSCs while p44/42 MAPK in gAD-MSCs (Elkhenany et al. 2016). gBM-MSCs tend to undergo low whole genome methylation as compared to fibroblasts as bone marrow requires cells that have high proliferation ability (Schopet al. 2009).

### 8.2.4 Reproductive Cyclicity and gMSCs

Reproductive cyclicity too may affect MSCs proliferation. Anestral and cyclic endometrium-derived MSCs (End-MSCs) at passage 4 had 40.6 h and 53 h PDT, respectively (Tamadon et al. 2017). MSCs proliferation in anestrus may be promoted by the available higher concentration of growth factors like basic fibroblast growth factor (bFGF) and transforming growth factor-β (TGF-β) in endometrial tissue (Tamada et al. 2000; Tamadon et al. 2017).

### 8.2.5 Effect of Donor Age and Health Status on gMSCs

Donor age and type of ailment may also affect MSCs characteristics. MSCs harvested from an osteoporotic goat tend to proliferate slowly and show reduced osteogenic potential. Such a reduced differentiation potential may be enhanced through specific scaffold (β-TCP) incorporation (Cao et al. 2012). Ageing tends to decrease MSCs proliferation and differentiation potential. A single gMSCs study, however, had failed to demonstrate decrease in such properties with ageing (Vertenten et al. 2009).

## 8.3 Potential In Vivo Applications

Goat has been utilized as the translational model animal for human. As such gMSCs have been evaluated in various models of different conditions of bone and cartilage, among others for their therapeutic effects (Gugjoo et al. 2020). To avoid animal sufferings, substitutes like organ culture system are being developed. One of the studies had shown comparable results in organ culture system and in vivo conditions with respect to the gAD-MSCs spatial and temporal behaviour (Peeters et al. 2015). The current chapter, however, focuses on the in vivo therapeutic applications of gMSCs.

MSCs local implantation makes them available to the tissue directly in sufficient concentrations, while distant sources like intravenous routes may be required in large quantities for effective utilization. Additionally, intravenous application of gMSCs may activate disseminated intravascular coagulation (DIC) cascade and lead to hypercoagulable state, but the condition is transient (Liao et al. 2017). This may be due to the MSCs ability to express pro-coagulant factors, which increases as the cells are culture expanded (Tatsumi et al. 2013; Liao et al. 2017) and can be controlled by heparin therapy (Liao et al. 2017). The potential therapeutic applications of gMSCs are discussed below.

### 8.3.1 Cutaneous Wounds

Cutaneous healing encompasses complex system of organized biological and molecular events. These events may sometime become disorganized or fail to occur as seen in diabetes or chronic infections (Wu et al. 2007). Extensive tissue damages too fail to heal or may show delayed healing. This compromises the quality of life and also decreases production of the animal. MSCs tend to resolve and enhance the injured tissue healing (Phinney and Prockop 2007). gWJ-MSCs on local implantation tend to be viable up to day 12 and might promote epithelialization by as early as day 7. MSCs tend to promote and improve healing through their anti-inflammatory activities and secrete paracrine factors that promote granulation tissue formation with minimal to nil scar tissue formation (Table 8.2) (Azari et al. 2008,2011). MSCs applications in delayed/non-union wound may be conducted to confirm their utility in such obstinate cases.

### 8.3.2 Osteochondral Defects

Goat knee cartilage thickness (0.7–1.5 mm) approximates human knee (2.2–2.5 mm) cartilage, and as such it is being utilized as translational model animal for MSCs chondrogenic studies (Frisbie et al. 2006). Even antiangiogenesis factor-transfected MSCs have been studied for their therapeutic effects in avascular tissues including hyaline cartilage (Table 8.3) (Sun et al. 2009). The transfected MSCs were equally viable and had comparable chondrogenic potential to that of non-transfected gMSCs (Jeng et al. 2010).

**Table 8.2** In vivo preclinical experimental mesenchymal stem cell studies on cutaneous wounds in goat

| Model type | Number of animals | Study period | Biomaterial used | Cell dose/ assembly | Evaluation criteria | Overall result | References |
|---|---|---|---|---|---|---|---|
| Bilateral cutaneous wound on thoracic cavity ($3 \times 3$ cm) | 5 (right wound treated with MSCs; left wound control) | 5 weeks | WJ-MSCs | $9 \times 10^6$ | Morphometric evaluation, immunohistochemical staining (α-smooth muscle actin) | % wound contraction, epithelialization, and wound healing were significantly higher in cell-treated wounds as compared to control | Azari et al. (2008) |
| Cutaneous wound, 3 cm linear full-thickness skin incision | 4 (4 wounds, 2 on each side of thoracic cavity, left wound treated with MSCs; right wounds with PBS) | 7, 12 days | WJ-MSCs | $3 \times 10^6$ | Histology | Transplanted cells were alive in wounds at day 12. At day 7 re-epithelialization was complete in cell-treated wounds while incomplete in control. Less inflammation and thinner granulation tissue with minimum scar tissue formation in cell-treated group | Azari et al. (2011) |

**Table 8.3** In vivo chondrogenic preclinical experimental mesenchymal stem cell studies in goat

| Model type | Number of animals | Model defect size/study period | Biomaterial used | Cell dose | Evaluation criteria | Overall result | References |
|---|---|---|---|---|---|---|---|
| Full-thickness meniscal repair (0.75 mm) | 32 (menisci repaired with two vertical sutures, $n = 8$; 0.5 mL exogenous fibrin clot, $n = 8$; fibrin clot + BM-MSCs, $n = 8$; left untreated, $n = 8$) | 4 months | Autologous BM-MSCs, fibrin clot | $20 \times 10^6$ | Gross, tensile strength, and histology | Statistically non-significant differences in healing between suture repaired and fibrin clot repaired groups. Addition of MSCs did not enhance menisci repair | Port et al. (1996) |
| Chronic model of excision of the medial meniscus and resection of the anterior cruciate ligament | 24 (6 animals cell treated/hyaluronan and 3 animals control for 12 weeks evaluation; 9 animals cell treated/hyaluronan and 6 animals for 26 weeks evaluation) | 12 and 26 weeks | BM-MSCs + hyaluronan | $10 \times 10^6$ loaded in hyaluronan | Histochemistry | Marked regeneration of the medial meniscus. Degeneration of the articular cartilage, osteophytic remodelling, and subchondral sclerosis were reduced in cell-treated group compared to control at 12 weeks. However, later the cell treated and control had severe osteoarthritis | Murphy et al. (2003) |

| | | | | | | | |
|---|---|---|---|---|---|---|---|
| Mandibular condyle osteochondral defect model | 50 (12 in each group) (group I: NELL-1-modified BM-MSCs/PLGA, group II: BM-MSCs/PLGA; group III: PLGA alone; group IV: control) | 3 mm diameter × 5 mm depth/6 weeks and 24 weeks | NELL-1 transfect autologous BM-MSCs and poly-lactic-co-polyglycolic acid scaffold | $3 \times 10^6$ cell seeded on PLGA scaffold | Macroscopic, histology, and immunohistochemistry, micro-CT | Group I showed rapid and vigorous healing leading to fibrocartilage formation at 6 weeks. At 24 weeks, complete repair of native articular cartilage and subchondral bone at 24 weeks. In group II: Repaired completely filled the defect with fibrocartilage at 24 weeks, but the cartilage was less well-organized group I. In groups III and IV, the defects were poorly repaired, and no cartilage in the group IV or only small portion of cartilage in the group III was formed | Zhu et al. (2011) |

(continued)

**Table 8.3** (continued)

| Model type | Number of animals | Model defect size/study period | Biomaterial used | Cell dose | Evaluation criteria | Overall result | References |
|---|---|---|---|---|---|---|---|
| Full-thickness chondral defect in medial femoral condyles | 8 (16 knees) chondron and cell treatment vs. microfracture | 5 mm/6 months | Chondron (chondrocytes in own matrix) and BM-MSCs | 10% chondron/90% BM-MSC combination at a concentration of $1 \times 10^6$ cells/mL | Macroscopic and microscopic scoring, biochemical analysis, and histological and immunohistochemical analyses | Combination of BM-MSCs and chondrons leads to significantly better microscopic, macroscopic, and biochemical cartilage regeneration compared to microfracture treatment | Bekkers et al. (2013) |
| Osteochondral defects created in medial condyles and trochlear grooves | 8 (group I: Scaffold seeded with cultured AD-MSCs; group II: Scaffold seeded with SVF cells; group III: Acellular scaffold) | 5 mm × 3 mm/1 month and 4 months | Stromal vascular fraction (SVF), AD-MSCs along with collagen type I/III scaffold | $5 \times 10^6$ (SVF) and $5 \times 10^5$ (AD-MSCs) seeded on scaffold | Macroscopy, immunohistochemistry, biomechanical analysis, micro-CT analysis, and biochemistry | Cell-treated groups had more extensive collagen type II, hyaline-like cartilage, and higher elastic moduli, and their glycosaminoglycan content in the cartilaginous layer that approached native tissue values. In control, lesser regenerative effect was seen. No difference in healing was seen between SVF-treated and AD-MSC-treated animals | Jurgens et al. (2013) |

| | | | | | | | |
|---|---|---|---|---|---|---|---|
| Chronic full-thickness chondral defect in medial femoral condyles | 18 (36 defects) group I: Bone marrow stimulation and BM-MSCs; group II: Bone marrow stimulation; group III: Control | 5 mm/6 months | BM-MSCs | $1 \times 10^7$ cells after 2 weeks after bone marrow stimulation for three consecutive weeks | Macroscopic, histology, biochemical assays (glycosaminoglycans), and gene expressions (aggrecan, collagen II and Sox9) | Hyaline-like tissue with higher glycosaminoglycans and chondrogenic gene expression in group I compared to group II that had fibrocartilage. Lowest healing in control | Nam et al. (2013) |
| Osteochondral defect (bilateral, 6 mm diameter and 12 mm depth) | *n*-12 (TEO cultured in double-chambered bioreactor without mechanical stimulation of stirring, $n = 4$; stimulation of stirring, $n = 4$; no TEO implantation, control, $n = 4$) | 12 and 24 weeks | Autologous BM-MSCs, β-TCP | – | Histology and immunohistochemistry, gross morphology | Reparative effects of TEO cultured in bioreactor were better than control, and mechanical stimulation could further improve the effects | Pei et al. (2014) |

(continued)

**Table 8.3** (continued)

| Model type | Number of animals | Model defect size/study period | Biomaterial used | Cell dose | Evaluation criteria | Overall result | References |
|---|---|---|---|---|---|---|---|
| Full-thickness femoral condyle cartilage defects | 6 (microfracture and cell/scaffold groups) | 6.5 mm diameter/6 and 9 months | Human WJ-MSCs seeded in an acellular cartilage extracellular matrix (ACECM)-oriented scaffold | $1 \times 10^6$ cells seeded on ACECM | Analysis of inflammatory response, magnetic resonance imaging, gross morphology, histology, immunohistochemical and immunofluorescent staining, biomechanical testing, and biochemical quantitative analyses | No significant differences between the two groups in immuno-inflammatory parameters. MRI demonstrated higher-quality cartilage and complete subchondral bone at defect sites in the cell-treated group at 9 months. Histological revealed extracellular cartilage, cartilage lacuna, and collagen type II levels were higher in cell-treated group compared to the microfracture, while the cell-treated group exhibited a higher elasticity modulus | Zhang et al. (2018) |

Various sources like bone marrow- (Murphy et al. 2003; Bekkers et al. 2013; Nam et al. 2013), adipose tissue- (Jurgens et al. 2013), and Wharton's Jelly- (Zhang et al. 2018) derived MSCs have been evaluated in osteochondral defect models. The cells along with various scaffolds like hyaluronan, collagen I/III, natural cartilage ingredients, and poly lactic co-glycolic acid (PLGA) have been implanted to prevent loss of cells through adverse effects of local environment. An overall improvement in osteochondral defect healing have been demonstrated to occur with MSCs and scaffold implantations (Table 8.3). The improved healing has been characterized by enhanced secretion of cartilage-specific matrices like collagen type II and glycosaminoglycan. The regenerated hyaline-like cartilage had biomechanical factors almost comparable to native tissue (Jurgens et al. 2013). These cells tend to retard degeneration of the articular cartilage and prevent osteophyte formation and sclerosis of subchondral bone (Murphy et al. 2003). Even mechanical stimulation of these cells seeded in scaffolds may further enhance their repair potential (Pei et al. 2014).

The femoral condyle defect-healing potential of MSCs (BM-MSCs or xenogenic WJ-MSCs) seeded over cartilage extracellular matrix may be better than commonly used surgical techniques (microfracture and bone marrow stimulation). The healing had been evaluated utilizing International Cartilage Repair Society (ICRS) and O'Driscoll scores, in addition to the expression profile of cartilage-specific genes (Bekkers et al. 2013; Nam et al. 2013; Zhang et al. 2018). However, no differences in inflammatory response at the defect site had been demonstrated among the cellular implantation- and surgical techniques-treated animals (Zhang et al. 2018). One of the studies had failed to demonstrate any significant improvement in meniscal repair with gBM-MSCs and fibrin clot (Port et al. 1996).

Transfected MSCs may further enhance osteochondral defect healing (Zhu et al. 2011; Wei et al. 2019). Among various factors, transfection of gBM-MSCs by NELL-like molecule-1 (Nell-1) seeded on poly lactic co-glycolic acid (PLGA) had repaired goat mandibular condyle defects by fibrocartilage at 6 weeks and by 24 weeks with native articular cartilage. Simple BM-MSCs and PLGA assembly had repaired the defect by fibrocartilage even at 24 weeks (Zhu et al. 2011). Nell-1 is a novel growth factor that specifically target cells committed to osteochondral lineage. Further studies in the area are desired.

### 8.3.3 Bone Defects

Bone tissue engineering is carried out extensively in humans (Costa-Pinto et al. 2011). Goat is increasingly being used as human orthopaedic research animal model (Proffen et al. 2012). Goat and human long bones have approximate size and macrostructure, in addition to the biomechanics and biochemistry (Anderson et al. 1999; Van Der Donk et al. 2001; Sousa et al. 2015). Furthermore, goat osteoporosis effectively mimics human condition (Cao et al. 2012).

Various goat bone defect models have been created to study MSCs therapeutic potential (Table 8.4). Variable critical size defects have been created that varied

**Table 8.4** In vivo preclinical experimental mesenchymal stem cell studies on osteogenesis in goat

| Model type | Number of animals | Model defect size/study period | Biomaterial used | Cell dose | Evaluation criteria | Overall result | References |
|---|---|---|---|---|---|---|---|
| Tibial bone defects (2.6 cm) | 26 (BCB + transfected cells, *n* = 9; β-gal-transfected cells + BCB, *n* = 6; BCB + simple cells, *n* = 8, BCB, *n* = 3) | 8 and 26 weeks | Biphasic calcined bone(BCB), BMP-2-transfected BM-MSCs, simple BM-MSCs | $2 \times 10^8$ | Histology, biomechanical examination | 5/8 defects were completely healed, while 3/8 had partial healed in BCB + transfected cell-treated group. In other experimental groups, slight or no healing was reported | Dai et al. (2005) |
| Cranial suture distraction (coronal) (0.4 mm/day for 8 days) | 10 (MSCs, *n* = 5; PBS, *n* = 5) | 4 weeks | MSCs | $5 \times 10^9$ | SEM and histology | The new bone formation at the edge of suture was superior in cell-treated group than control | Shen et al. (2006) |
| Bilateral early stage osteonecrosis of femoral head (ligation of lateral medial circumflex arteries and delivery of liquid nitrogen to femoral head, 6-mm defect filled with biomaterials) | 23 (right and left femoral defect-treated BMP-2- and β-gal-transfected cells, respectively, *n* = 20; control, *n* = 3) | 16 weeks | Bone morphogenetic protein-2 (BMP-2)- or β-galactosidase (β-gal)-transfected BM-MSCs, β-tricalcium phosphate (β-TCP) | $1 \times 10^7$ | Radiography, bone mineral density, histology, and biomechanical testing | The femoral heads in BMP-2-transfected cell-treated group had normal density and surface, while those in β-gal-treated cell group had low density and irregular surface. Histologically, lamellar bone was formed in BMP-2 cell-treated group, while in β-gal cell-treated group residual fibrous tissue was formed. The new bone formed in BMP-2 group had significantly higher bone as compared to β-gal. The maximum compressive strength and Young's modulus of repaired tissue in BMP-2 treated cell group were similar to normal bone and significantly higher as compared to β-gal cell-treated group. In non-treated group, femoral heads collapsed at 16 weeks | Tang et al. (2007) |

| | | | | | | | |
|---|---|---|---|---|---|---|---|
| Segmental femoral defect (2 cm, nail stabilized) | 6 (HASi, $n = 4$; tissue-engineered HASi, $n = 2$) | 4 months | BM-MSCs, triphasic ceramic-coated hydroxyapatite (HASi) | $1 \times 10^5$/cm$^2$ of HASi | Radiography, micro-CT, histology, histomorphometry, scanning electron microscopy, and inductive couple plasma spectrometry | Good osteointegration and osteoconduction were observed in TE HASi and HASi. Tissue-engineered HASi showed better and faster performance evident by lamellar bone organization of newly formed bone with the degradation of material. In HASi only group, immature interwoven bony bridges still intermingled with scattered small remnants of material | Nair et al. (2009) |
| Bilateral segmental tibial defect (30 mm) | 6 (β-TCP seeded BM-MSCs cultured in dynamic perfusion reactor implanted in left tibia; β-TCP seeded BM-MSCs cultured static implanted in right tibia) | 24 weeks | Autologous BM-MSCs | $2 \times 10^6$/mL | Radiography, histology and micro-CT | Significant osteogenic capacity of experimental group compared to control. Segmental bone defect repair by dynamic perfusion reactor culture MSCs outweigh static mode | Wang et al. (2010) |
| Tibial unicortical defect (6 mm) (four defects per tibia) | 8 (4 defects in each tibia; PLA, PLA + cell; PLA + cells + β-TCP; control) | 2, 4, 8, and 12 weeks | BM-MSCs, β-TCP, methacrylate-endcapped poly(D,L-lactide-Co-έ-caprolactone) | – | Radiography, histology, histomorphology, and immunohistochemistry | Cell survival and proliferation in polymer-substituted bone defects was demonstrated. The incorporation of β-TCP was associated with less expansion and growth of MSCs | Vertenten et al. (2009) |
| Non-critical-sized unicortical tibial defect model (bilateral, 4 holes, 6 mm) | 8 (autologous BM-MSCs, Pluronic® F127 hydrogel; tubular sintered hydroxyapatite) | 2 and 4 weeks, 6 and 8 weeks | Autologous BM-MSCs, Pluronic® F127 hydrogel, tubular sintered hydroxyapatite | $4 \times 10^5$ cultured on carriers | Clinical, radiology, and histology | Faster bone repair in defects filled with hydrogel plus cultiSpher-S carriers in comparison to control. No new bone formation was originating from hydroxyapatite carriers. However, the hydrogels stimulated the natural repair process | Lippens et al. (2010) |

(continued)

**Table 8.4** (continued)

| Model type | Number of animals | Model defect size/study period | Biomaterial used | Cell dose | Evaluation criteria | Overall result | References |
|---|---|---|---|---|---|---|---|
| Segmental tibial defect (25 mm) | 32 (nHACP/CF + cells, $n = 8$; autograft, $n = 8$; nHACP/CF, $n = 8$; control, $n = 8$) | 8 weeks | Autologous BM-MSCs, nano-hydroxyapatite/collagen/poly (L-lactic acid), PLLA/chitin fibres (nHACP/CF) | $2 \times 10^6$ | Radiography, histology, and biomechanics | Cell along with scaffold and autograft groups had defect repaired by 8 weeks | Liu et al. (2010) |
| Medial femoral condyle defect of osteoporotic defect (oestrogen deficiency induced, 0.8-cm diameter and 8-mm deep cylindrical defects) | 16 (sham, $n = 6$; β-TCP treated, $n = 10$; BM-MSCs/β-TCP, $n = 16$) | 16 weeks after defect creation | Autologous BM-MSCs (oestrogen deficient donor), β-TCP | – | Radiography, CT, micro-CT and histomorphometry | MSCs/ β-TCP treated animals had successful repair of bone defects | Cao et al. (2012) |
| Maxillary sinus augmentation (bilateral) | 9 (cells/CPC, $n = 6$; CPC alone, $n = 6$; autogenous graft, $n = 6$) | 3 months | BM-MSCs, calcium phosphate cement(CPC) | $2 \times 10^7$ | Micro-CT, histology and histomorphometry | Group treated with cells/CPC promoted early bone formation and mineralization and maximally maintain volume and height of augmented maxillary sinus as compared to other groups. Osteointegration was significantly higher in cells/CPC treated group as compared to others | Zou et al. (2012) |

| | | | | | | | |
|---|---|---|---|---|---|---|---|
| Critical sized femur defects (bilateral, 20 mm) | 12 (right defects treated with SAP/DBM/BMA, $n = 12$; left with DBM/BMA, $n = 12$) | 8 and 16 weeks | Bone marrow aspirate (BMA), nano-scale self-assembly peptide (SAP) modified demineralized bone matrix (DBM) | – | Radiography, CT and 3D image processing techniques | Volume of new bone formed in SAP/DBM/BMA assembly implanted bone defects was significantly higher as compared to DBM/MSCs | Li et al. (2015) |
| Iliac crest defect (4 defects bilaterally) | 10 (one defect received either autologous bone graft or was left empty ($n = 5$). In the other 3 locations, the experimental groups ($n = 10$) were tested: (1) biphasic calcium phosphate (BCP) control, (2) BCP with alginate, pBMP-2 and allogeneic MSCs and (3) BCP with alginate and pBMP-2 | 3–12 weeks | Allogeneic BM-MSCs, bicalcium phosphate (BCP) | $2 \times 10^6$ | Histomorphometry, fluorescent microscopy, immunohistochemistry | Fluorochrome incorporation evaluated at 3, 6 and 9 weeks and histomorphometry at 12 weeks after implantation revealed clear differences between the groups, with pBMP-2 combined with MSCs being the most effective. The BMP-2 was demonstrated in a variety of bone-residing cells through immunohistochemistry. Further analysis indicated that multinucleated giant cells might have an important role in transgene expression. The model appeared robust, showed no neighbouring effects and demonstrated effectivity of combined cell and gene therapy | Loozen et al. (2015) |
| Distraction osteogenesis (mandibular) | 12 (MSCs, $n = 6$; control, $n = 6$) | 30 days | BM-MSCs | $1.5 \times 10^6$ (two injections at day 10 and day 20 after surgery) | Energy dispersive X-ray, scanning electron microscopy (SEM) and histology | An increased calcification and significant trabecular bone size and limited osteoid formation in cell-treated group as compared to control | El Hadidi et al. (2016) |

(continued)

**Table 8.4** (continued)

| Model type | Number of animals | Model defect size/study period | Biomaterial used | Cell dose | Evaluation criteria | Overall result | References |
|---|---|---|---|---|---|---|---|
| Maxillary sinus floor elevation | 18 ($n$ = 6 each group); group A (SGDs-CPC compound), group B (CPC alone), and group C (autogenous bone obtained from an iliac crest) | 3 months | Deciduous teeth MSCs (DT-MSCs), calcium phosphate cement (CPC) | | Computed tomography (CT), histology, and histomorphometric analysis | At 3 months, SGDs-CPC compound had promoted bone formation and mineralization. The areas of new bone formation in elevated sinuses were DT-MSCs/CPC group, which was significantly higher than the CPC-alone group or in the autogenous bone group. Immunohistochemical staining revealed that GFP and OCN were both expressed in the new bone tissue, which suggested that the implanted DT-MSCs might have contributed to new bone formation on the elevated sinus floor. DT-MSCs can promote new bone formation and maturation in the goat maxillary sinus, and the tissue-engineered bone composite of SGDs and CPC might be a potential substitute for existing maxillary sinus floor elevation methods | Zhao et al. (2017) |

among studies even on the same bone type. Critical size defect for ilium had been 17 mm (Anderson et al. 1999), femoral defect 20 mm (Li et al. 2015), and tibial defects 25 mm (Liu et al. 2010) and 30 mm (Wang et al. 2010). Various bone tissue-engineered constructs along with MSCs have been employed in goats for critical size defects of mandible, femur, and tibia (Table 8.4). Mostly BM-MSCs have been utilized along with various kinds of scaffolds (Table 8.4). The cellular scaffolds had led to the early and improved bone repair both in quality and quantity as compared to the acellular scaffolds. The lamellar bone organization and its osteo-integration in the cell-treated defects had been better than those treated without cells. In vitro cell-scaffold compatibility does not guarantee in vivo regenerative effects (Vertenten et al. 2009). Thus, it is important to confirm in vivo applicability of MSCs and/or scaffolds.

gBM-MSCs genetic engineering with human bone morphogenetic protein (hBMP) gene (Tang et al. 2007; Loozen et al. 2015) and beta-galactosidase (gal)-gene (Tang et al. 2007) have been demonstrated to further promote healing of bone. The healed bone tissue formed had higher volume with better mechanical strength as compared to that with normal MSCs. BMP-2 transduced gBM-MSCs, however, may be more pro-healing as compared to BM-MSCs transfected with beta-galactosidase (Tang et al. 2007). Bone remains under stress that may determine the healing outcome. On that principle, dynamically cultured MSCs and/scaffolds and distraction osteogenesis had favoured bone healing (Shen et al. 2006; Cao et al. 2012; Gardel et al. 2013).

Bone tissue-engineered cellular scaffolds have favoured early and better bone regeneration. To achieve evidence-based medicine and for definitive applications of these constructs, more uniform and extensive studies are desired.

### 8.3.4 Intervertebral Disc Disease

Goats carry larger size of lumbar intervertebral discs (~5 mm disc height) and have disc mechanical properties and biochemical composition resembling to that of human. This makes goats more suitable animal model for human (Table 8.5) (Beckstein et al. 2008). Goat preliminary ex vivo studies on MSCs and/scaffold have shown favourable role in intervertebral disc diseases (Gullbrand et al. 2017). Few in vivo studies conducted have been pro to the utilization of MSCs along with scaffolds. The healing may be observed as early as 3 weeks of time (Xu et al. 2017). Among various factors, basement membrane molecules too determine nucleus pulposus chondrogenesis and cartilage regeneration (Toh et al. 2013).

### 8.3.5 Periodontal Tissue

The cellular grafting of periodontal defects may further promote periodontal tissue regeneration against the normal grafting. Among various tissue-regenerated structures are cementum, bone, and periodontal ligament (Table 8.5) (Marei et al. 2009). The preliminary reports demand in-depth studies for any definitive outcome.

**Table 8.5** In vivo preclinical experimental mesenchymal stem cell studies on different conditions in goat

| Model type | Number of animals | Study period | Biomaterial used | Cell dose/ assembly | Evaluation criteria | Overall result | References |
|---|---|---|---|---|---|---|---|
| Annulus fibrosus defect (30 discs, T1-L5, 1 × 1 cm) | $n = 6$ (30 discs in each group) Sham group, injury group, and treatment group | 3, 6 and 12 weeks | BM-MSCs, platelet-rich plasma, gelatine sponge | – | HE and mason trichrome staining, Alcian blue periodic acid Schiff and collagen II staining | Repaired annulus fibrosus in cell-treated group had cell density and morphology of cells closer to sham group. Chondrocytes participated in repair process and collagen II progressively increased | Xu et al. (2017) |
| Canine tooth extraction (bilateral, titanium fixture) | $n = 5$ (control: Poly (DL-Lactide-co-glycolide), $n = 5$; treatment: BM-MSCs, poly (DL-Lactide-co-glycolide) | 10 days and 1 month | Autologous BM-MSCs, poly (DL-Lactide-co-glycolide) | $9.4 \times 10^6$ per scaffold | Radiography, histology, histophotometric, and scanning electron microscopy | Experimental side showed periodontal-like tissue with newly formed bone at 10 days and 1 month. Control showed connective tissue regeneration at 10 days but not at 1 month | Marei et al. (2009) |
| Bronchopleural fistula (upper tracheal lobectomy) | 9 (BM-MSCs, $n = 4$; bronchoscopic fibrin glue, $n = 5$) | 28 day | BM-MSCs, bronchoscopic fibrin glue | $2 \times 10^6$/mL (5 mL injected) | Histopathology, CT, and MRI | Bronchopleural closure occurred by extraluminal fibroblast proliferation and collagenous matrix development in cell-treated group. In the other group, the fistula was still present. Even 40% (2/5) animals died | Petrella et al. (2014) |

### 8.3.6 Cardiovascular System

Goat has also been used to study the effect of MSCs in cardiovascular affections relevant to the humans. Autologous gBM-MSCs and/or scaffold in coronary artery ligation model has been evaluated. The assembly could prevent left ventricular chamber dilatation and had improved myocardial perfusion and contraction (Liao et al. 2006). It is a preliminary study that demands further evaluation.

### 8.3.7 Bronchopleural Fistula

Lung cancer remains one of the leading causes of death (Wakelee et al. 2007). Surgical resection in limited lung disease is preferable option but tends to develop bronchopleural fistula (BPF) causing higher mortality (Sonobe et al. 2000). In an induced bronchopleural fistula, implantation of the gBM-MSCs had led to the healing of fistula in about a month (Table 8.5) (Petrella et al. 2014), demanding further studies.

### 8.3.8 Reproduction

Nuclear Transfer (NT) technology is used to develop clones. The current efficiency remains low, and one of the important limiting factors is the nucleus of donor cell type (Kwong et al. 2014). gMSCs as donor karyoplast appear promising to improve such efficiency as these cells can be passaged for extended periods (Ren et al. 2014). Ex vivo studies have demonstrated gBM-MSCs as donor karyoplast to produce cloned caprine embryos. These cells had better developed up to the hatched blastocyst stage than that with ear fibroblasts. Compared to somatic cells, better proliferation rate, growth capacity, transfection efficiency and convergence, and cleavage had been demonstrated with MSCs (Kwong et al. 2014; Ren et al. 2014). Contrarily, one of the studies had failed to show any difference in fusion and cleavage rates between gBM-MSCs and somatic cells (Kwong et al. 2014).

Apart from the NT, MSCs have been evaluated for their effect on the growth and activation of pre-antral follicles. A significant reduction in the percentage of morphologically normal follicles cultured along with BM-MSCs had been demonstrated as compared to the fresh control. However, a significantly higher rate of morphologically normal pre-antral follicles had been reported when cultured in BM-MSCs as compared to those cultured normally. The follicular diameter, however, had remained comparable in all the follicles irrespective of the culture conditions. In conclusion, BM-MSCs had no influence on follicular or oocyte development but had promoted survival of pre-antral follicles (Arrivabene Neves et al. 2019).

### 8.3.9 Mastitis

Stem cells have been harnessed from mammary gland and as such are considered as potential targets to improve milk production (Martignani et al. 2009; Capuco et al.

2012; Martignani et al. 2014). MSCs due to their characteristic features have potential to regenerate mammary tissue. A single MSCs study on goat chronic mastitis had demonstrated their potential to reduce fibrosis and in maintaining normal tissue architecture and ability to secrete milk (Costa et al. 2019).

### 8.3.10 Urinary System

Goats may act as a model for stress urinary incontinence disorder of human (Burdzinska et al. 2017) and as such applicability of gMSCs in such ailments may be evaluated.

## 8.4 Conclusions

gMSCs have been harvested from various sources including foetal membranes. MSCs characteristic properties like proliferation and differentiation potential have been evaluated in relation to tissue sources. The cells have been characterized as per the standard protocols. Bone marrow, adipose tissue, and foetal adnexa-derived MSCs have been mainly evaluated under experimental studies. A wide array of gMSCs research mainly focussed on human translational studies has been conducted. Preliminary therapeutic literature on gMSCs has variably favoured their in vivo applications. Different biomaterials like cellularized scaffolds and genetically engineered cells together with growth factors have also been evaluated. Currently, the studies remain inconclusive.

## References

Akram T, Hakim M, Mir BA, Hussain SS, Shabir N, Andrabi M, Dar PA, Ganaia NA, Shabir N (2017) Comparative efficiency of goat mesenchymal stem cell isolation from bone marrow and bone chip. Small Rum Res 153:87–94

Anderson ML, Dhert WJ, De Bruijn JD, Dalmeijer RA, Leenders H, Van Blitterswijk CA, Verbout AJ (1999) Critical size defect in the goat's os ilium. A model to evaluate bone grafts and substitutes. Clinical Orthop Rel Res 364:231–239

Arrivabene Neves C, dos Santos Silva L, Sousa de Carvalho CE, Carvalho MS, Sarmento JLR, Cavalcante TV, Arrivabene M, Arrivabene Neves T, de Sousa Bezerra MÉ, Gomes Júnior AL, Gomes da Silva CM, Martins de Carvalho MA (2019) Culture of goat preantral follicles in situ associated with mesenchymal stem cell from bone marrow. Zygote 28:65–71

Azari O, Babaei H, Molaei MM, Nematollahi-Mahani SN, Layasi T (2008) The use of Wharton's jelly-derived mesenchymal stem cells to accelerate second-intention cutaneous wound healing in goat. Ir J Vet Surg 3(3):15–26

Azari O, Babaei H, Derakhshanfar A, Nematollahi-Mahani SN, Poursahebi R, Moshrefi M (2011) Effects of transplanted mesenchymal stem cells isolated from Wharton's jelly of caprine umbilical cord on cutaneous wound healing; histopathological evaluation. Vet Res Commun 35(4):211–222

Beckstein JC, Sen S, Schaer TP, Vresilovic EJ, Elliott DM (2008) Comparison of animal discs used in disc research to human lumbar disc: axial compression mechanics and glycosaminoglycan content. Spine 33:E166–E173

Bekkers JE, Tsuchida AI, van Rijen MH, Vonk LA, Dhert WJ, Creemers LB, Saris DB (2013) Single-stage cell-based cartilage regeneration using a combination of chondrons and mesenchymal stromal cells: comparison with microfracture. Am J Sports Med 41(9):2158–2166
Burdzinska A, Dybowski B, Zarychta-Wisniewska W, Kulesza A, Zagozdzon R, Gajewski Z, Paczek L (2017) The anatomy of caprine female urethra and characteristics of muscle and bone marrow derived caprine cells for autologous cell therapy testing. Anat Rec 300(3):577–588
Cao L, Liu G, Gan Y, Fan Q, Yang F, Zhang X, Tang T, Dai K (2012) The use of autologous enriched bone marrow MSCs to enhance osteoporotic bone defect repair in long-term estrogen deficient goats. Biomaterials 33(20):5076–5084
Capuco AV, Choudhary RK, Daniels KM, Li RW, Evock-Clover CM (2012) Bovine mammary stem cells: cell biology meets production agriculture. Animal 6(3):382–393
Costa CRM, Feitosa MLT, Rocha AR, Bezerra DO, Leite YKC, Argolo Neto NM et al (2019) Adipose stem cells in reparative goat mastitis mammary gland. PLoS One 14(10):e0223751
Costa-Pinto AR, Reis RL, Neves NM (2011) Scaffolds based bone tissue engineering: the role of chitosan. Tissue Eng Part B Rev 17:331–347
da Silva Filho OF, Argôlo Neto NM, de Carvalho MAM, de Carvalho YK, das Neves Diniz A, da Silva Moura L, Ambrósio CE, Monteiro JM, de Almeida HM, Miglino XMA, de Jesus Rosa Pereira A, Macedo JKV, da Rocha AR, Feitos MLT, Ribeiro Alves F (2014) Isolation and characterization of mesenchymal progenitors derived from the bone marrow of goats native from northeastern Brazil. Acta Cir Bras 29(8):478–484
Dai KR, Xu XL, Tang TT, Zhu ZA, Yu CF, Lou JR, Zhang XL (2005) Repairing of goat Tibial bone defects with BMP-2 gene-modified tissue-engineered bone. Calcif Tissue Int 77:55–61
El Hadidi YN, El Kassaby M, El Fatah Ahmed SA, Khamis N (2016) The effect of mesenchymal stem cells application on the distracted bone microstructure. (An experimental study). J Oral Maxillofac Surg 74(7):1463.e1–1463.e11
Elkhenany H, Amelse L, Lafont A, Bourdo S, Caldwell M, Neilsen N, Dervishi E, Derek O, Biris AS, Anderson D, Dhar M (2015) Graphene supports in vitro proliferation and osteogenic differentiation of goat adult mesenchymal stem cells: potential for bone tissue engineering. J Appl Toxicol 35(4):367–374
Elkhenany H, Amelse L, Caldwell M, Abdelwahed R, Dhar M (2016) Impact of the source and serial passaging of goat mesenchymal stem cells on osteogenic differentiation potential: implications for bone tissue engineering. J Anim Sci Biotechnol 7:16
Eslaminejad MB, Nazarian H, Falahi F, Taghiyar L, Daneshzadeh MT (2009) Ex vivo expansion and differentiation of mesenchymal stem cells from goat bone marrow. Ir J Basic Med Sci 12(2):70–79
Frisbie DD, Cross MW, McIlwraith CW (2006) A comparative study of articular cartilage thickness in the stifle of animal species used in human pre-clinical studies compared to articular cartilage thickness in the human. Vet Comp Orthop Traumatol 19(3):142–146
Gardel LS, Correia-Gomes C, Serra LA, Gomes ME, Reis RL (2013) A novel bidirectional continuous perfusion bioreactor for the culture of large-sized bone tissue-engineered constructs. J Biomed Mat Res Part B 101B:1377–1386
Ghaneialvar H, Soltani L, Rahmani HR, Lotfi AS, Soleimani M (2018) Characterization and classification of mesenchymal stem cells in several species using surface markers for cell therapy purposes. Ind J Clin Biochem 33(1):46–52
Gugjoo, MB, Amarpal, Fazili MR, Shah RA, Mir MS, Sharma GT (2020). Goat mesenchymal stem cell basic research and potential applications. Small Rum Res. 183:106045.
Gullbrand SE, Schaer TP, Agarwal P, Bendigo JR, Dodge GR, Chen W, Elliott DM, Mauck RL, Malhotra NR, Smith LJ (2017) Translation of an injectable triple-interpenetrating-network hydrogel for intervertebral disc regeneration in a goat model. Acta Biomater 15(60):201–209
Jeng L, Olsen BR, Spector M (2010) Engineering Endostatin-producing cartilaginous constructs for cartilage repair using nonviral transfection of chondrocyte-seeded and Mesenchymal-stem-cell-seeded collagen scaffolds. Tissue Eng Part A 16(10):3011–3021
Jurgens WJFM, Kroeze RJ, Zandieh-Doulabi B, van Dijk A, Renders GAP, Smit T, van Milligen FJ, Ritt MJ, Helder MN (2013) One-step surgical procedure for the treatment of Osteochondral

defects with adipose-derived stem cells in a Caprine knee defect: a pilot study. BioRes Open Access 2(4):315–325

Knippenberg M, Helder M, Doulabi BZ, Semeins CM, Wuisman PIJM, Klein-Nulend J (2005) Adipose tissue-derived Mesenchymal stem cells acquire bone cell-like responsiveness to fluid shear stress on Osteogenic stimulation. Tissue Eng 11(11/12):1780–1787

Knippenberg M, Helder M, Doulabi BZ, Semeins CM, Wuisman PIJM, Klein-Nulend J (2006) Osteogenesis versus chondrogenesis by BMP-2 and BMP-7 in adipose stem cells. Biochem Biophys Res Commun 342:902–908

Kumar K, Agarwal P, Das K, Milli B, Madhusoodan AP, Kumar A, Bag S (2016) Isolation and characterization of mesenchymal stem cells from caprine umbilical cord tissue matrix. Tissue Cell 48(6):653–658

Kwong PJ, Nam HY, Wan Khadijah WE, Kamarul T, Abdullah RB (2014) Comparison of in vitro developmental competence of cloned Caprine embryos using donor Karyoplasts from adult bone marrow mesenchymal stem cells vs ear fibroblast cells. Reprod Domest Anim 49:249–253

Li Z, Hou T, Moyuan D, Luo F, Wu X, Xing J, Chang Z, Xu J (2015) The Osteogenetic efficacy of goat bone marrow-enriched self-assembly peptide/demineralized bone matrix *in vitro* and *in vivo*. Tissue Eng 21(7–8):1398

Li PZ, Yan GY, Han L, Pang J, Zhong BS, Zhang GM, Wang F, Zhang YL (2017) Overexpression of STRA8, BOULE, and DAZL genes promotes goat bone marrow-derived mesenchymal stem cells in vitro trans-differentiation toward putative male germ cells. Reprod Sci 24(2):300–312

Liao B, Deng L, Wang F (2006) Effects of bone marrow mesenchymal stem cells enriched by small intestinal submucosal films on cardiac function and compensatory circulation after myocardial infarction in goats. Zhongguo Xiu Fu Chong Jian Wai Ke Za Zhi 20(12):1248–1252

Liao L, Shi B, Chang H, Su X, Zhang L, Bi C, Shuai Y, Du X, Deng Z, Jin Y (2017) Heparin improves BMSC cell therapy: anticoagulant treatment by heparin improves the safety and therapeutic effect of bone marrow-derived mesenchymal stem cell cytotherapy. Theranostics 7(1):106–116

Lippens E, Vertenten G, Gironès J, Declercq H, Saunders J, Luyten J, Duchateau L, Schacht E, Vlaminck L, Gasthuys F, Cornelissen M (2010) Evaluation of bone regeneration with an injectable, in situ polymerizable Pluronic F127 hydrogel derivative combined with autologous mesenchymal stem cells in a goat tibia defect model. Tissue Eng Part A 16(2):617–627

Liu X, Li X, Fan Y, Zhang G, Li D, Dong W, Sha Z, Yu X, Feng Q, Cui F, Watari F (2010) Repairing goat tibia segmental bone defect using scaffold cultured with mesenchymal stem cells. J Biomed Mat Res Part B Appl Biomat 94:44–52

Loozen LD, van der Helm YJM, Oner FC, Dhert WJA, Kruyt MC, Alblas J (2015) Bone morphogenetic protein-2 nonviral gene therapy in a goat iliac crest model for bone formation. Tissue Eng 21(9):10

Marei MK, Saad MM, El-Ashwah AM, El-Backly RM, Al-Khodary MA (2009) Experimental formation of periodontal structure around titanium implants utilizing bone marrow mesenchymal stem cells: a pilot study. J Oral Implantol 35(3):106–129

Martignani E, Eirew P, Eaves C, Baratta M (2009) Functional identification of bovine mammary epithelial stem/progenitor cells. Vet Res Commun 33:101–103. https://doi.org/10.1007/s11259-009-9254-z

Martignani E, Cravero D, Miretti S, Accornero P, Baratta M (2014) Bovine mammary stem cells: new perspective for dairy science. Vet Q 34:52–58

Martins GR, Marinho RC, Bezerra-Junior RQ, Câmara LMC, Albuquerque-Pinto LC, Teixeira MFS (2017) Isolation, culture and characterization of multipotent mesenchymal stem cells from goat umbilical cord blood. Pesquisa Vet Bras 37(6):643–649

Mohamad-Fauzi N, Ross PJ, Maga EA, Murray JD (2015) Impact of source tissue and ex vivo expansion on the characterization of goat mesenchymal stem cells. J Anim Sci Biotechnol 6:1

Murphy JM, Fink DJ, Hunziker EB, Barry FP (2003) Stem cell therapy in a Caprine model of osteoarthritis. Arthritis Rheumatol 48(12):3464–3474

Nair MB, Varma HK, Menon KV, Shenoy SJ, John A (2009) Tissue regeneration and repair of goat segmental femur defect with bioactive triphasic ceramic-coated hydroxyapatite scaffold. Biomed Mat Res 91A:855–865

Nam HY, Karunanithi P, Poh Loo WC, Naveen SV, Chen HC, Hussin P, Chan L, Kamarul T (2013) The effects of staged intra-articular injection of cultured autologous mesenchymal stromal cells on the repair of damaged cartilage: a pilot study in caprine model. Arthr Res Ther 15:R129

Peeters M, van Rijn S, Vergroesen P-PA, Paul CPL, Noske DP, Vandertop WP, Wurdinger T, Helder MN (2015) Bioluminescence-mediated longitudinal monitoring of adipose-derived stem cells in a large mammal *ex vivo* organ culture. Sci Rep 5:13960

Pei Y, Fan J-J, Zhang X-Q, Zhang Z-Y, Yu M (2014) Repairing the Osteochondral defect in goat with the tissue-engineered Osteochondral graft Preconstructed in a double-chamber stirring bioreactor. BioMed Res Int 219203:11

Petrella F, Toffalorio F, Brizzola S, Martino De Pas T, Rizzo S, Barberis M, Pelicci P, Spaggiari L, Acocella F (2014) Stem cell transplantation effectively occludes Bronchopleural fistula in an animal model. Ann Thorac Surg 97:480–483

Phinney DG, Prockop DJ (2007) Concise review: Mesenchymal stem/multipotent stromal cells: the state of transdifferentiation and modes of tissue repair—current view. Stem Cells 25:2896–2902

Port J, Jackson DW, Lee TQ, Simon TM (1996) Meniscal repair supplemented with exogenous fibrin clot and autogenous cultured marrow cells in the goat model. Am J Sports Med 24(4):547–555

Pratheesh MD, Gade NE, Katiyar AN, Dubey PK, Sharma B, Saikumar G, Amarpal, Sharma GT (2013) Isolation, culture and characterization of caprine mesenchymal stem cells derived from amniotic fluid. Res Vet Sci 94(2):313–319

Pratheesh MD, Gade NE, Dubey PK, Nath A, Sivanarayanan TB, Madhu DN, Sharma B, Amarpal, Saikumar G, Sharma GT (2014) Molecular characterization and xenogenic application of Wharton's jelly derived caprine mesenchymal stem cells. Vet Res Commun 38(2):139–148

Pratheesh MD, Gade NE, Nath A, Dubey PK, Sivanarayanan TB, Madhu DN, Sreekumar TR, Amarpal, Saikumar G, Sharma GT (2017) Evaluation of persistence and distribution of intra-dermally administered PKH26 labelled goat bone marrow derived mesenchymal stem cells in cutaneous wound healing model. Cytotechnology 69(6):841–849

Proffen BL, McElfresh M, Fleming BC, Murray MM (2012) A comparative anatomical study of the human knee and six animal species. Knee 19:493–499

Qiu P, Bai Y, Liu C, He X, Cao H, Li M, Zhu H, Hua J (2012) A dose-dependent function of follicular fluid on the proliferation and differentiation of umbilical cord mesenchymal stem cells (MSCs) of goat. Histochem Cell Biol 138:593–603

Ren Y, Wu H, Ma Y, Yuan J, Liang H, Liu D (2014) Potential of adipose-derived Mesenchymal stem cells and skeletal muscle-derived satellite cells for somatic cell nuclear transfer mediated Transgenesis in Arbas cashmere goats. PLoS One 9(4):e93583

Schop D, Janssen FW, van Rijn LDS, Fernandes H, Bloem RM, de Bruijn JD, van Dijkhuizen-Radersma R (2009) Growth, metabolism, and growth inhibitors of Mesenchymal stem cells. Tissue Eng 15(8):1877–1886

Shen Q-R, Zou S-J, Chen U-X, Xou H-X, Hu J (2006) Effect of Mesenchymal stem cells transplantation on cranial suture distraction osteogenesis in growing goats. West China J Stomatol 24(2):153–155

Somal A, Bhat IA, Baiju I, Pandey S, Panda BSK, Thakur N, Sarkar M, Chandra V, Saikumar G, Sharma GT (2016) A comparative study of growth kinetics, in vitro differentiation potential and molecular characterization of fetal adnexa derived Caprine Mesenchymal stem cells. PLoS One 11(6):e0156821

Somal A, Bhat IA, Baiju I, Singh AP, Panda BSK, Desingu PA, Pandey S, Bharti MK, Pal A, Saikumar G, Chandra V, Sharma GT (2017) Impact of cryopreservation on caprine fetal adnexa derived stem cells and its evaluation for growth kinetics, phenotypic characterization and wound healing potential in xenogenic rat model. J Cell Physiol 232(8):2186–2200

Sonobe M, Nakagawa M, Ichinose M, Ikegami N, Nagasawa M, Shindo T (2000) Analysis of risk factors in bronchopleural fistula after pulmonary resection for primary lung cancer. Eur J Cardiothorac Surg 18:519–523

Sorensen AL, Jacobsen BM, Reiner AH, Andersen IS, Collas P (2010) Promoter DNA methylation patterns of differentiated cells are largely programmed at the progenitor stage. Mol Biol Cell 21:2066–2077

Sousa CP, Dias IR, Lopez-Peña M, Camassa JA, Lourenço PJ, Judas FM, Gomes ME, Reis RL (2015) Bone turnover markers for early detection of fracture healing disturbances: a review of the scientific literature. An Acad Bras Cienc 87:1049–1061

Sun X-D, Jeng L, Bolliet C, Olsen BR, Spector M (2009) Non-viral endostatin plasmid transfection of mesenchymal stem cells via collagen scaffolds. Biomaterials 30:1222–1231

Tamada H, Yoh C, Inaba T, Takano H, Kawate N, Sawada T (2000) Epidermal growth factor (EGF) in the goat uterus: immunohistochemical localization of EGF and EGF receptor and effect of EGF on uterine activity in vivo. Theriogenol 54(1):159–169

Tamadon A, Mehrabani D, Zarezadeh Y, Rahmanifar F, Dianatpour M, Zare S (2017) Caprine endometrial Mesenchymal stromal stem cell: multilineage potential, characterization, and growth kinetics in breeding and anestrous stages. Vet Med Int 2017:5052801

Tang TT, Lu B, Yue B, Xie XH, Xie YZ, Dai KR, Lu JX, Lou JR (2007) Treatment of osteonecrosis of the femoral head with hBMP-2-gene-modified tissue-engineered bone in goats. J Bone Joint Surg Br 89(1):127–129

Tatsumi K, Ohashi K, Matsubara Y, Kohori A, Ohno T, Kakidachi H, Horii A, Kanegae K, Utoh R, Iwata T, Okano T (2013) Tissue factor triggers procoagulation in transplanted mesenchymal stem cells leading to thromboembolism. Biochem Biophys Res Commun 431:203–209

Toh WS, Foldager CB, Olsen BR, Spector M (2013) Basement membrane molecule expression attendant to chondrogenesis by nucleus Pulposus cells and Mesenchymal stem cells. J Orthop Res 31:1136–1143

Tripathi AK, Ramani UV, Ahir VB, Rank DN, Joshi CG (2010) A modified enrichment protocol for adult caprine skeletal muscle stem cell. Cytotechnology 62:483–488

Van Der Donk S, Buma P, Aspenberg P, Schreurs BW (2001) Similarity of bone in growth in rats and goats: a bone chamber study. Comp Med 51:336–340

Vertenten G, Lippens E, Gironès J, Gorski T, Declercq H, Saunders J, Van den Broeck W, Chiers K, Duchateau L, Schacht E, Cornelissen M, Gasthuys F, Vlaminck L (2009) Evaluation of an injectable, photopolymerizable, and three-dimensional scaffold based on methacrylate-endcapped poly(D,L-lactide-co-epsilon-caprolactone) combined with autologous mesenchymal stem cells in a goat tibial unicortical defect model. Tissue Eng Part A 15(7):1501–1511

Wakelee HA, Chang ET, Gomez SL, Keegan TH, Feskanich D, Clarke CA, Holmberg L, Yong LC, Kolonel LN, Gould MK, West DW (2007) Lung cancer incidence in never smokers. J Clin Oncol 25:472–478

Wang C, Wang Z, Li A, Bai F, Lu J, Xu S, Li D (2010) Repair of segmental bone-defect of goat's tibia using a dynamic perfusion culture tissue engineering bone. J Biomed Mat Res A 92:1145–1153

Wang X, Wang Z, Wang Q, Wang H, Liang H, Liu D (2017) Epigenetic modification differences between fetal fibroblast cells and mesenchymal stem cells of the Arbas cashmere goat. Res Vet Sci 114:363–369

Wang Q, Wang X, Lai D, Deng J, Hou Z, Liang H, Liu D (2018) BIX-01294 promotes the differentiation of adipose mesenchymal stem cells into adipocytes and neural cells in Arbas cashmere goats. Res Vet Sci 119:9–18

Wei X, Liu B, Liu G, Yang F, Cao F, Dou X (2019) Mesenchymal stem cell-loaded porous tantalum integrated with biomimetic 3D collagen-based scaffold to repair large osteochondral defects in goats. Stem Cell Res Ther 10:72

Wu Y, Chen L, Scott PG, Tredget EE (2007) Mesenchymal stem cells enhance wound healing through differentiation and angiogenesis. Stem Cells 25:2648–2654

Xu X, Hu J, Lu H (2017) Histological observation of a gelatin sponge transplant loaded with bone marrow-derived mesenchymal stem cells combined with platelet-rich plasma in repairing an annulus defect. PLoS One 12(2):e0171500

Zhang Y, Liu S, Guo W, Wang M, Hao C, Gao S, Zhang X, Li X, Chen M, Jing X, Wang Z, Peng J, Lu S, Guo Q (2018) Human umbilical cord Wharton's jelly mesenchymal stem cells combined with an acellular cartilage extracellular matrix scaffold improve cartilage repair compared with microfracture in a Caprine model. Osteoarthr Cartil 30077(18):S1063–S4584

Zhang, X.L., Jiang, X.Q., 2006. Osteoblastic differentiation of goat bone marrow stromal cells after AdBMP-2 mediated transduction in vitro. Shanghai Kou Qiang Yi Xue 15(6):610–3.

Zhang YL, Li PZ, Pang J, Wan YJ, Zhang GM, Fan YX, Wang ZY, Tao NH (2019) Induction of goat bone marrow mesenchymal stem cells into putative male germ cells using mRNA for STRA8, BOULE and DAZL. Cytotechnology 71(2):563–572

Zhao W, Lu JY, Hao YM, Cao CH, Zou DR (2017) Maxillary sinus floor elevation with a tissue-engineered bone composite of deciduous tooth stem cells and calcium phosphate cement in goats. J Tissue Eng Regen Med 11(1):66–76

Zhu S, Zhang B, Man C, Ma Y, Hu J (2011) NELL like molecule-1 modifed bone marrow mesenchymal stem cells/poly-lactic-co-glycolic acid composite improves repair of large osteochondral defects in mandibular condyle. Osteoarthr Cartil 19:743–750

Zou D, Guo L, Lu J, Zhang X, Wei J, Liu C, Zhang Z, Jiang X (2012) Engineering of bone using porous calcium phosphate cement and bone marrow stromal cells for maxillary sinus augmentation with simultaneous implant placement in goats. Tissue Eng 18(13,14):1464–1478

# Cattle/Buffalo Mesenchymal Stem Cell Basic Research and Potential Applications

9

M. B. Gugjoo, Amar Pal, M. R. Fazili, R. A. Shah, and G. T. Sharma

**Abstract**

Cattle/buffalo has an important position in the economics of the livestock industry, worldwide. These animals are main contributors of the meat and milk, and any relevant ailment like reproductive inefficiency and mastitis ipso facto pose a major challenge to the veterinarian. To address these health issues and other relevant conditions, stem cells especially mesenchymal stem cells are being evaluated. MSCs in large ruminants have been harvested from various tissue including fetal membranes and are well characterized. Slim assimilation of knowledge in their basic physiological properties limits their definitive applications. MSCs applications in large ruminants have mostly been evaluated for production-related traits. In this chapter potential applications of cattle and buffalo MSCs, in addition to their characteristics, have been discussed.

M. B. Gugjoo (✉) · M. R. Fazili
Division of Veterinary Clinical Complex, FVSc & AH, SKUAST-Kashmir, Jammu and Kashmir, India

A. Pal
Division of Surgery, Indian Veterinary Research Institute, Izatnagar, UP, India

R. A. Shah
Division of Animal Biotechnology, FVSc & AH, SKUAST-K, Srinagar, Jammu and Kashmir, India

G. T. Sharma
Division of Physiology and Climatology, Indian Veterinary Research Institute, Izatnagar, UP, India


M. B. Gugjoo, A. Pal (eds.), *Mesenchymal Stem Cell in Veterinary Sciences*, https://doi.org/10.1007/978-981-15-6037-8_9

## 9.1 Introduction

Among various domestic animals, cattle/buffalo has a crucial position in the economics of the livestock industry, worldwide. These animals mainly contribute through meat and milk production, in addition to hides and other biological materials (Jean and Anderson 2014; Aghamohammadi et al. 2018; Das et al. 2018). Medical conditions like endometritis, mastitis, lameness, fracture, and others can badly affect their production and reproductive efficiency (Izquierdo et al. 2017). With the biotechnological advancement, cattle- or buffalo-MSCs (c-/buf-MSCs) are being studied for improved production, reproduction, and/or genetic improvement (Shibiru et al. 2017). Additionally, bovines can offer suitable translational experimental animal model for human medicine (Bosnakovski et al. 2004). c-/buf-MSCs potential applications are being evaluated mostly in relation to reproduction and/ or production. The current chapter throws light on the c-/buf-MSCs characteristics and their potential applications.

## 9.2 c/buf-MSCs Sources and Their Characterization

Various sources have been utilized to harvest c-/buf-MSCs. These cells are being evaluated and characterized as per the standard criteria laid down for humans. Cattle- and buffalo-MSCs characterization markers including differentiation potential have been detailed in Tables 9.1 and 9.2, respectively (Gugjoo et al. 2018). Plastic adherent c-/buf-MSCs exhibit fibroblastic morphology. All these studies have not demonstrated all the laid down characterization principles (Donofrio et al. 2008). Even some of the studies have failed to report typical surface marker expression of these MSCs as shown in Table 9.1. In addition to the trilineage, the cells have been studied for their extended differentiation potential including germ cell-, hepatocyte-, islet cell-, myocyte-, and neurocyte-like cells (Cardoso et al. 2012a; Lu et al. 2014a, b, c; Xiong et al. 2014; Peng et al. 2017; Deng et al. 2018).

### 9.2.1 c-/buf-MSCs Morphology

Plastic adherent c-/buf-MSCs show variable morphologies. These cells may exhibit variable morphologies in early passages followed by mostly fibroblastic in extended passages as has been reported for cattle amniotic fluid MSCs (cAF-MSCs). The early passage morphology includes fibroblastic, and flat and circular type (Corradetti et al. 2013) or fusiform (Yang et al. 2016). These fibroblasts tend to show higher mean population doubling potential and are better able to migrate and home to distant tissues. The higher migration potential and homing was confirmed by their ability to show greater CD44 expression (Rossi et al. 2014). Buffalo amniotic membrane MSCs (bufAm-MSCs) too had shown mixed cell morphologies in early passages, which had changed to mostly fibroblastic subsequently (Dev et al. 2012). The time of cell plastic adherence may also determine their ability to appear fibroblasts.

**Table 9.1** Cattle mesenchymal stem cell sources and their characterization

| Positive markers | Negative markers | Differentiation | References |
|---|---|---|---|
| *Adipose tissue MSCs (AD-MSCs)* | | | |
| Vimentin, CD49d, CD13 | CD34 | Adipogenic | Ren et al. (2010) |
| β-Integrin, CD44, and CD73 | – | Adipogenic, osteogenic | Lu et al. (2014c) |
| CD29, CD73, CD90, CD105 | CD34, CD45 | Adipogenic, chondrogenic, osteogenic | da Silva et al. (2016a) |
| CD70, CD90 | CD34, CD45 | Adipogenic | Cebo (2017) |
| CD29, CD44, Vimentin | CD34, MHC-II | Adipogenic, chondrogenic, osteogenic | Queiroz et al. (n.d.) |
| CD73, Oct4 and Nanog | CD34 | – | Cahuascanco et al. (2019) |
| *Amniotic fluid MSCs (AF-MSCs)* | | | |
| CD29, CD44, CD73, CD166, MHC-I, Oct-4, C-Myc | CD14, CD34, CD45, CD90, MHC-II | Adipogenic, chondrogenic, neurogenic | Corradetti et al. (2013) |
| CD44, CD73, CD166, Oct-4 | CD34, CD45 | Adipogenic, osteogenic, neurogenic | Gao et al. (2014) |
| CD90, CD105, CD34 | CD14, CD73, CD34, Oct-4, Sox2, Cytokeratin, E-Cadherin | Adipogenic, osteogenic, chondrogenic | Rossi et al. (2014) |
| CD29, CD73, CD90, CD105 | CD34, CD44, CD45 | Adipogenic, chondrogenic, osteogenic | da Silva et al. (2016a) |
| β-Integrin, CD44, CD73, CD166, Oct-4 | CD34, CD45 | Adipogenic, osteogenic, chondrogenic | Chang et al. (2018) |
| *Amniotic membrane MSCs (Am-MSCs)* | | | |
| CD29, CD44, CD73, CD105, CD166, MHC-I, Oct-4, C-Myc | CD14, CD34, CD45, MHC-II | Adipogenic, chondrogenic, neurogenic | Corradetti et al. (2013) |
| Nanog, Oct-4, Sox-2 | – | Osteogenic | Mann et al. (2013) |
| CD44, Nanog, Oct-4 | MHC-II | Adipogenic, chondrogenic, osteogenic | Campos et al. (2017) |
| *Bone marrow MSCs (BM-MSCs)* | | | |
| CD29, CD44, and CD166 | CD14, CD31, CD34, CD45, CD90, CD105, CD140a, and CD117 | Adipogenic, osteogenic, neuroectodermal | Kato et al. (2004) |
| CD73, CD90, CD105 | CD34, CD45 | Adipogenic, chondrogenic, osteogenic | Cortes et al. (2013) |
| CD29, CD73, Oct-4, Nanog | CD34, CD45 | Hepatogenic, neurogenic | Dueñas et al. (2014) |

(continued)

**Table 9.1** (continued)

| Positive markers | Negative markers | Differentiation | References |
|---|---|---|---|
| CD29, CD44, CD73 | CD31, CD34, CD45 | Osteoblasts, adipocytes, hepatic, and islet-like cells | Lu et al. (2014a) |
| CD29, CD90, CD44, CD105, Oct-4, Sox2 Nanog | CD45 | Adipogenic, chondrogenic, osteogenic | Lee et al. (2015) |
| CD73, Oct4 and Nanog | CD34 | – | Cahuascanco et al. (2019) |
| *Dermal MSCs (D-MSCs)* | | | |
| CD29, CD44, CD73, CD90 | CD34, CD106, CD166, Oct-4 | Adipogenic, osteogenic, neurogenic | Sun et al. (2014) |
| *Endometrium MSCs (End-MSCs)* | | | |
| CD44, CD117, STAT-3, Oct-4, Sox-2 | Nanog | Chondrogenic, osteogenic | Cabezas et al. (2014) |
| CD44, CD29, Vimentin | CD34 (low expression), MHC-II | Adipogenic, osteogenic | de Moraes et al. (2016) |
| OCT4, SOX2, CD44, c-KIT | NANOG | Adipogenic, osteogenic, chondrogenic | Lara et al. (2017a, b) |
| *Liver MSCs (Li-MSCs)* | | | |
| CD29, CD44, CD73, CD90, CD106, CD166 | CD34, CD45, BLA-DR | Adipogenic, osteogenic, chondrogenic | Lu et al. (2014b) |
| | *Lung MSCs (L-MSCs)* | | |
| CD29, CD44, CD73, CD166 | CD34, CD45, BOLA-DRα | Osteogenic, neural lineages | Hu et al. (2015) |
| *Placental MSCs (Pl-MSCs)* | | | |
| CD166, CD73, β-integrin, Oct-4 | CD45 | Islet like cells (insulin, glucagon, pax-4, Nkx6.1, pax-6, Fox) | Peng et al. (2017) |
| *Tendon MSCs (T-MSCs)* | | | |
| CD44, Tenascin, Collagen I, Collagen III | Collagen II | Adipogenic, chondrogenic, osteogenic | Yang et al. (2016) |
| *Umbilical cord MSCs (UC-MSCs)* | | | |
| Oct-4, CD73 | – | Adipogenic, chondrogenic, osteogenic | Raoufi et al. (2011) |
| CD29, CD44, CD73, CD90, CD166 | CD34, CD45, BLA-DR | Adipogenic, osteogenic, hepatocyte-, Islet cell-, neurocyte-like cells, | Xiong et al. (2014) |
| CD44, Nanog, Oct-4 | MHC-II | Adipogenic, chondrogenic, osteogenic | Campos et al. (2017) |
| *Wharton's jelly MSCs (WJ-MSCs)* | | | |
| CD29, CD73, CD90, CD105, POUF51, ITSN1 | CD34, CD45 | Adipogenic, chondrogenic, osteogenic, neural-like cells | Cardoso et al. (2012b) |

(continued)

**Table 9.1** (continued)

| Positive markers | Negative markers | Differentiation | References |
|---|---|---|---|
| CD105, CD29, CD73, CD90 | – | Adipogenic, chondrogenic, osteogenic, astrocyte | Cardoso et al. (2017) |
| CD29, CD73, CD90, CD105 | CD34, CD44, CD45 | Adipogenic, chondrogenic, osteogenic | da Silva et al. (2016a) |
| *Yolk sac MSCs (YS-MSCs)* | | | |
| CD90, CD105, and CD79 | CD45, CD44, and CD79 | Adipogenic, chondrogenic, osteogenic | Mançanares et al. (2015) |

**Table 9.2** Buffalo mesenchymal stem cell sources and their characterization

| Positive markers | Negative markers | Differentiation | References |
|---|---|---|---|
| *Adipose tissue MSCs* | | | |
| Oct-4, CD44 | – | Adipogenic, chondrogenic, osteogenic | Hepsibha et al. (2011) |
| CD105, CD90, CD73 | CD34, CD45, CD79A | Adipogenic, osteogenic, chondrogenic | Sampaio et al. (2015) |
| *Amniotic membrane MSCs* | | | |
| Oct-4, Sox-2, Nanog | – | Neurogenic | Dev et al. (2012) |
| CD29, CD44, CD105, OCT4, SOX2, NANOG, TERT | CD34 | Adipogenic, chondrogenic, osteogenic | Ghosh et al. (2015) |
| CD29, CD44, Nanog, Oct-4, Sox-2 | CD34 | Adipogenic, chondrogenic, osteogenic | Sadeesh et al. (2016) |
| CD44, CD90, CD105, CD73, CD29, CD166, OCT4, SOX2, NANOG, REX-1, SSEA-1, SSEA-4 and TRA-1-81, Cytokeratin 18 (epithelial) | CD34, CD45 | Adipogenic, osteogenic, chondrogenic, and neural lineages (FABP4, Ost, ACAN, COL2A1, Nestin, and β III-tubulin) | Deng et al. (2018) |
| *Bone marrow MSCs* | | | |
| CD73, CD90, CD105, Nanog, Oct-4, STRO-1, Sox2 | CD31, CD34 | Adipogenic, chondrogenic, osteogenic | Gade et al. (2013) |
| *Umbilical cord MSCs* | | | |
| OCT4, NANOG and SOX2 | – | Adipogenic, osteogenic | Singh et al. (2013) |
| *Wharton's jelly MSCs* | | | |
| CD-73, CD-105 and CD-90, Stro-1, Oct-4, Nanog, Sox-2 | CD34 | Adipogenic, chondrogenic, osteogenic | Sreekumar et al. (2014) |

*AD-MSCs* adipose tissue MSCs, *BM-MSCs* bone marrow MSCs, *d-MSCs* dermis MSCs, *Am-MSCs* amniotic membrane MSCs, *UC-MSCs* umbilical cord MSCs, *WJ-MSCs* Wharton's jelly MSCs

Earlier the cells attach to plastic, earlier the fibroblastic cells appear as hasd been demonstrated for early gestation-derived bufAm-MSCs (Deng et al. 2018)

### 9.2.2 Culture Characteristics

Cattle and buffalo MSCs tend to have similar culture characteristics. Only difference may remain in their ability to survive for longer periods. Buffalo adipose tissue (AD)-MSCs had been proliferated for more passages in comparison to cattle (cAD)-MSCs, yielding higher cell concentration in the former sources (Sampaio et al. 2015). Such differences may arise with respect to tissue sources, their physiological state, and age of donor.

### 9.2.3 Extended Passaging and MSCs Proliferation

MSCs in general fail to survive indefinitely. Their colony-forming unit (CFU) potential along the path of extended passaging decreases (Lu et al. 2014a, b, c). MSCs may show differences in the proliferation potential and self-renewal properties based on the sources. Cattle bone marrow MSCs (cBM-MSCs) had shown CFU potential as 63.6 ± 7.57% at passage 3, 59.0 ± 4.38% at passage 9, 32.6 ± 3.88% at passage 15, and 19.5 ± 3.27% at passage 25 (Lu et al. 2014a, b, c). Likewise, population doubling time (PDT) of bufAm-MSCs had remained 43.9 ± 20 h. at passage 3, 45.9 ± 30 h. at passage 6, 46.7 ± 30 h. at passage 9, and 60.0 ± 50 h, at passage 20 (Deng et al. 2018). However, cattle bone marrow MSCs (cBM-MSCs) population doubling had been maintained up to passage 7 (Cortes et al. 2013).

MSCs proliferation potential and viability may be maintained up to as low as passage 10 (Corradetti et al. 2013) to the maximum of passage 60 (Cardoso et al. 2012b). Cattle tendon-derived MSCs (cT-MSCs), dermal MSCs (cDer-MSCs), and lung MSCs (cL-MSCs) had been maintained up to passages 42, 19, and 25 and afterward these cells had undergone senescence (Sun et al. 2014; Hu et al. 2015; Yang et al. 2016). Morphologically senescent cells appear vacuolated and show karyopyknosis while growth kinetics show much reduced or no growth at all.

Comparison of cAF-MSCs and cAm-MSCs had shown higher proliferation capacity of former cell lines up to the passage 10. Thereafter both cell lines had reduced viability (Corradetti et al. 2013). Another study on cAF-MSCs however, had been able to successfully culture them up to passage 34 (Gao et al. 2014). Similarly, cattle Wharton's jelly (cWJ)-MSCs and cattle bone marrow (cBM)-MSCs had comparable proliferation up to passage 10 after which both the cell lines had become senescent (Cortes et al. 2013; Xiong et al. 2014). Thus, MSCs proliferation is mainly affected by tissue source and culture environment. To keep the record of cellular activity, evaluation of karyotypic features is desired (Colleoni et al. 2005).

### 9.2.4 Culture Modification and MSCs Characteristics

MSCs may also be cultured in serum-free media. cWJ-MSCs cultured in serum-free media had remained viable and had stable karyotype up to passage 60. Under 3-D culture system, cMSCs may undergo morphological changes and tend to differentiate (Cardoso et al. 2012a, b). These systems offer uniform culture environment and thus can potentially give rise to a uniform batch of cells. However, further studies are required to standardize these culture techniques.

### 9.2.5 Gestational Phase and cMSCs Proliferation

cAF-MSCs had shown differential proliferation potential as per the gestational phase being higher in first trimester in comparison to other stages (Rossi et al. 2014). Estral cyclity had non-significant effect on cattle endometrium (cEnd)-MSCs as the cells at different stages had almost comparable cellular properties up to passage 3 (de Moraes et al. 2016). The late luteal phase MSCs, however, may be more amenable to mesodermal lineages (Cabezas et al. 2014).

### 9.2.6 Effect of Age and Health Status on MSCs

MSCs proliferation potential tends to decrease with age. cBM-MSCs from young donor had demonstrated vigorous proliferation in comparison to the adult source (Jiwu et al. 2008). Health status also affects the MSCs culture characteristics as infections like endometritis had reduced proliferation potential of cEnd-MSCs (Lara et al. 2017a, b), while tumor growth (adenomyotic uterus) had enhanced their pluripotency expression profile (Nanog and Oct-4) (Łupicka et al. 2015).

## 9.3 Potential Applications of c-/buf-MSCs

c-/buf-MSCs are seen as potential candidates to improve production and reduce clinical problems in animals. Therapeutic applications of MSCs in large ruminants have lagged far behind as compared to equine or pet animals. It is due to the lack of interest in high-end therapeutics for such species as these animals are mainly reared for economic benefits. In vivo research is preferably conducted in lab animals and in advanced stage large animals are employed. The large animals being ontogenically closer to humans are considered to provide better translational models for them. Considering the importance of large ruminants in relation to production purpose, currently MSCs applications ipso facto have been prioritized in the field (Gugjoo et al. 2018). However, the potential goes beyond as these animals have clinical relevance to veterinary and human medicine. It, therefore, becomes imperative to carry out therapeutic studies on c-/buf-MSCs. Below are described the potential applications of c-/buf-MSCs.

### 9.3.1 Reproductive System

#### 9.3.1.1 MSCs as Donor in Nuclear Transfer Technique

Nuclear transfer technique (NTT) is aimed at improving the reproductive efficiency though the current success rate is very low (0–5%). Donor cell plays a key role in the success. In comparison to fibroblasts, MSCs as donor cell have shown better fusion rate but little success in cleavage rate. The overall blastocyst formation efficiency has remained good (da Silva et al. 2016b; Nazari et al. 2016; Sadeesh et al. 2016). MSCs fusion rate may vary with the donor tissue source as cAF-MSCs had shown better fusion rate almost 30–35% higher than the cAD-MSCs (85.89% vs. 50.06%). However, fusion rate of MSCs from different sources had remained higher than fibroblasts (44.85%). In contrast the cleavage rate had been better for fibroblasts, and an overall comparable percent of calves had been born (da Silva et al. 2016a, b).

Apart from MSCs source, oocyte quality too plays an important role in the success of nuclear transfer (NT). cBM-MSCs form only 7% blastocysts to preserved oocyte and 39% blastocysts to fresh oocytes (Kato et al. 2004), showing that recipient oocyte too has bearing on development of blastocysts. The cleavage rate of bufAm-MSCs had been lower than that of in vitro fertilized (IVF) embryos, though comparable percent of blastocysts were formed (Sadeesh et al. 2016).

Cell size may affect fusion rate as better fusion rate with larger cAF-MSCs/cWJ-MSCs had been demonstrated in comparison to smaller cAD-MSCs/ fibroblasts. The larger size of cell favors better contact with the zona pellucida and/or cell membrane of oocyte (da Silva et al. 2016a, b). However, larger-sized cAF-MSCs may lead to large offspring syndrome (larger-sized offspring). This larger-sized offspring may not survive long due to dystocia and/or fetal abnormalities that tend to develop with limited available space in dam uterus (da Silva et al. 2016a, b). Furthermore, short survival of offspring may also develop due to the epigenetically dysregulated expression of imprinted genes (Farin et al. 2006; Opiela et al. 2017).

Contrary to these reports, cBM-MSCs had failed to show any significant difference in the developmental rates at any stage in comparison to the fibroblasts (Colleoni et al. 2005). Thus, further studies are desired to evaluate the possible options to improve the success rate of the fusion and cleavage with the MSCs.

#### 9.3.1.2 MSCs as Supplementation Medium

MSCs may provide secretory factors that can replace granulosa cells. Granulosa cells supplement embryo culture media for their growth. cAD-MSCs ($10^3/10^4$ cells/mL) supplementation to in vitro fertilized embryos had led to their better blastocyst formation rates. The higher cell concentration ($10^4$ cells/mL) and their conditioned medium were more effective than lower ($10^3$ cells/mL) ones (Miranda et al. 2016).

#### 9.3.1.3 MSCs and Genital Affections

The cells may be effective in male/female genital affections due to their trophic actions including immunomodulatory actions. Amniotic fluid MSCs injection into the ovary had induced estrus in bilateral ovarian dystrophic cattle. Fifty percent of animals (4 out of 8 animals) had come into heat. Out of those four animals, 50% (2 out of 4 animals) had ovulated, formed corpus luteum (CL), and attained pregnancy. It could be inferred that cAF-MSCs injection may bring animals into heat and subsequently CL formation can occur but may also be affected by other factors (Chang et al. 2018). Endometritis reduce cEnd-MSCs proliferation potential (Lara et al. 2017a, b), and as such extrinsic implantation of MSCs may be effective way to control endometritis due to their immunomodulatory or anti-inflammatory properties. The cellular implantation may be made through per vaginal route where cells may migrate and home to endometrium as has been demonstrated in horse (Mambelli et al. 2013; Falomo et al. 2015).

MSCs may be utilized to improve and/ or maintain elite bull germplasm. In vitro cMSCs differentiated into germ cell-like cells may be implanted into the testicles. It has been demonstrated that ram testis had supported germ cell-like cells growth and survival. The cells had even migrated and homed to the basement membrane of seminiferous tubules. These cells had formed spermatogonia-like cell colonies similar to testicular native spermatogonia (Ghasemzadeh-Hasankolaei et al. 2015). In lieu cBM-MSCs trans-differentiated to germ cell-like cells may be evaluated for their utility.

### 9.3.2 Oxidative Stress

In livestock sector, elite animals are preferred for their higher production potential, which creates an environment of stress (Jóźwik et al. 2012). Stress makes these tissue cells more amenable to early aging. MSCs exhibit dynamic antioxidant action through endogenous production of hemeoxygenase-1 (HO-1) and glutathione and restricting cellular production of reactive oxygen species (Vanella et al. 2012; Calió et al. 2014; Ayatollahi et al. 2014). These features also attribute to MSCs tolerance toward the oxidative stress (Brandl et al. 2011). Due to these features, MSCs may prevent telomere shortening in cells preventing their senescence (Estrada et al. 2013).

### 9.3.3 Bovine Mastitis

Mastitis that causes decrease in milk production has lot more associated problems as it makes human health amenable to diseases through bad quality milk and through antibiotic resistance. Currently, no ideal treatment is available to combat this problem in productive cattle. Mammary stem cells are being isolated and characterized from mammary gland of bovines in addition to other species (Shackleton et al. 2006; Martignani et al. 2009; Martignani et al. 2014). Isolation of different stem cell types from mammary gland is currently under process (Thomas et al. 2011; Rauner and Barash 2012; Martignani et al. 2014).

MSCs can prove useful in two ways: on one way the cells may secrete specific proteins and on other way their antimicrobial effects may prevent mastitis. Co-culture of mammary epithelial cells and cUC-MSCs had enhanced quantity of β-casein and κ casein, in addition to the expression of other proteins in mammary epithelial cells. The possible pathway followed in the mammary epithelial cells might have been JAK/ STAT as IGF-1 had inhibited their expression profile (Zhao et al. 2017).

Fetal cBM-MSCs and cAD-MSCs conditioned medium carries antibacterial activity. The conditioned medium had decreased about 30% of the growth of *Staphylococcus aureus* under in vitro conditions (Cahuascanco et al. 2019). Transfection of c-/buf-MSCs with bovine lactoferricin (LFcinB) gene may endow them with the properties to express antibacterial peptide LfcinB (Sharma et al. 2017). Such LFcinB-cloned MSCs if injected into the mammary gland may prove fruitful through their higher antibacterial properties especially against *Staphylococcus aureus* and *Escherichia coli* (Sharma et al. 2017). Thus, extensive studies are desired to confirm the feasibility and efficacy of using these transfected MSCs in clinical settings.

### 9.3.4 Cultured Meat Production

Due to growing human population demands and impediments to animal-rearing (greenhouse gases emission) and slaughter policies (ethics), lab-grown meat is being aimed. Cultured meat like burger and meat balls have been produced in 2013 and 2016, respectively (Post 2012; Arshad et al. 2017). The main focus remains to produce cost-effective meat that has acceptable taste and appearance. MSCs that can trans-differentiate into muscle cell-like cells and adipocytes may be cultured on edible scaffold (Post 2012; Bhat et al. 2015; Arshad et al. 2017) and, thus, need further studies.

### 9.3.5 Wound Healing

Cattle/ buffalo teat fistula and fibrosis have huge economic impact through loss of milk and potential to develop mastitis. MSCs applications in such conditions are desired. In our own unpublished study, allogeneic MSCs applications in post-surgical repair of teat fistula promote early healing of wounds. Thus, MSCs applications appear promising in such cases. The possible beneficial effects may be through MSCs differentiation into various types of epithelial cells (Wu et al. 2007; Borena et al. 2009) and/or through vasculogenesis (Borena et al. 2009).

### 9.3.6 Bovine Spongiform Encephalopathy (BSE) and Nerve Injuries

BM-MSCs/ peripheral blood (PB)-MSCs neurogenic potential is affected in scrapie (Mediano et al. 2015), although marrow infectivity in BSE remains undiagnosed on the spleen, lymph nodes and brain assays (Sohn et al. 2009). Thus, there is need to confirm if the said potential is also reduced in BSE and whether these cells can be used to treat infected animals.

Stem cells may trans-differentiate into the neural cell-like cells or secrete trophic factors (Hokari et al. 2008; Saito et al. 2008; Cardoso et al. 2012a; Xiong et al. 2014), though the direct neuronal differentiation is questionable (Hokari et al. 2008; Wu et al. 2003; Dobkin et al. 2006). Commonly affected peripheral nerves like radial, obturator, and sciatic nerves make large ruminants lame and/or recumbent. Thus, MSCs studies may be undertaken to evaluate their therapeutic potential in these peripheral nerve injuries.

### 9.3.7 Degenerative Joint Diseases (DJD) and Laminitis

MSCs are extensively studied for their therapeutic effects in osteoarthritis. Cattle chronic osteoarthritis (OA) or degenerative joint disease (DJD) tends to cause male infertility (impotentia coeundi) in bulls (Heinola et al. 2013; Wolfe 2018). MSCs may be effective in OA; however, currently the therapy has not been able to provide desired results and may be due to the inhibitory effects of locally available inflammatory mediators. The inflammatory environment tends to enhance expression of adhesion molecule preventing cell migration and decrease their secretion of matrix (Zayed et al. 2016; Reesink et al. 2017).

Cattle and horse laminitis carry almost similar pathogenesis (Nilsson 1982) and as MSCs have been demonstrated to be play effective role in equine acute laminitis (Dryden et al. 2013; Morrison et al. 2014), studies in cattle may also be undertaken.

### 9.3.8 Miscellaneous Conditions

c-/buf-MSCs potentially trans-differentiate into myocyte-, osteocyte-, tenocyte-, hepatocyte-, and islet cell-like cells and as such may be employed to treat associated ailments. Additionally, MSCs may also be evaluated for their antiviral effect (Cardoso et al. 2012a, b).

## 9.4 Conclusions

Cattle and buffalo MSCs have been harvested from numerous sources and are well characterized. MSCs properties may vary with respect to the source and may be affected with respect to the age and health status. Literature on cattle/ buffalo MSCs applications mainly focuses on their production and reproduction-related traits. The results of these studies however are currently inconclusive. MSCs potential applications go beyond these reproductive and/ production-related traits and may be studied in such areas as well.

## References


Aghamohammadi M, Haine D, Kelton DF, Barkema HW, Hogeveen H, Keefe GP, Dufour S (2018) Herd-level mastitis-associated costs on Canadian dairy farms. Front Vet Sci 5:100

Arshad MS, Javed M, Sohaib M, Saeed F, Imran A, Amjad Z (2017) Tissue engineering approaches to develop cultured meat from cells: a mini review. Cogent Food Agric 3:1320814

Ayatollahi M, Hesami Z, Jamshidzadeh A, Gramizadeh B (2014) Antioxidant effects of bone marrow mesenchymal stem cell against carbon tetrachloride-induced oxidative damage in rat livers. Int J Organ Transplant Med 5:166–173

Bhat ZF, Kumar S, Fayaz H (2015) *In vitro* meat production: challenges and benefits over conventional meat production. J Integr Agric 14(2):241–248

Borena BM, Pawde AM, Pal A, Aithal HP, Kinjavdekar P, Singh R, Kumar D (2009) Autologous bone marrow-derived cells for healing excisional dermal wounds of rabbits. Vet Rec 165(19):563–568

Bosnakovski D, Mizuno M, Kim G, Ishiguro T, Okumura M, Iwanaga T, Kadosawa T, Fujinaga T (2004) Chondrogenic differentiation of bovine bone marrow mesenchymal stem cells in pellet cultural system. Exp Hematol 32(5):502–509

Brandl A, Meyer M, Bechmann V, Nerlich M, Angele P (2011) Oxidative stress induces senescence in human mesenchymal stem cells. Exp Cell Res 317:1541–1547

Cabezas J, Lara E, Pacha P, Rojas D, Veraguas D, Saravia F, Rodríguez-Alvarez L, Castro FO (2014) The endometrium of cycling cows contains populations of putative mesenchymal progenitor cells. Reprod Domest Anim 49(4):550–559

Cahuascanco B, Bahamonde J, Huaman O, Jervis M, Cortez J, Palomino J, Escobar A, Retama P, Torres CG, Peralta OA (2019) Bovine fetal mesenchymal stem cells exert antiproliferative effect against mastitis causing pathogen Staphylococcus aureus. Vet Res 50:25

Calió ML, Marinho DS, Ko GM, Ribeiro RR, Carbonel AF, Oyama LMM, Ormanji T, Guirao P, Calió PL, Reis LA, de Jesus Simões M, Lisbôa-Nascimento T, Ferreira AT, Bertoncini CR (2014) Transplantation of bone marrow mesenchymal stem cells decreases oxidative stress, apoptosis, and hippocampal damage in brain of a spontaneous stroke model. Free Radic Biol Med 70:141–154

Campos LL, Landim-Alvarenga FC, Ikeda TL, Monteiro BA, Maia L, Freitas-Dell'Aqua CP, De Vita B (2017) Isolation, culture, characterization and cryopreservation of stem cells derived from amniotic mesenchymal layer and umbilical cord tissue of bovine fetuses. Pesqui Vet Bras 37(3):278–286

Cardoso TC, Ferrari HF, Garcia AF, Novais JB, Silva-Frade C, Ferrarezi MC, Andrade AL, Gamiero R (2012a) Isolation and characterization of Wharton's jelly-derived multipotent mesenchymal stromal cells obtained from bovine umbilical cord and maintained in defined serum free three-dimensional system. BMC Biotechnol 12:18

Cardoso TC, Novais JB, Antello TF, Silva-Frade C, Ferrarezi MC, Ferrari HF, Gameiro R, Flores EF (2012b) Susceptibility of neuron-like cells derived from bovine Wharton's jelly to bovine herpesvirus type 5 infections. BMC Vet Res 8:242

Cardoso TC, Okamura LH, Baptistella JC, Gameiro R, Ferreira HL, Marinho M, Flores EF (2017) Isolation, characterization and immunomodulatory-associated gene transcription of Wharton's jelly-derived multipotent mesenchymal stromal cells at different trimesters of cow pregnancy. Cell Tissue Res 367(2):243–256

Cebo D (2017) Characterization of bovine adipose-derived stem cells. Int J Sci Technol Res 6(5):16–18

Chang L-B, Peng S-Y, Chou C-J, Chen Y-J, Shiu J-S, Tu P-A, Gao S-X, Chen Y-C, Lin T-K, Wu S-C (2018) Therapeutic potential of amniotic fluid stem cells to treat bilateral ovarian dystrophy in dairy cows in a subtropical region. Reprod Domest Anim 53(4):1024–1024

Colleoni S, Donofrio G, Lagutina I, Duchi R, Galli C, Lazzari G (2005) Establishment, differentiation, electroporation, viral transduction, and nuclear transfer of bovine and porcine mesenchymal stem cells. Cloning Stem Cells 7(3):154–166

Corradetti B, Meucci A, Bizzaro D, Cremonesi F, Lange Consiglio A (2013) Mesenchymal stem cells from amnion and amniotic fluid in the bovine. Reproduction 145:391–400
Cortes Y, Ojeda M, Araya D, Dueñas F, Fernández MS, Peralta OA (2013) Isolation and multilineage differentiation of bone marrow mesenchymal stem cells from abattoir-derived bovine fetuses. BMC Vet Res 9:133
da Silva CG, Martins CF, Cardoso TC, da Cunha ER, Bessler HC, MacManus CM, Pivato I, Bao SN (2016a) Isolation and characterization of Mesenchymal stem cells derived from bovine Wharton's jelly and their potential use in cloning by nuclear transfer. Ciência Rural 46(10):1830–1837
da Silva CG, Martins CF, Cardoso TC, da Cunha ER, Bessler HC, Martins GHL, Pivato I, Bao SN (2016b) Production of bovine embryos and calves cloned by nuclear transfer using Mesenchymal stem cells from amniotic fluid and adipose tissue. Cell Reprogram 18(2):127–135
Das D, Panda SK, Jena B, Sahoo AK (2018) Economic impact of subclinical and clinical mastitis in Odisha. India Int J Curr Microbiol App Sci 7(3):3651–3654
de Moraes CN, Maia L, Dias MC, Dell'Aqua CP, da Mota LS, Chapwanya A, Landim-Alvarenga FD, Oba E (2016) Bovine endometrial cells: a source of mesenchymal stem/progenitor cells. Cell Biol Int 40(12):1332–1339
Deng Y, Huang G, Zou L, Nong T, Yang X, Cui J, Wei Y, Yang S, Shi D (2018) Isolation and characterization of buffalo (*bubalus bubalis*) amniotic mesenchymal stem cells derived from amnion from the first trimester pregnancy. J Vet Med Sci 80(4):710–719
Dev K, Giri S, Kumar A, Yadav A, Singh B, Gautam S (2012) Derivation, characterization and differentiation of buffalo (Bubalus bubalis) amniotic fluid derived stem cells. Reprod Domest Anim 47(5):704–711
Dobkin BH, Curt A, Guest J (2006) Cellular transplants in China: observational study from the largest human experiment in chronic spinal cord injury. Neurorehabil Neural Repair 20(1):5–13
Donofrio G, Franceschi V, Capocefalo A, Cavirani S, Sheldon IM (2008) Bovine endometrial stromal cells display osteogenic properties. Reprod Biol Endocrinol 6:65
Dryden VC, Morrison S, Bras R, Morrell SA (2013) Using stem cells in clinical cases. J Equine Vet 33(10):872–873
Dueñas F, Becerra V, Cortes Y, Vidal S, Sáenz L, Palomino J, De Los Reyes M, Peralta OA (2014) Hepatogenic and neurogenic differentiation of bone marrow mesenchymal stem cells from abattoir-derived bovine fetuses. BMC Vet Res 10:154
Estrada JC, Torres Y, Benguría A, Dopazo A, Roche E, Carrera-Quintanar L, Pérez RA, Enríquez JA, Torres R, Ramírez JC, Samper E, Bernad A (2013) Human mesenchymal stem cell-replicative senescence and oxidative stress are closely linked to aneuploidy. Cell Death Discov 4:e691
Falomo ME, Ferroni L, Tocco I, Gardin C, Zavan B (2015) Immunomodulatory role of adipose-derived stem cells on equine endometriosis. BioMed Res Int 2015:1
Farin PW, Piedrahita JA, Farin CE (2006) Errors in development of fetuses and placentas from in vitro produced bovine embryos. Theriogenology 65:178–191
Gade NE, Pratheesh MD, Nath A, Dubey PK, AmarPal, Sharma B, Saikumar G, Sharma GT (2013) Molecular and cellular characterization of buffalo bone marrow-derived mesenchymal stem cells. Reprod Domest Anim 48(3):358–367
Gao Y, Zhu Z, Zhao Y, Hua J, Ma Y, Guan W (2014) Multi-lineage potential research of bovine amniotic fluid mesenchymal stem cells. Int J Mol Sci 15:3698–3710
Ghasemzadeh-Hasankolaei M, Baghaban Eslaminejad M, Sedighi-Gilani M (2015) Derivation of male germ cells from ram bone marrow mesenchymal stem cells by three different methods and evaluation of their fate after transplantation into the testis. Vitro Cell Dev Biol Anim 52(1):49–61
Ghosh K, Kumar R, Singh J, Gahlawat SK, Kumar D, Selokar NL, Yadav SP, Gulati BR, Yadav PS (2015) Buffalo (Bubalus bubalis) term amniotic-membrane-derived cells exhibited mesenchymal stem cells characteristics in vitro. Vitro Cell Dev Biol Anim 51(9):915–921
Gugjoo MB, Amarpal, Fazili MR, Shah RA, Sharma GT (2018) Mesenchymal stem cell: Basic research and potential applications in cattle and buffalo. J Cell Physiol 234(6):8618–8635

Heinola T, Grauw JC, Virkki L, Kontinen A, Raulo SM, Sukura A, Kontinen YT (2013) Bovine chronic osteoarthritis causes minimal change in synovial fluid. J Comp Pathol 148(Suppl.4):335–344

Hepsibha P, Meenambigai TV, Mangalagowri A, Palanisamy A, Stalin A, Nithya S, Kumanan K (2011) Multipotent differentiation potential of buffalo adipose tissue derived mesenchymal stem cells. Asian J Anim Vet Adv 6:772–788

Hokari M, Kuroda S, Shichinohe H, Yano S, Hida K, Iwasaki Y (2008) Bone marrow stromal cells protect and repair damaged neurons through multiple mechanisms. J Neurosci Res 86:1024–1035

Hu P, Pu Y, Li X, Zhu Z, Zhao Y, Guan W, Ma Y (2015) Isolation, *in vitro* culture and identification of a new type of mesenchymal stem cell derived from fetal bovine lung tissues. Mol Med Rep 12:3331–3338

Izquierdo AC, Guerra Liera JE, Cervantes RE, Inzunza Castro JF, Abel Villa Mancera E, Crispín RH, Juárez Mosqueda ML, Vázquez AG, Pérez JO, Aparicio PS, Rodriguez Denis BE (2017) Production of milk and bovine mastitis. J Adv Dairy Res 5:174

Jean GS, Anderson DE (2014) Decision analysis for fracture management in cattle. Vet Clin North Am Food Anim Pract 30(1):1–10

Jiwu F, Xuejin B, Jing Z (2008) A comparative study of calf and adult bone marrow mesenchymal stem cell in vitro. Sci Agric Sin 41(7):2128–2135

Jóźwik A, Krzyżewski J, Strzałkowska N, Poławska E, Bagnicka E, Wierzbicka A, Niemczuk K, Lipińska P, Horbańczuk JO (2012) Relations between the oxidative status, mastitis, milk quality, and disorders of reproductive functions in dairy cows—a review. Anim Sci Paper Rep 30:297–307

Kato Y, Imabayashi H, Mori T, Tani T, Taniguchi M, Higashi M, Matsumoto M, Umezawa A, Tsunoda Y (2004) Nuclear transfer of adult bone marrow mesenchymal stem cells: developmental totipotency of tissue-specific stem cells from an adult mammal. Biol Reprod 70(2):415–418

Lara E, Rivera N, Rojas D, Rodríguez-Alvarez LL, Castro FO (2017a) Characterization of mesenchymal stem cells in bovine endometrium during follicular phase of oestrous cycle. Reprod Domest Anim 52(5):707–714

Lara E, Velásquez A, Cabezas J, Rivera N, Pacha P, Rodríguez-Alvarez L, Saravia F, Castro FO (2017b) Endometritis and in vitro PGE2 challenge modify properties of cattle endometrial mesenchymal stem cells and their transcriptomic profile. Stem Cells Int. 2017:1

Lee SH, Cha S-H, Kim C-L, Lillehoj HS, Song J-Y, Lee K-W (2015) Enhanced adipogenic differentiation of bovine bone marrow derived mesenchymal stem cells. J Appl Anim Res 43(1):15–21

Lu T, Huang Y, Wang H, Ma Y, Guan W (2014a) Multi-lineage potential research of bone marrow-derived stromal cells (BMSCs) from cattle. Appl Biochem Biotechnol 172:21–35

Lu T, Hu P, Su X, Li C, Ma Y, Guan W (2014b) Isolation and characterization of mesenchymal stem cells derived from fetal bovine liver. Cell Tissue Bank 15(3):439–450

Lu T, Xiong H, Wang K, Wang S, Ma Y, Guan W (2014c) Isolation and characterization of adipose-derived mesenchymal stem cells (ADSCs) from cattle. Appl Biochem Biotechnol 174(2):719–728

Łupicka M, Socha B, Szczepańska A, Korzekwa A (2015) Expression of pluripotency markers in the bovine uterus with adenomyosis. Reprod Biol Endocrinol 13:110

Mambelli LI, Winter GH, Kerkis A, Malschitzky E, Mattos RC, Kerkis I (2013) A novel strategy of mesenchymal stem cells delivery in the uterus of mares with endometrosis. Theriogenology 79(5):744–750

Mançanares CA, Oliveira VC, Oliveira LJ, Carvalho AF, Sampaio RV, Mançanares AC, Souza AF, Perecin F, Meirelles FV, Miglino MA, Ambrósio CE (2015) Isolation and characterization of mesenchymal stem cells from the yolk sacs of bovine embryos. Theriogenology 84(6):887–898

Mann A, Yadav RP, Singh J, Kumar D, Singh B, Yadav PS (2013) Culture, characterization and differentiation of cells from buffalo (Bubalus bubalis) amnion. Cytotechnology 65(1):23–30

Martignani E, Eirew P, Eaves C, Baratta M (2009) Functional identification of bovine mammary epithelial stem/progenitor cells. Vet Res Commun 33:101–103

Martignani E, Cravero D, Miretti S, Accornero P, Baratta M (2014) Bovine mammary stem cells: new perspective for dairy science. Vet Q 34:52–58

Mediano DR, Sanz-Rubio D, Bolea R, Marín B, Vázquez FJ, Remacha AR, López-Pérez Ó, Fernández-Borges N, Castilla J, Zaragoza P, Badiola JJ, Rodellar C, Martín-Burriel I (2015) Characterization of mesenchymal stem cells in sheep naturally infected with scrapie. J Gen Virol 96:3715–3726

Miranda MS, Nascimento HS, Costa MP, Costa NN, Brito KN, Lopes CT, Santos SS, Cordeiro MS, Ohashi OM (2016) Increasing of blastocyst rate and gene expression in co-culture of bovine embryos with adult adipose tissue-derived mesenchymal stem cells. J Assist Reprod Genet 33(10):1395–1403

Morrison S, Dryden VC, Bras R, Morrell S (2014) How to use stem cells in clinical laminitis cases. How to choose regenerative medicine and therapies. In: American association of equine practitioners: proceedings, vol 60. American Association of Equine Practitioners, Lexington, pp 499–503

Nazari H, Shirazi A, Shams-Esfandabadi N, Afzali A, Ahmadi E (2016) The effect of amniotic membrane stem cells as donor nucleus on gene expression in reconstructed bovine oocytes. Int J Dev Biol 60:95–102

Nilsson SA (1982) Laminitis in cattel in comparison with equine laminitis. In: Forth international symposium on disorders of ruminant digit, Paris

Opiela J, Samiec M, Romanek J (2017) In vitro development and cytological quality of inter-species (porcine/bovine) cloned embryos are affected by trichostatin A-dependent epigenomic modulation of adult mesenchymal stem cells. Theriogenology 97:27e33

Peng S-Y, Chou C-W, Kuo Y-H, Shen P-C, Shaw S-WS (2017) Potential differentiation of islet like cells from pregnant cow derived placental stem cells. Taiwan J Obstet Gynecol 56:306–311

Post MJ (2012) Cultured meat from stem cells: challenges and prospects. Meat Sci 92(3):297–301

Queiroz CM, Moraes CN, Maia L, Kandim-Alvaranga FC, Ferriera JCP (n.d.) Isolation, culture and characterization of mesenchymal stem cells from adipose tissue. Vet Zootec

Raoufi MF, Tajik P, Dehghan MM, Eini F, Barin A (2011) Isolation and differentiation of mesenchymal stem cells from bovine umbilical cord blood. Reprod Domest Anim 46:95–99

Rauner G, Barash I (2012) Cell hierarchy and lineage commitment in the bovine mammary gland. PLoS One 7:e30113

Reesink HL, Sutton RM, Shurer CR, Peterson RP, Tan JS, Jin S, Paszek MJ, Nixon AJ (2017) Galectin-1 and galectin-3 expression in equine mesenchymal stromal cells (MSCs), synovial fibroblasts and chondrocytes, and the effect of inflammation on MSC motility. Stem Cell Res Ther 8:243

Ren Y, Wu H, Ma Y, Cang M, Wang R, Liu D (2010) Isolation, cultivation and identification of adipose-derived stem cell in bovines. Chin J Biotechnol 26(12):1645–1651

Rossi B, Merlo B, Colleoni S, Iacono E, Tazzari PL, Ricci F, Lazzari G, Galli C (2014) Isolation and in vitro characterization of bovine amniotic fluid derived stem cells at different trimesters of pregnancy. Stem Cell Rev Rep 10(5):712–724

Sadeesh EM, Shah F, Yadav PS (2016) Differential developmental competence and gene expression patterns in buffalo (Bubalus bubalis) nuclear transfer embryos reconstructed with fetal fibroblasts and amnion mesenchymal stem cells. Cytotechnology 68(5):1827–1848

Saito F, Nakatani T, Iwase M, Maeda Y, Hirakawa A, Murao Y, Suzuki Y, Onodera R, Fukushima M, Ide C (2008) Spinal cord injury treatment with intrathecal autologous bone marrow stromal cell transplantation: the first clinical trial case report. J Trauma 64:53–59

Sampaio RV, Chiaratti MR, Santos DCN, Bressan FF, Sangalli JR, Sá ALA, Silva TVG, Costa NN, Cordeiro MS, Santos SSD, Ambrosio CE, Adona PR, Meirelles FV, Miranda MS, Ohashi OM (2015) Generation of bovine (*Bos indicus*) and buffalo (*Bubalus bubalis*) adipose tissue derived stem cells: isolation, characterization, and multipotentiality. Genet Mol Res 14(1):53–62

Shackleton M, Vaillant F, Simpson KJ, Stingl J, Smyth GK, Asselin-Labat ML, Wu L, Lindeman GJ, Visvader JE (2006) Generation of a functional mammary gland from a single stem cell. Nature 439:84–88

Sharma N, Huynh DL, Kim SW, Ghosh M, Sodhi SS, Singh AK, Kim NE et al (2017) A PiggyBac mediated approach for lactoferricin gene transfer in bovine mammary epithelial stem cells for management of bovine mastitis. Oncotarget 8(61):104272

Shibiru B, Abriham K, Ashebr A (2017) Review on reproductive biotechnology and its role in dairy cattle production and health. Reprod Opin 9(3):60–70

Singh J, Mann A, Kumar D, Duhan JS, Yadav PS (2013) Cultured buffalo umbilical cord matrix cells exhibit characteristics of multipotent mesenchymal stem cells. In Vitro Cell Dev Biol Anim 49:408–416

Sohn HJ, Lee YH, Green RB, Spencer YI, Hawkins SAC, Stack MJ, Konold T, Wells GAH, Matthews D, Cho IS, Joo YS (2009) Bone marrow infectivity in cattle exposed to the bovine spongiform encephalopathy agent. Vet Rec 164:272–274

Sreekumar TR, Ansari MM, Chandra VG, Sharma T (2014) Isolation and characterization of buffalo Wharton's jelly derived mesenchymal stem cells. J Stem Cell Res Ther 4:207

Sun T, Yu C, Gao Y, Zhao C, Hua J, Cai L, Guan W, Ma Y (2014) Establishment and biological characterization of a dermal mesenchymal stem cells line from bovine. Biosci Rep 34(2):139–146. https://doi.org/10.1042/BSR20130095

Thomas E, Zeps N, Cregan M, Hartmann P, Martin T (2011) 14-3-3sigma (sigma) regulates proliferation and differentiation of multipotent p63-positive cells isolated from human breast milk. Cell Cycle 10:278–284

Vanella L, Sanford C Jr, Kim DH, Abraham NG, Ebraheim N (2012) Oxidative stress and heme oxygenase-1 regulated human mesenchymal stem cells differentiation. Int J Hyperth 2012:1

Wolfe DF (2018) Abnormalities of the bull–occurrence, diagnosis and treatment of abnormalities of the bull, including structural soundness. Animal 12:s148–s157

Wu S, Suzuki Y, Ejiri Y, Toru N, Hongliang B, Masaaki K, Kazuya K, Ohta M, Chou H, Ide C (2003) Bone marrow stromal cells enhance differentiation of cocultured neurosphere cells and promote regeneration of injured spinal cord. J Neurosci Res 72:343–351

Wu Y, Chen L, Scott PG, Tredget E (2007) Mesenchymal stem cells enhance wound healing through differentiation and angiogenesis. Stem Cells 25:2648–2659

Xiong H, Bai C, Wu S, Gao Y, Lu T, Hu Q, Guan W, Ma Y (2014) Biological characterization of mesenchymal stem cells from bovine umbilical cord. Anim Cell Syst 1(1):59–67

Yang J, Zhao Q, Wang K, Liu H, Ma C, Huang H, Liu Y (2016) Isolation and biological characterization of tendon-derived stem cells from fetal bovine. Vitro Cell Dev Biol Anim 52(8):846–856

Zayed MN, Schumacher J, Misk N, Dhar MS (2016) Effects of pro-inflammatory cytokines on chondrogenesis of equine mesenchymal stromal cells derived from bone marrow or synovial fluid. Vet J 217:26–32

Zhao L, Wang Z, Zhang J, Yang J, Gao XF, Wu B, Zhao G, Bao S, Hu S, Liu P, Li X (2017) Characterization of the single-cell derived bovine induced pluripotent stem cells. Tissue Cell 49(5):521–527

# Cat Mesenchymal Stem Cell Characteristics and Potential Applications

# 10

M. B. Gugjoo and Amar Pal

**Abstract**

Cat mesenchymal stem cells (fMSCs) have been culture expanded and characterized from various tissue sources, though mainly from adipose tissues. The cells have been evaluated for their therapeutic effects in various experimental models and clinical trials relevant to veterinary and human practice. The studies have potentially favored MSCs applications; however, limited available studies currently fail to provide any conclusive result(s). The current chapter discusses fMSCs characteristics and their potential applications.

## 10.1 Introduction

Cat is increasingly being kept as a companion animal, and as such their clinical practice is increasing by each passing day. Regenerative medicine involving MSCs is gradually being employed in feline practice. Cat and human genome are highly linked, and as such cat MSCs experimental/clinical studies may provide proof of therapeutics for human medicine (Munoz et al. 2012), in addition to the therapeutics in veterinary practice.

M. B. Gugjoo (✉)
Division of Veterinary Clinical Complex, FVSc & AH, SKUAST-Kashmir, Jammu and Kashmir, India

A. Pal
Division of Surgery, Indian Veterinary Research Institute, Izatnagar, UP, India



M. B. Gugjoo, A. Pal (eds.), *Mesenchymal Stem Cell in Veterinary Sciences*, https://doi.org/10.1007/978-981-15-6037-8_10

## 10.2 fMSCs In Vitro Characteristics

An initial report on feline mesenchymal stem cells was reported by Martin and co-workers in 2002 that had isolated and characterized the cells from bone marrow (Martin et al. 2002). Since then many reports have been published on their isolation and culture expansion. The cells have been isolated from various although limited tissue types as detailed in Table 10.1. Most of the studies have focused on the adipose tissue derived MSCs. Cat MSCs (fMSCs) are being characterized based on the standard criteria laid down by the International Society for Cellular Therapy (ISCT). fMSCs are plastic adherent with spindle-shaped fibroblast-like appearance (Quimby et al. 2011; Arzi et al. 2015; Sato et al. 2016; Quimby and Borjesson 2018). These cells variably express various surface markers and lack hematopoietic markers. Among various sources, some of the studies had shown MSCs expression of the hematopoietic (CD34) and other markers like MHC-II, while others had reported MSCs failure to express CD90 (Iacono et al. 2012; Mumaw et al. 2015; Vidane et al. 2016). The cells show trilineage differentiation (adipogenic, chondrogenic, and osteogenic lineages) (Table 10.1).

The cells have been culture expanded in growth medium containing the fetal bovine serum (FBS). A study that utilized combination of FBS (10%) and autologous plasma had shown similar fPB-MSCs characteristics as that of cells being cultured under FBS. Currently, studies on FBS alternatives to culture expand fMSCs is lacking. fMSCs tend to undergo early senescence with extensive passaging. Cat peripheral blood (fPB)-MSCs had steadily expanded up to 5–6 passages, but almost all the cells had stopped proliferation at passages 7–9. Among 15 cultures, only 01 culture had proliferated up to passage 10. The cellular proliferation, however, had remained unaffected by age, sex, and breed (Sato et al. 2016). The population doubling levels of cat adipose tissue (fAD)-MSCs demonstrated at different passages were 2.08 ± 0.02 (P1), 3.95 ± 0.07 (P2), 5.88 ± 0.15 (P3), 7.14 ± 0.15 (P4), 8.00 ± 0.16 (P5), and 8.78 ± 0.19 (P6). From P6 onward, minimal increase in doubling levels had occurred. As such the population doubling time had significantly increased to an average of 95.7 h. at P4 and 141.3 h. at P5 from 62.1 h. at P3 (Kim et al. 2017). Similarly, another study had shown that fAD-MSCs population doubling time at passage 5 had increased significantly. The decreased expression of pluripotency markers and of surface markers was also seen after extended culture, significantly at P5. The musculoskeletal (especially osteogenic and chondrogenic) differentiation potential too had reduced with progressive culturing in these cells (Lee et al. 2018). fAD-MSCs at P2 had been characterized by their normal metaphase and karyotype (Villatoro et al. 2018).

To improve cellular proliferation and self-renewal, culture modifications are being worked out. Culture addition of recombinant acid ceramidase (rAC) had improved growth of cat bone marrow (fBM)-MSCs by twofold at 1 week. Even addition of rAC to chondrogenic medium had improved cellular chondrogenic potential, synergistic with TGF-β1 (Simonaro et al. 2013). rAC might have reduced the cell stress that arises with the harvesting procedure and during their ex vivo expansion. The possible beneficial effect of the rAC treatment in chondrogenic

**Table 10.1** Cat mesenchymal stem cell sources and characterization

| Positive markers | Negative markers | Differentiation | References |
|---|---|---|---|
| *Adipose tissue MSCs (AD-MSCs)* | | | |
| CD44, CD90, CD105 | CD4, MHC-II (low expression) | Adipogenic, chondrogenic, and osteogenic | Webb et al. (2011) |
| CD44, CD90 | CD4, MHC-II | Adipogenic, chondrogenic, and osteogenic | Quimby et al. (2013) |
| CD44, CD90, CD105 | CD14, CD34, CD45 | Adipogenic, chondrogenic, osteogenic, and myogenic | Kono et al. (2014) |
| CD105, CD44, CD90 | CD18, MHC-II | – | Arzi et al. (2015) |
| CD90, CD105, CD146, Oct-4, Nanog, Sox-2, and Klf4 | CD73, CD14, CD45, CD271, HLA-DR, Sox-2 | Adipogenic, chondrogenic, osteogenic, and neurogenic | Gómez et al. (2015) |
| CD29, CD44, CD73, CD90, CD105, CD45 | MHCII, CD79α, CD34 | Adipogenic, chondrogenic, and osteogenic | Mumaw et al. (2015) |
| CD90, CD105, CD44 | MHC-II | Adipogenic, chondrogenic, and osteogenic | Parys et al. (2016) |
| CD29, CD44, CD90, CD105, CD166, and MHC-I, Oct3/4, Nanog, and SSEA-4 | CD14, CD34, CD45, CD73, Sox2 | Adipogenic, chondrogenic, and osteogenic | Kim et al. (2017) |
| CD29, CD44, CD73, CD90, MHC-I, Stro-1 | CD34, CD45, MHC-II | Adipogenic, chondrogenic, and osteogenic | Villatoro et al. (2018) |
| CD44, CD90, CD105 | MHC-II | Adipogenic, chondrogenic, and osteogenic | |
| *Amniotic membrane MSCs (Am-MSCs)* | | | |
| CD90, CD44, CD105 | CD34, CD14, CD45, CD73 | Adipogenic, chondrogenic, and osteogenic | Iacono et al. (2012) |
| CD73, CD90 | CD34, CD45, CD79 | Adipogenic, chondrogenic, and osteogenic | Vidane et al. (2014) |
| CD73 and CD90, CD105 (low expression), OCT4, SOX2. | CD30, CD45 | – | Cardoso et al. (2016) |
| CD73 (89.5%) and CD90 (91.9%), CD79 (29.1%), CD34 (47%), | CD45 | – | Vidane et al. (2016) |

(continued)

**Table 10.1** (continued)

| Positive markers | Negative markers | Differentiation | References |
|---|---|---|---|
| *Bone marrow MSCs (BM-MSCs)* | | | |
| CD44, CD90, CD105 | CD4, MHC-II | Adipogenic, chondrogenic, and osteogenic | Quimby et al. (2011) |
| CD44, CD90, CD105 | CD4, MHC-II (low expression) | Adipogenic, chondrogenic, and osteogenic | Webb et al. (2011) |
| CD29, CD44, CD105, Stro-1 | CD14, CD34, CD45, MHC-II | Adipogenic, osteogenic, and neurogenic | Munoz et al. (2012) |
| CD29, CD44, CD105, CD45 (30%), MHC-II (39%) | CD73, CD90, CD79α, and CD34, | Adipogenic, chondrogenic, and osteogenic | Mumaw et al. (2015) |
| *Peripheral blood MSCs (PB-MSCs)* | | | |
| CD44, CD90 | CD4, MHC-II | Adipogenic, chondrogenic, and osteogenic | Sato et al. (2016) |

*AD-MSCs* adipose tissue MSCs, *Am-MSCs* amniotic membrane MSCs, *BM-MSCs* bone marrow MSCs, *PB-MSCs* peripheral blood MSCs

differentiation process may occur through better TGF-β pathway signaling, in addition to the effects of sphingolipid changes on chondrogenic differentiation. Further, stress induced with serum-free chondro-differentiation of MSCs may also be controlled (Simonaro et al. 2013).

The cells isolated from various tissues and at variable locations may have variable culture characteristics. fBM-MSCs and fAD-MSCs had similar surface markers expression, but fAD-MSCs had proliferated significantly faster as compared to BM-MSCs. Thus, fBM-MSC and fAD-MSC were phenotypically similar, but higher intrinsic proliferation rate had made fAD-MSC culture proliferation easier (Webb et al. 2011). In relation to variable donor tissue locations, abdominal fAD-MSCs harbored higher percentage of cells expressing positive surface markers (CD90 and CD105) than those of subcutaneous fAD-MSCs. Although comparable pluripotency marker (Oct-4 and Klf4) expression had been reported, except for Nanog that had been highly expressed in subcutaneous fAD-MSCs. The cells had shown an equivalent in vitro multilineage differentiation including the neuron-like cells (Gómez et al. 2015).

### 10.2.1 MSCs from Infected Donor

fAD-MSCs can contract latent Feline Foamy Virus (FFV, retrovirus) infection that induces formation of syncytial cell and subsequent apoptosis in them. FFV in cultured cells had led to their morphologic and proliferative alterations. FFV-infected fAD-MSC lines cultured in growth media had formed multinucleated giant cells and had topped proliferation at passages 3–5. Incorporation of antiviral tenofovir (reverse transcriptase inhibitor) at 5 or 10 mM had prevented such alterations in morphology and the proliferative capacity of FFV-infected fAD-MSCs. The effect on differentiation potential, however, remains to be evaluated (Arzi et al. 2015).

## 10.3 MSCs Applications

MSCs culture expansion is aimed to achieve enough concentration for their in vivo applications. MSCs have been applied either through intravenous and intraperitoneal or through local routes. All these routes have been found to be safe and without any potential adverse effects. For better concentration of MSCs in cerebrospinal fluid (CSF), intraventricular route of implantation was demonstrated to be superior as compared to the subdural and brain parenchyma route (Glage et al. 2011). The mild reactions in the form of local lymph node swelling and transient lethargy and inactivity had been encountered (Webb and Webb 2015; Trzil et al. 2016; Parys et al. 2016). The intravenous applications of MSCs have usually been followed by coagulation cascade being prevented by heparin. Below are the in vivo fMSCs experimental (Table 10.2) and clinical (Table 10.3) studies.

### 10.3.1 Chronic Enteropathy

Feline chronic enteropathy especially idiopathic inflammatory bowel disease (IBD) requires lifelong medication including immunosuppressive and chemotherapeutic drugs. IBD compromises gastrointestinal (GI) mucosal immunity and tolerance to intestinal antigens (Janeczko et al. 2008; Trepanier 2009). MSCs due to their immunomodulatory features are considered to play the therapeutic role.

In a feline chronic enteropathy of 02 months study, two intravenous doses of allogeneic AD-MSCs had improved the clinical symptoms based on owners report. Five out of seven cases (in blinded study) had significant improvement in clinical signs, while those in placebo (normal saline) treated cases had no improvement and/or symptoms worsened with time. Owner (aware about the treatment) had mixed response. One had reported marked improvement in condition; other conveyed no change while third case follow-up report had been unavailable (Webb and Webb 2015). Thus, further studies are desired to actually confirm MSCs efficacy in these ailments.

### 10.3.2 Chronic Kidney Disease

Kidney ailments are one of the common clinical conditions with life-threatening potential in domestic cats (Lawson et al. 2015). Currently, an effective and specific treatment to restore the affected kidney to normalcy is unavailable (Vidane et al. 2016).

**Table 10.2** In vivo preclinical experimental mesenchymal stem cell studies in cats

| Model type | Number of animals | Study period | Biomaterial used | Cell dose/ assembly | Evaluation criteria | Overall result | References |
|---|---|---|---|---|---|---|---|
| Chronic allergic asthma (Bermuda grass allergen induced) | 9 (MSCs, $n = 5$; placebo PBS, $n = 4$) | 8 months and 12 months | Allogeneic, AD-MSCs | $1.44 \times 10^7$ MSCs/ infusion (bimonthly intravenous injected 6 doses) | Bronchoalveolar lavage fluid cytology, pulmonary mechanics and clinical scoring to assess airway hyper-responsiveness AHR; and thoracic computed tomography (CT) | No differences between treatment groups or over time with respect to airway eosinophilia or AHR have been seen. Significantly lower lung attenuation and bronchial wall thickening scores were noted in CT images of MSC-treated animals compared to placebo-treated cats at month 8 of the study, although the effect was not sustained at month 12. No differences were noted between groups in the immunologic assays | Trzil et al. (2014) |
| Allergic asthma (Bermuda grass allergen induced) | 15 (MSCs treated, 5; placebo-treated 4; healthy cat, 6) | 9 months after treatment | Allogeneic AD-MSCs | $2 \times 10^6$, $4 \times 10^6$, $4.7 \times 10^6$, $1 \times 10^7$ and $1 \times 10^7$ in each cat (at 0, 14, 28, 98, 130 days) | Airway eosinophilia, pulmonary mechanics, thoracic computed tomography, and several immunologic assays | At early points (days 12, 26, 47, 108, and 133), airway eosinophil percentage was not affected by MSC administration. By month 9, eosinophil percentages in all MSC-treated cats decreased to normal reference intervals. Diminished airway hyper-responsiveness was noted in all MSC-treated compared with placebo-treated cats at day 133. Lung attenuation and bronchial wall thickening scores were significantly reduced in MSC-treated vs. untreated asthmatic cats, consistent with decreased airway remodeling at month 9 | Trzil et al. (2016) |

| | | | | | | | |
|---|---|---|---|---|---|---|---|
| Ischemic kidney injury (IKI) | 18 (AD-MSCs, $n = 5$; BM-MSCs, $n = 5$, fibroblasts, $n = 5$; reference, $n = 3$) | 6 days | Allogeneic AD-MSCs, BM-MSCs, fibroblasts | $4 \times 10^6$ (intravenous after 1 h of injury per animal in each group) | Serum creatinine and blood urea nitrogen concentrations were measured. Urine-specific gravity, urine protein to urine creatinine ratio, and glomerular filtration rate. International Renal Interest Society (IRIS) Grade | Time, but not treatment, had a significant effect on renal function. No difference was noted in % of cats with IRIS AKI. Significantly fewer mitotic figures were observed in ischemic kidneys that received bone marrow-derived MSCs vs. fibroblasts. No differences in smooth muscle actin staining were noted. This study did not support the use of allogeneic MSCs in AKI in the regimen described here. Type of renal injury, MSC dose, allogeneicity, duration, and route or timing of administration could influence the efficacy of MSCs | Rosselli et al. (2016) |
| Cardiomyopathy (L-isoproterenol induced for 10 days) | 22 (cardiac stem cells + atrial biopsy, $n = 7$; atrial biopsy + sham, $n = 5$; bone marrow + BM-MSCs, $n = 4$; bone marrow aspiration + sham, $n = 6$) | 66 days after induction of cardiomyopathy | Autologous cardiac stem cells (CSCs), MSCs | $1 \times 10^6$ (CSCs), $1 \times 10^6$ (MSCs); intracoronary injected | Echocardiography, invasive hemodynamics, and immunohistochemistry | Fractional shortening significantly improved with both CSC and MSC as compared to shams. MSC was superior to CSC in improving left ventricle end-diastolic (LVED) volume and ejection fraction. LVED pressure was less in MSC as compared to CSC and sham. LV BrdU + myocytes were higher in MSC than CSC and sham. CSC and MSC therapies improve cardiac function and attenuate pathological remodeling. However, MSC appear to confer additional benefit | Taghavi et al. (2015) |

**Table 10.3** In vivo clinical mesenchymal stem cell studies in cats

| Clinical case | Number of animals | Study period | Biomaterial used | Cell dose/assembly | Evaluation criteria | Overall result | References |
|---|---|---|---|---|---|---|---|
| Chronic enteropathy | 14: 7 treated with allogeneic MSCs (blinded study); 4 placebo; 3 treated with allogeneic MSCs (nonblind study) | 2 weeks and 2 months | Allogeneic AD-MSCs | $2 \times 10^6$/kg b wt (two doses repeated after 2 weeks) | Clinical evaluation based on questionnaire | Owners of 5/7 fMSC-treated cats reported significant improvement or complete resolution of clinical signs, while the owner of the remaining two cats reported modest but persistent improvement. Owners of placebo-treated cats reported no change or worsening of clinical signs. Of the owners not blinded to the treatment, one reported marked improvement, one reported no change, and one was lost to follow-up | Webb and Webb (2015) |
| Chronic kidney disease (CKD) | 6 (control healthy, $n = 2$; CKD, $n = 4$) | 7 and 30 days | Autologous AD-MSCs, BM-MSCs | Unilateral intrarenal injection | Minimum database and glomerular filtration rate (GFR) via nuclear scintigraphy were determined pre-injection, at 7 days and at 30 days post-injection | Intrarenal injection did not induce immediate or long-term adverse effects. Two cats with CKD that received aMSC experienced modest improvement in GFR and a mild decrease in serum creatinine concentration | Quimby et al. (2011) |
| Chronic kidney disease (CKD) | 16 (pilot study 1: 6; pilot study 2: 5; pilot study 3: 5) | 0, 2, 4, 6, and 8 weeks | Allogeneic AD-MSCs | $2 \times 10^6$ (cryopreserved cells), $4 \times 10^6$ (cryopreserved cells), $4 \times 10^6$ (cryopreserved adipose tissue) | Serum biochemistry, complete blood count, urinalysis, urine protein, glomerular filtration rate, and urinary cytokine concentrations | Administration of cryopreserved aMSCs was associated with significant adverse effects and no discernible clinically relevant improvement in renal functional parameters. Administration of aMSCs cultured from cryopreserved adipose was not associated with adverse effects but was also not associated with improvement in renal functional parameters | Quimby et al. (2013) |

| | | | | | | | |
|---|---|---|---|---|---|---|---|
| Chronic kidney disease (CKD) (randomized, placebo-controlled) | 8 MSCs treated, $n = 6$; placebo control, $n = 2$) | 0 and 8 weeks | Allogeneic AD-MSCs | $2 \times 10^6$ per kg b wt (3 doses at 2, 4, and 6 weeks) | Complete blood counts, chemistry and urinalysis were performed at weeks 0, 2, 4, 6, and 8. Glomerular filtration rate (GFR) via nuclear scintigraphy and urine protein:creatinine ratio (UPC) | Six cats received three doses of allogeneic MSC culture expanded from cryopreserved adipose without adverse effects. No significant change in serum creatinine, blood urea nitrogen, potassium, phosphorus, GFR by nuclear scintigraphy, UPC, or packed cell volume was seen in cats treated with MSCs. While administration of MSC culture expanded from cryopreserved adipose was not associated with adverse effects, significant improvement in renal function was not observed immediately after administration. Long-term follow-up is necessary to determine whether MSC administration affects disease progression in cats with CKD | Quimby et al. (2016) |
| Chronic kidney disease (CKD) | 10 (1healthy cat, intrarenal infusion of cells; 9 CKD cats intravenous infused cells) | 60 days after initial injection | Allogeneic amniotic membrane MSCs | $2 \times 10^6$ (CKD cats two doses 21 days apart), healthy cat ($5 \times 10^5$ bilateral intrarenal infusion) | Ultrasonography and lab analysis | CDK cats treated with MSCs registered a significant improvement of renal function (decrease in serum creatinine and urine protein concentrations and increase in urine-specific gravity). The kidney architecture and morphology did not change following the treatment. The feline AMSCs have a renoprotective effect and improve renal function in cats with naturally occurring CKD, stabilizing the clinical condition and disease progression | Vidane et al. (2016) |

**Table 10.3** (continued)

| Clinical case | Number of animals | Study period | Biomaterial used | Cell dose/assembly | Evaluation criteria | Overall result | References |
|---|---|---|---|---|---|---|---|
| Feline chronic gingivostomatitis | 9 cats | 6 and 24 weeks | Autologous AD-MSCs | $20 \times 10^6$ (2 doses 30 days apart, intravenous) | Histology and immunohistochemistry, blood immune cell subsets, serum protein, and cytokine levels | 7/9 cats completed the study. Five cats responded to treatment by either complete clinical remission ($n = 3$) or substantial clinical improvement ($n = 2$). Two cats were nonresponders. Cats that responded to treatment also exhibited systemic immunomodulation demonstrated by decreased numbers of circulating CD8+ T cells, a normalization of the CD4/CD8 ratio, decreased neutrophil counts, and interferon-g and interleukin (IL)-1β concentration, and a temporary increase in serum IL-6 and tumor necrosis factor-α concentration. No clinical recurrence has occurred following complete clinical remission (follow-up of 6–24 months) | Arzi et al. (2016) |
| Feline eosinophilic keratitis (FEK) (unilateral) | 5 | 1, 3, 6, and 11 months | Allogeneic AD-MSCs | $2 \times 10^6$ (2 doses 2 months apart subconjunctivally) | Ocular surface integrity, cytology, PCR, and biochemistry | Clinical signs showed a significant change during the follow-up with resolution of the corneal and conjunctiva lesions, and there were no signs of regression or worsening | Villatoro et al. (2018) |
| L1-L5 compressive fracture (treated by hemilaminectomy) | 1 | 3 months | Autologous BM-MSCs, collagen I gel, and physiotherapy | $7 \times 10^8$ | Clinical evaluation | Seven days after surgery and cell transplantation, the examination revealed a progressive recovery of the panniculus reflexes and of the responses to superficial and deep pain stimuli despite the low proprioceptive and hyperreflexic ataxic hind limbs. Physiotherapy protocols were applied for clinical rehabilitation after surgery. The cat's first steps, 3-min weight-bearing, and intestine and urinary bladder partial reestablishment were observed 75 days post-surgery | Penha et al. (2012) |

In an ischemic kidney injury model, comparative efficacy of allogeneic AD-MSCs vs. BM-MSCs vs. fibroblasts had shown that restoration in renal function occurs only with time in either case and as such MSCs application in ischemic kidney injury was not supported. It can be inferred that further studies that consider all the criteria like renal injury type, cell dose and source, and route or timing of administration should be considered in order to draw any conclusion (Rosselli et al. 2016).

In a clinical chronic kidney disease (CKD) study, intravenous infusion of amniotic membrane MSCs had shown improved clinical condition although without any change in the kidney architecture. The intrarenal implantation of MSCs in CKD cats had failed to yield desired result at 1 month post-injection (Quimby et al. 2011). Intra-renal infusion of the cells may be associated with the anesthetic requirement and has associated stress response in a healthy cat. The intravenous injection of MSCs for CKD may be preferable route (Vidane et al. 2016). In another CKD clinical study, intravenous implantation of MSCs had failed to improve the clinical signs. Administration of cryopreserved cells, however, had elicited the adverse reaction, while cells harvested from cryopreserved adipose tissue were devoid of such adverse reactions in the cases (Quimby et al. 2013). Same authors in another study had demonstrated no immediate improvement with allogeneic MSCs. However, it was concluded that improvement that may occur later in the follow-up requiring long duration studies (Quimby et al. 2016).

### 10.3.3 Cat Asthma

Feline asthma exhibits airway eosinophilia, airway hyper-responsiveness (AHR) and finally remodeling (Norris Reinero et al. 2004). Current therapeutics in such cases involves glucocorticoids and bronchodilators, but these drugs fail to provide consistent results (Murdoch and Lloyd 2010; Freishtat et al. 2010; Cocayne et al. 2011; Durrani et al. 2011). To control such ailment and provide relief to the patients, advanced therapeutic is desired.

In an induced chronic asthma model (Bermuda grass allergen) of cat, allogeneic AD-MSCs and placebo-treated cases had similar AHR and eosinophilia. At 8 months, lung attenuation and bronchial wall thickening were significantly lowered in MSC-treated animals in comparison to placebo-treated cats. The effect, however, was not sustained at month 12 (Trzil et al. 2014). In another similar model, asthmatic cats injected with multiple doses of allogeneic AD-MSCs had shown an improvement in clinical response (lowered lung attenuation bronchial wall thickening). The improvement was not demonstrated earlier (up to 133 days), but later on at 9 months much improved clinical response had been demonstrated. It was concluded that MSCs tend to show delayed effect and as such delayed response may be observed (Trzil et al. 2016). However, to confirm sustenance of these improved results further investigations are required.

### 10.3.4 Chronic Gingivostomatitis

Feline chronic gingivostomatitis (FCGS) is an idiopathic, oral inflammatory disease that affects 0.7%–10% of cat population (Arzi et al. 2010; Alpsoy et al. 2015; Jennings et al. 2015). It is generally treated by extraction of all the teeth. Even about 30% of cats do not respond to full mouth extraction demanding continuous medication (Jennings et al. 2015). Thus, MSCs due to their immunomodulatory and/anti-inflammatory role are being evaluated for FCGS.

In a feline chronic gingivostomatitis study, intravenous implantation of two doses of MSCs had improved the clinical symptoms in more than 50% (5 out of 7) of the cases. Among responders some had shown clinical remission (3/5), while others (2/5) had significant clinical improvement. In responsive cases the systemic immunomodulation was seen (Arzi et al. 2016).

### 10.3.5 Feline Eosinophilic Keratitis (FEK)

In FEK cornea undergoes superficial vascularization, yellow-white plaques, stromal infiltration, and edema (Allgoewer et al. 2001). The etiological agent is currently unknown, but disease appears to be immune mediated (Maggs 2008) and as such anti-inflammatory or immunosuppressants are being prescribed for its control (Dean and Meunier 2013). Since these compromise body's immunity and lead to the other ailments (Villatoro et al. 2015), other treatment options are being evaluated. Topical cyclosporine to some extent appears useful but leads to corneal irritation and chemosis (Allgoewer et al. 2001). MSCs due to their immunomodulation and/anti-inflammatory properties appear suitable therapeutic option.

In a clinical feline eosinophilic keratitis (FEK) study, two doses of allogeneic MSCs applied subconjunctivally had led to the significant improvement in its symptoms. The clinical signs like corneal and conjunctiva lesions were resolved without any sign of regression or worsening (Villatoro et al. 2018).

### 10.3.6 Neurological Ailments

fMSCs have been demonstrated to increase the tyrosine hydroxylase expression of PC12 cell line being used for neurological studies (Jin et al. 2007). This could possibly be through MSCs secreted nerve growth factor (NGF) and brain-derived nerve factor (BDNF) in combination with dopamine produced by PC12 cells (Zhou et al. 1994; Li et al. 1997; Zhou et al. 1998). Red fluorescent protein (RFP)-transfected cat umbilical cord blood (UCB)-MSCs under neurogenic differentiation medium were differentiated into neuron-like cells. Immunocytochemistry had demonstrated that differentiated cells were $\beta$ III-tubulin-positive (6.85%), neurofilament light (NF-L)-positive (3.37%), and neurofilament medium (NF-M)-positive (7.04%) (Jin et al. 2008). Thus, UCB-MSCs can act as an alternative source.

In a cat lumbar compressive fracture (surgically rectified by hemilaminectomy), implantation of MSCs along with physiotherapy had improved the clinical symptoms. The initial observable positive responses were demonstrated at day 7 with the return of panniculus reflex. After 75 days, a partial improvement in weight bearing along with intestine and urinary bladder tone was recorded (Penha et al. 2012). These positive results demand further extensive investigations of MSCs application in such conditions (Table10.2).

### 10.3.7 Cardiomyopathy

In cats, hypertrophic cardiomyopathy is common. In an isoproterenol induced hypertrophic model, intracoronary injection of cardiac stem cells (CSCs) and BM-MSCs had improved fractional shortening in comparison to sham. MSCs had led to superior improvement in left ventricle end-diastolic volume and ejection fraction as compared to CSCs (Taghavi et al. 2015). Such a preliminary study shows promising results with the use of MSCs though further extensive studies are desired for any conclusion.

## 10.4 Conclusions

fMSCs have been harvested from various sources mainly from adipose tissues. The cells have been culture expanded and characterized as per the ISCT. The cells exhibit limited self-renewal properties in currently available media as their viability decreases and become more senescent at early culture passages (p6). fAD-MSCs are more prone to FFV that leads to their morphologic and proliferative alterations. The cellular applications have been made in various experimental models and or clinical conditions. The results of these studies have variably been pro-therapeutic. In case of kidney failure, however, little to no success with fMSCs has been demonstrated. These studies are very limited with lack of sufficient data to be considered sufficient for any conclusion(s).

## References

Allgoewer I, Schäffer EH, Stockhaus C, Vögtlin A (2001) Feline eosinophilic conjunctivitis. Vet Ophthalmol 4:69–74

Alpsoy E, Akman-Karakas A, Uzun S (2015) Geographic variations in epidemiology of two autoimmune bullous diseases: pemphigus and bullous pemphigoid. Arch Dermatol Res 307:291–298

Arzi B, Murphy B, Cox DP et al (2010) Presence and quantification of mast cells in the gingival of cats with tooth resorption, periodontitis and chronic stomatitis. Arch Oral Biol 55:148–154

Arzi B, Kol A, Murphy B, Walker NJ, Wood JA, Clark K, Verstraete FJM, Borjesson DL (2015) Feline foamy virus adversely affects feline mesenchymal stem cell culture and expansion: implications for animal model development. Stem Cells Dev 24(7):814–823

Arzi B, Mills-Ko E, Verstraete FJM, Kol A, Walker NJ, Badgley MR, Fazel N, Murphy WJ, Vapniarsky N, Borjesson DL (2016) Therapeutic efficacy of fresh, autologous mesenchymal stem cells for severe refractory gingivostomatitis in cats. Stem Cells Transl Med 5:75–86

Cardoso MT, Pinheiro AO, Vidane AS, Casals JB, de Oliveira VC, Gonçalves NJN, Martins DS, Ambrósio CE (2016) Characterization of teratogenic potential and gene expression in canine and feline amniotic membrane-derived stem cells. Reprod Domest Anim 51(Suppl. 3):1–7

Cocayne CG, Reinero CR, DeClue AE (2011) Subclinical airway inflammation despite high-dose oral corticosteroid therapy in cats with lower airway disease. J Feline Med Surg 13:558–563

Dean E, Meunier V (2013) Feline eosinophilic keratoconjunctivitis: a retrospective study of 45 cases (56 eyes). J Feline Med Surg 15:661–666

Durrani SR, Viswanathan RK, Busse WW (2011) What effect does asthma treatment have on airway remodeling? Current perspectives. J Allergy Clin Immunol 128:439–448

Freishtat RJ, Nagaraju K, Jusko W, Hoffman EP (2010) Glucocorticoid efficacy in asthma: is improved tissue remodeling upstream of anti-inflammation. J Investig Med 58:19–22

Gómez MC, Qin Q, Biancardi MN, Galiguis J, Dumas C, MacLean RA, Wang G, Pope E (2015) Characterization and multilineage differentiation of domestic and black-footed cat mesenchymal stromal/stem cells from abdominal and subcutaneous adipose tissue. Cell Reprogram 17(5):376. https://doi.org/10.1089/cell.2015.0040

Glage S, Klinge PM, Miller MC, Wallrapp C, Geigle P, Hedrich HJ, Brinker T (2011) Therapeutic concentrations of glucagon-like peptide-1 in cerebrospinal fluid following cell-based delivery into the cerebral ventricles of cats. Fluids Barriers CNS 8:18

Iacono E, Cunto M, Zambelli D, Ricci F, Tazzari PL, Merlo B (2012) Could fetal fluid and membranes be an alternative source for mesenchymal stem cells (MSCs) in the feline species? A preliminary study. Vet Res Commun 36:107–118

Janeczko S, Atwater D, Bogel E et al (2008) The relationship of mucosal bacteria to duodenal histopathology, cytokine mRNA, and clinical disease activity in cats with inflammatory bowel disease. Vet Microbiol 128:178–193

Jennings MW, Lewis JR, Soltero-Rivera MM et al (2015) Effect of tooth extraction on stomatitis in cats: 95 cases (2000-2013). J Am Vet Med Assoc 246:654–660

Jin G-Z, Yin X-J, Yu X-F, Cho S-J, Lee H-S, Lee H-J, Kong I-K (2007) Enhanced tyrosine hydroxylase expression in PC12 cells co-cultured with feline mesenchymal stem cells. J Vet Sci 8(4):377–382

Jin G-Z, Yin X-J, Yu X-F, Cho S-J, Choi E-G, Lee Y-S, Jeon J-T, Yee S-T, Kong I-K (2008) Generation of neuronal-like cells from umbilical cord blood-derived mesenchymal stem cells of a RFP-transgenic cloned cat. J Vet Med Sci 70(7):723–726

Kim H-R, Lee J, Byeon JS, Gu N-Y, Lee J, Cho I-S, Cha S-H (2017) Extensive characterization of feline intra-abdominal adipose-derived mesenchymal stem cells. J Vet Sci 18(3):299–306

Kono S, Kazama T, Kano K, Harada K, Uechi M, Matsumoto T (2014) Phenotypic and functional properties of feline dedifferentiated fat cells and adipose-derived stem cells. Vet J 199:88–96

Lawson J, Elliott J, Wheeler-Jones C, Syme H, Jepson R (2015) Renal fibrosis in feline chronic kidney disease: known mediators and mechanisms of injury. Vet J 203:18–26

Lee B-Y, Li Q, Song W-J, Chae H-K, Kweon K, Ahn J-O, Youn H-Y (2018) Altered properties of feline adipose-derived mesenchymal stem cells during continuous *in vitro* cultivation. J Vet Med Sci 80(6):930–938

Li XM, Qi J, Juorio AV, Boulton AA (1997) Reciprocal regulation of the content of aromatic L-amino acid decarboxylase and tyrosine hydroxylase mRNA by NGF in PC12 cells. J Neurosci Res 47:449–454

Maggs DJ (2008) Cornea and sclera. In: Maggs DJ, Miller PE, Ofri R, Slatter DH (eds) Slatter's fundamentals of veterinary ophthalmology, 4th edn. Elsevier Saunders, St. Louis, pp 175–202

Martin DR, Cox NR, Hathcock TL et al (2002) Isolation and characterization of multipotential mesenchymal stem cells from feline bone marrow. Exp Hematol 30:879–886

Mumaw JL, Schmiedt CW, Breidling S et al (2015) Feline mesenchymal stem cells and supernatant inhibit reactive oxygen species production in cultured feline neutrophils. Res Vet Sci 103:60–69

Munoz JL, Greco SG, Patel SA, Sherman LS, Bhatt S, Bhatt RS, Shrensel JA, Guan Y-Z, Xie G, Ye J-H, Rameshwar P, Siegel A (2012) Feline bone marrow-derived mesenchymal stromal cells (MSCs) show similar phenotype and functions with regards to neuronal differentiation as human MSCs. Differentiation 84(2):214–222

Murdoch JR, Lloyd CM (2010) Chronic inflammation and asthma. Mutat Res 690:24–39

Norris Reinero CR, Decile KC, Berghaus RD et al (2004) An experimental model of allergic asthma in cats sensitized to house dust mite or Bermuda grass allergen. Int Arch Allergy Immunol 135:117–131

Parys M, Nelson N, Koehl K, Miller R, Kaneene JB, Kruger JM, Yuzbasiyan-Gurkan V (2016) Safety of intraperitoneal injection of adipose tissue-derived autologous mesenchymal stem cells in cats. J Vet Intern Med 30:157–163

Penha EM, Aguiar PHP, Barrouin-Melo SM, de Lima RS, da Silveira ACC, Otelo ARS, Pinheiro CMB, Ribeiro-dos-Santos R, Soares MBP (2012) Clinical neurofunctional rehabilitation of a cat with spinal cord injury after hemilaminectomy and autologous stem cell transplantation. IntJStem Cells 5(2):146–150

Quimby JM, Borjesson DL (2018) Mesenchymal stem cell therapy in cats current knowledge and future potential. J Feline Med Surg 20:208–216

Quimby JM, Webb TL, Gibbons DS, Dow SW (2011) Evaluation of intrarenal mesenchymal stem cell injection for treatment of chronic kidney disease in cats: a pilot study. J Feline Med Surg 13:418–426

Quimby JM, Webb TL, Habenicht LM, Dow SW (2013) Safety and efficacy of intravenous infusion of allogeneic cryopreserved mesenchymal stem cells for treatment of chronic kidney disease in cats: results of three sequential pilot studies. Stem Cell Res Ther 4:48

Quimby JM, Webb TL, Randall E, Marolf A, Valdes-Martinez A, Dow SW (2016) Assessment of intravenous adipose derived allogeneic mesenchymal stem cells for the treatment of feline chronic kidney disease: a randomized, placebo-controlled clinical trial in eight cats. J Feline Med Surg 18(2):165–171

Rosselli DD, Mumaw JL, Dickerson V, Brown CA, Brown SA, Schmiedt CW (2016) Efficacy of allogeneic mesenchymal stem cell administration in a model of acute ischemic kidney injury in cats. Res Vet Sci 108:18–24

Sato K, Yamawaki-Ogata A, Kanemoto I, Usui A, Narita Y (2016) Isolation and characterisation of peripheral blood-derived feline mesenchymal stem cells. Vet J 216:183–188

Simonaro CM, Sachot S, Ge Y, He X, DeAngelis VA, Eliyahu E, Leong D, Sun HB, Mason JB, Haskins ME, Richardson DW, Schuchman EH (2013) Acid ceramidase maintains the chondrogenic phenotype of expanded primary chondrocytes and improves the chondrogenic differentiation of bone marrow-derived mesenchymal stem cells. PLoS One 8(4):e62715

Taghavi S, Sharp TE, Duran JM, Makarewich CA, Berretta RM, Starosta T, Kubo H, Barbe M (2015) Autologous c-kit+ mesenchymal stem cell injections provide superior therapeutic benefit as compared to c-kit+ cardiac-derived stem cells in a feline model of isoproterenol-induced cardiomyopathy. Clin Transl Sci 8(5):425–431

Trepanier L (2009) Idiopathic inflammatory bowel disease in cats. Rational treatment selection. J Feline Med Surg 11:32–38

Trzil JE, Masseau I, Webb TL, Chang C-H, Dodam JR, Cohn LA, Liu H, Quimby JM, Dow SW, Reinero CR (2014) Long term evaluation of mesenchymal stem cell therapy in a feline model of chronic allergic asthma. Clin Exp Allergy 44(12):1546–1557

Trzil JE, Masseau I, Webb TL, Chang C-H, Dodam JR, Liu H, Quimby JM, Dow SW, Reinero CR (2016) Intravenous adipose-derived mesenchymal stem cell therapy for the treatment of feline asthma: a pilot study. J Feline Med Surg 18:981. https://doi.org/10.1177/1098612X15604351

Vidane AS, Souza AF, Sampaio RV, Bressan FF, Pieri NC, Martins DS, Meirelles FV, Miglino MA, Ambrósio CE (2014) Cat amniotic membrane multipotent cells are nontumorigenic and are safe for use in cell transplantation. Stem Cells Cloning: Adv Appl 7:71–78

Vidane AS, Pinheiro AO, Casals JB, Passarelli D, Hage MCFNS, Bueno RS, Martins DS, Ambrósio CE (2016) Transplantation of amniotic membrane-derived multipotent cells ameliorates and delays the progression of chronic kidney disease in cats. Reprod Domest Anim 51(Suppl. 3):1–11

Villatoro AJ, Fernández V, Claros S, Rico-Llanos GA, Becerra J, Andrades JA (2015) Use of adipose-derived mesenchymal stem cells in keratoconjunctivitis sicca in a canine model. Biomed Res Int 2015:1

Villatoro AJ, Claros S, Fernández V, Alcoholad C, Fariñas F, Moreno A, Becerra J, Andrades JA (2018) Safety and efficacy of the mesenchymal stem cell in feline eosinophilic keratitis treatment. BMC Vet Res 14:116

Webb TL, Webb CB (2015) Stem cell therapy in cats with chronic enteropathy: a proof-of-concept study. J Feline Med Surg 17:901. https://doi.org/10.1177/1098612X14561105

Webb TL, Quimby JM, Dow SW (2011) In vitro comparison of feline bone marrow derived and adipose tissue-derived mesenchymal stem cells. J Feline Med Surg 14(2):165–168

Zhou J, Bradford HF, Stern GM (1994) The stimulatory effect of brain-derived neurotrophic factor on dopaminergic phenotype expression of embryonic rat cortical neurons *in vitro*. Brain Res Dev Brain Res 81:318–324

Zhou J, Pliego-Rivero B, Bradford HF, Stern GM, Jauniaux ER (1998) Induction of tyrosine hydroxylase gene expression in human foetal cerebral cortex. Neurosci Lett 252:215–217

# Dog Mesenchymal Stem Cell Basic Research and Potential Applications

# 11

M. B. Gugjoo, Amar Pal, and G. T. Sharma

**Abstract**

Mesenchymal stem cells (MSCs) being available in numerous adult tissues and foetal membranes have been harvested from almost all these sources in dog. Due to the presence of miniscule number of MSCs in these tissues, culture expansion followed by their characterization becomes imperative. These characterized cells have been made subject in preclinical experimental and clinical settings in dogs. The studies have variably but undeniably supported their therapeutic effects, although currently inconclusive. Number of dog clinical ailments compeer to that of human. Therapeutic benefits of dMSCs in these ailments may provide proof-of-therapeutic principle for human medicine. The current chapter details dMSCs culture characteristics and potential applications.

## 11.1 Introduction

The population of pets like dog is increasing in the world. These pets are considered as family members and owners spent many dollars to keep them healthy. Dog, *Canis familiaris*, suffers from numerous ailments that lack standard treatment. Many of

M. B. Gugjoo (✉)
Division of Veterinary Clinical Complex, FVSc & AH, SKUAST-Kashmir, Jammu and Kashmir, India

A. Pal
Division of Surgery, Indian Veterinary Research Institute, Izatnagar, UP, India

G. T. Sharma
Division of Physiology and Climatology, Indian Veterinary Research Institute, Izatnagar, UP, India



M. B. Gugjoo, A. Pal (eds.), *Mesenchymal Stem Cell in Veterinary Sciences*, https://doi.org/10.1007/978-981-15-6037-8_11

their ailments have patho-physiological resemblance to that of human ailments, involving genetic basis as well (Starkey et al. 2005). This makes dog a suitable model animal for human. Dog offers many advantages including similarity in lifespan. This allows long-term studies on the safety and efficacy of therapeutic agents (Baird et al. 2015). To address issues of these enigmatic ailments and/or to promote early healing, regenerative medicine involving stem cell therapy is currently being studied. Among various types of stem cells, MSCs mainly contribute to the therapeutics in dog.

## 11.2 In Vitro Studies

### 11.2.1 Sources of Dog MSCs

A diverse set of tissues have been evaluated for stem cell reserve in dogs. This includes adult issues and the foetal membranes (Table 11.1) (Gugjoo et al. 2019a). These cells have also been harvested from induced pluripotent stem cells (Chow et al. 2017). The current tissue harvesting procedures are invasive; efforts are being made to develop advanced harvesting procedures that reduce animal sufferings. For muscle tissue harvesting, minimally invasive technique of microbiopsy (leads to insignificant muscle damage) had been demonstrated as reasonably safe way to harvest MSCs (Ceusters et al. 2017). Foetal membrane tissues are limited in dogs as pets undergo elective castration in juvenile age (Sultana et al. 2018). dMSCs in compensation, may be harvested from ovaries and/or uterus of spayed animals (Sultana et al. 2018).

### 11.2.2 Cellular Variability with Respect to Sources

#### 11.2.2.1 Cell Concentration

The tissues harbour dMSCs in limited and variable concentrations. The cell yield per gram of tissue varies. At day 16 of culture, the highest cell yield had been obtained from periosteum ($19.4 \times 10^6$) followed in descending order by muscle ($3.37 \times 10^6$), adipose tissue ($2.33 \times 10^6$) and bone marrow ($1.45 \times 10^6$) (Kisiel et al. 2012). dMSCs tend to lose their self-renewal potential after certain passages. With extended passaging, these cells had shown reduced proliferation potential and appeared senescent. dMSCs had been demonstrated to maintain their consistent population doubling time (PDT) up to 20 passages, though other factors like cell source and culture environment may also affect such properties (Spencer et al. 2012; Kresic et al. 2017). One of the studies had evaluated MSCs from various sources and it was demonstrated that highest cell proliferation in all these sources was at passage 3. Among these tissue sources, average PDT was lowest for adipose tissue MSCs (15 h) followed in ascending order by placenta-MSCs (21.2 h), bone marrow MSCs (26.2 h) and umbilical cord MSCs (41.9 h) (Zhan et al. 2019). Early passage dMSCs may therefore be clinically more useful than those obtained after extended passaging.

**Table 11.1** Dog Mesenchymal stem cell (MSC) sources and their characterization

| Positive markers | Negative markers | Differentiation | References |
|---|---|---|---|
| *Adipose MSCs (AD-MSCs)* | | | |
| CD44, CD90, MHC-I, CD34 (partially) | CD14 and CD45 | Osteogenic | Kang et al. (2008) |
| OCT4, NANOG, SOX2 | Not available | Osteogenic, Chondrogenic and limited Adipogenic | Neupane et al. (2008) |
| CD44, CD90, CD105 | CD14, CD34, CD45 | Osteogenic, Adipogenic, neurogenic | Ryu et al. (2009) |
| CD44, CD29, CD90 | CD14, CD34, CD45, CD117, CD13, CD105, CD73 | Osteogenic, Adipogenic, Chondrogenic, myogenic | Vieira et al. (2010) |
| CD90, CD44, CD140a, CD117 | CD34, CD45 | Osteogenic, Adipogenic, myogenic | Martinello et al. (2011) |
| CD29, CD44, thy 1.1 | CD31, CD34, CD73, CD105 | Osteogenic, Adipogenic, Chondrogenic, myogenic, neurogenic | Oh et al. (2011) |
| OCT4, NANOG, SOX2 | – | Osteogenic, Adipogenic, chondrogenic | Guercio et al. (2012) |
| CD44, CD90 SOX2, OCT4, NANOG | CD34, CD45, CD146 | Osteogenic, Adipogenic | Kisiel et al. (2012) |
| CD29, CD44, CD 90, SSEA-1 (low), OCT3/4, SOX2, NANOG | CD34, CD45, SSEA-3, SSEA-4, TRA-1-60 and TRA-1-81 | Osteogenic, Adipogenic | Takemitsu et al. (2012) |
| CD44, CD73, CD90, CD105 | CD14, CD34, CD45 | Neurogenic | Ryu et al. (2012) |
| CD29, CD90 and STRO-1 | CD34, CD45 and MHC-II | Osteogenic, Adipogenic, Chondrogenic | Villatoro et al. (2015) |
| CD90, CD44, CD166, Oct4, Sox2 | CD34, CD45 | Primordial germ cell-like cells and male germ cells | Wei et al. (2016) |
| CD 90, CD 105 | CD 14, CD 45 | Osteogenic, Adipogenic, Chondrogenic, neurogenic | Blecker et al. (2017) |
| CD9, CD44, CD90, and CD105, NANOG, OCT4, and SOX2 | CD34, CD45 and STRO-1 | Osteogenic, Adipogenic, Chondrogenic | Bearden et al. (2017) |

(continued)

**Table 11.1** (continued)

| Positive markers | Negative markers | Differentiation | References |
|---|---|---|---|
| CD44, CD90 | CD34, CD45 | Osteogenic, Adipogenic, neurogenic | Kriston-Pál et al. (2017) |
| CD90, CD44, NANOG, TERT, SOX2, OCT4 | CD34, CD45 | Osteogenic, Chondrogenic, Adipogenic, neurogenic, endoderm, primordial germ cells | Trindade et al. (2017) |
| CD90, CD44, Sox2 and Nanog | CD45 | Osteogenic, Chondrogenic | Ayala-Cuellar et al. (2019) |
| CD73, CD90, CD44 | CD45, CD14, HLA-DR and CD34. | Osteogenic, Chondrogenic, Adipogenic | Bçrziòð et al. (2018) |
| CD90, OCT4, SOX9, RUNX2, PPARG | CD34, CD45, MCH-II | – | Enciso et al. (2018) |
| CD90, CD44 | CD45, CD11b | Osteogenic, Chondrogenic, Adipogenic | Sasaki et al. (2018) |
| CD44, CD90 | CD 34 and CD45 | Osteogenic, Chondrogenic | Shah et al. (2018) |
| CD29, CD44, CD73, CD90 and STRO-1 | CD11/18, CD34, CD45, MHC-II | Osteogenic, Adipogenic, Chondrogenic | Villatoro et al. (2018) |
| *Amniotic fluid MSCs* | | | |
| Oct-4, CD44, DLA-DRA1 and DLA-79 | CD34, CD45 | Osteogenic, Chondrogenic, Adipogenic, neurogenic | Uranio et al. (2011) |
| *Amniotic membrane MSCs* | | | |
| CD90, CD105 | CD3, CD11c, CD28, CD38, CD62L, CD34, CD45, CD41a | Osteogenic, Chondrogenic, Adipogenic, neurogenic | Park et al. (2012a) |
| Oct-4, CD44, CD184 and CD29 | CD34, CD45 | Osteogenic, Chondrogenic, Adipogenic, neurogenic | Uranio et al. (2011) |
| *Bone marrow MSCs* | | | |
| CD90, MHC-I | CD34, CD45, MHC-II | Neurogenic | Kamishina et al. (2006) |
| CD105, CD90 | CD45, CD34 | Osteogenic, Chondrogenic, Adipogenic | Csaki et al. (2007) |

(continued)

**Table 11.1** (continued)

| Positive markers | Negative markers | Differentiation | References |
|---|---|---|---|
| CD10, CD13, CD29, CD44, CD73/SH-3, CD90/ Thy-1 and CD106/ VCAM-1 | CD3, CD14, CD34, CD45 | – | Lee et al. (2011b) |
| CD 44 and CD 90 | CD34 | Osteogenic Chondrogenic, Adipogenic, neurogenic | Tharasanit et al. (2011) |
| CD44, STRO-1 | CD34, CD45 | Chondrogenic | Hodgkiss-Geere et al. (2012) |
| CD44, CD90. SOX2, OCT4, NANOG | CD34, CD45, CD146 | Osteogenic, Adipogenic | Kisiel et al. (2012) |
| CD44, CD73, CD90, CD105 | CD14, CD34, CD45 | Neurogenic | Ryu et al. (2012) |
| CD29, CD44, CD 90, OCT3/4, SOX2, NANOG | CD34, CD45, SSEA-3, SSEA-4, TRA-1-60, TRA-1-81 | Osteogenic, Adipogenic | Takemitsu et al. (2012) |
| CD44, CD105 | CD34, CD45 | – | Kazemi et al. (2017) |
| CD29, CD44, CD90 | CD45 and CD11b | Osteogenic, Adipogenic | Matsuda et al. (2017) |
| CD90, OCT4, SOX9, RUNX2, PPARG | CD34, CD45, MCH-II | – | Enciso et al. (2018) |
| CD 166, CD 29, CD 90, CD 105 and CD 44 | CD45 and CD 34 | Osteogenic, Chondrogenic | Li et al. (2018) |
| CD90, CD44 | CD45, CD11b | Osteogenic, Chondrogenic, Adipogenic | Sasaki et al. (2018) |
| CD9, CD44, CD90, and CD105, NANOG, OCT4, and SOX2 | CD34, CD45 and STRO-1 | Osteogenic, Adipogenic, Chondrogenic | Bearden et al. (2017) |
| *Corneal MSCs* | | | |
| CD90, CD73, CD105, N-cadherin, Pax6 | CD34 | Osteogenic, Chondrogenic, Adipogenic | Kafarnik et al. (2020) |
| *Dental pulp MSCs* | | | |
| CD90, CD44, CD146, SSEA-4 (low) | – | Osteogenic, Chondrogenic, Adipogenic | Khorsand et al. (2013) |

(continued)

**Table 11.1** (continued)

| Positive markers | Negative markers | Differentiation | References |
|---|---|---|---|
| *Endometrium MSCs* | | | |
| CD44, CD29, CD90, CD13, CD133, CD73, CD105, Oct4 | CD45, CD34, CD31 (low to nil) | Osteogenic, Chondrogenic, Adipogenic | De Cesaris et al. (2017) |
| *Limbal epithelium MSCs* | | | |
| CD90, CD44, CD49 | CD45, CD11b\c | Osteogenic, Chondrogenic, Adipogenic, keratinocyte | Sancak et al. (2014) |
| *Muscle MSCs* | | | |
| CD44, CD90, SOX2, OCT4, NANOG | CD34, CD45 | Osteogenic, Adipogenic | Kisiel et al. (2012) |
| *Olfactory epithelium MSCs* | | | |
| CD73, CD90, CD105 | CD34, CD45, CD79 | Osteogenic, Chondrogenic, Adipogenic, neurogenic | Pinheiro et al. (2016) |
| *Omental MSCs* | | | |
| CD90, CD44 | CD45, CD11b | Osteogenic, Chondrogenic, Adipogenic | Bahamondes et al. (2017) |
| *Ovarian MSCs* | | | |
| CD90, CD44, NANOG, TERT, SOX2, OCT4 | CD34, CD45 | Osteogenic, Chondrogenic, Adipogenic, neurogenic, endoderm, primordial germ cells | Trindade et al. (2017) |
| *Periodontal ligament MSCs* | | | |
| – | – | Adipogenic, Osteogenic | Sedigh et al. (2010) |
| *Periosteal MSCs* | | | |
| CD44, CD90, SOX2, OCT4, NANOG | CD34, CD45 | Osteogenic, Adipogenic | Kisiel et al. (2012) |
| *Synovial membrane MSCs* | | | |
| CD9, CD44, CD90, and CD105, NANOG, OCT4, and SOX2. | CD34, CD45 and STRO-1 | Osteogenic, Adipogenic, Chondrogenic | Bearden et al. (2017) |
| CD90, CD44 | CD45, CD11b | Osteogenic, Chondrogenic, Adipogenic | Sasaki et al. (2018) |
| *Peripheral blood MSCs* | | | |
| CD34(low) | CD34 | Osteocalcin | Huss et al. (2000) |

(continued)

**Table 11.1** (continued)

| Positive markers | Negative markers | Differentiation | References |
|---|---|---|---|
| *Placental MSCs* | | | |
| CD90, CD29, CD44, CD105, CD73, SOX2 (low) | CD45, CD34 and MHC-II, NANOG, OCT4 | Osteogenic, Chondrogenic, Adipogenic | Saulnier et al. (2016) |
| CD44, CD29, CD90 | MHC-II, CD45, CD34 | Osteogenic, Adipogenic, Chondrogenic | Cabon et al. (2019) |
| *Umbilical cord MSCs* | | | |
| CD29, CD33, CD44, CD105, CD184, Oct4 | CD4, CD8a, CD10, CD14, CD20, CD24, CD31, CD34, CD38, CD41a, CD45, CD49b, CD41/61, CD62p, CD73, CD90, CD133, HLA-DR | Osteogenic, Chondrogenic, neurogenic | Seo et al. (2009) |
| *Umbilical cord tissue MSCs* | | | |
| CD44, CD54, CD61, CD80, CD90, CD105, Flk1, HMGA2, Sox2, Nanog, Oct3/4 | CD4, CD8A, CD25, CD33, CD34 and CD117 | Chondrogenic | Lee et al. (2013) |
| Oct-4, CD44, CD184 and CD29 | CD34, CD45 | Osteogenic, Chondrogenic, Adipogenic, neurogenic | Uranio et al. (2011) |
| *Umbilical cord vein MSCs* | | | |
| CD44, CD29, CD90 | CD14, CD34, CD45, CD117 | Osteogenic, Adipogenic, Chondrogenic | Zucconi et al. (2010) |
| CD44, CD73, CD90, CD105 | CD14, CD34, CD45 | Neurogenic | Ryu et al. (2012) |
| *Wharton's jelly MSCs* | | | |
| CD90, CD105 | CD3, CD11c, CD28, CD38, CD62L, CD34, CD45,CD41a | Osteogenic, Chondrogenic, neurogenic | Seo et al. (2012) |
| CD44, CD73, CD90, CD105 | CD14, CD34, CD45 | Neurogenic | Ryu et al. (2012) |

*AD* Adipose tissue, *AF* Amniotic fluid, *AM* Amniotic membrane, *BM* Bone marrow, *DP* Dental pulp, *End* Endometrium, *LE* Limbal epithelium, *Li* liver, *MSC* Mesenchymal stem cell, *Mu* Muscle, *OE* Olfactory epithelium, *Om* Omentum, *Ov* Ovary, *PL* Periodontal ligament, *Pe* Periosteum, *PB* Peripheral blood, *Pl* Placenta, *Sy* synovium, *UCB* Umbilical cord blood, *UCT* Umbilical cord tissue, *UCV* Umbilical cord vein, *WJ* Wharton's jelly

### 11.2.3 Characterization of dMSCs

The characterization criteria followed in dMSCs is similar to that of human including plastic adherence, surface marker expression and differentiation potential (Dominici et al. 2006). dMSCs show plastic adherence and appear morphologically as spindle-shaped fibroblasts, although morphologies resembling slender and/

elongated to cuboidal shapes too have been demonstrated (Huss et al. 2000; Csaki et al. 2007; de Bakker et al. 2013).

#### 11.2.3.1 Expression Profile of dMSCs

dMSCs in general express various surface markers as per the laid down criteria except for few surface markers as detailed in Table 11.1. dMSCs show plasticity and variably express pluripotency markers depending upon their tissue source as enlisted in Table 11.1. dBM-MSCs and dAD-MSCs had comparable expression of Oct2/4 and Sox2 although Nanog expression was 2.5 times higher in the latter as compared to the former cells (Takemitsu et al. 2012). Such an expression remains comparable in foetal adnexa MSCs at various stages of gestation (Uranio et al. 2014). One study on placental MSCs, however, had failed to demonstrate expression profile of some of the pluripotency markers like Oct3/4 and Nanog (Saulnier et al. 2016). The basis of these differences needs to be understood as it may have bearing on in vivo therapeutic results.

#### 11.2.3.2 dMSCs Differentiation Potential

dMSCs undergo recommended tri-lineage differentiation involving osteogenic, adipogenic and chondrogenic. In two studies MSCs adipogenic differentiation had failed to occur (Neupane et al. 2008; Seo et al. 2009) but was effectuated in presence of rabbit serum (Neupane et al. 2008). Since dMSCs express pluripotency markers, extended differentiation potential into neural cell-, myocyte-, keratinocyte-, endothelial cell- and primordial germ cell-like cells too has been established (Table 11.1).

### 11.2.4 Comparative Source Based dMSC Characteristics

dMSCs from various sources show variable characteristics. Dog endometrium-(dEnd-) MSCs had been demonstrated to have lower replicative potential as compared to dAD-MSCs (De Cesaris et al. 2017). Surface marker expression may not be similar in all the cases. Among various surface markers consistent expression of CD13 and CD105 has been demonstrated from different tissue sources. Other important surface marker (CD73) has also been consistently expressed barring a single study (Vieira et al. 2010). A study on dog peripheral blood MSCs (dPB-MSCs) had reported partial expression of haematopoietic markers (CD34) and could be attributed to their dynamic shift in blood (Huss et al. 2000). Overall expression of dMSCs derived from bone marrow, synovium and adipose tissue had been comparable except for CD105 expression that was more in adipose tissue (AD-) MSCs and synovial (Sy-) MSCs.

Osteogenesis had been significantly higher in Sy-MSCs and BM-MSCs while pellet size in chondrogenic medium was significantly more in AD-MSCs and Sy-MSCs (Bearden et al. 2017). Other study had shown higher chondrogenic potential of dSy-MSCs than dBM-MSCs and MSCs derived from inguinal and infrapatellar fat pad (Sasaki et al. 2018). Between dBM-MSCs and dLi-MSCs, chondrogenic potential had been higher in former while liver specific genes were better expressed in the latter (Malagola et al. 2016). Among various sources like adipose tissue, bone marrow, umbilical cord and Wharton's Jelly, dAD-MSCs had higher potential to proliferate. Vascular endothelial growth factor expression (VEGF) had been higher in dBM-MSCs. At early phase dAD-MSCs and dog umbilical cord blood- (dUCB-) MSCs had greater ability of osteogenesis, however, comparable osteogenesis was demonstrated in all these cell lines (Kang et al. 2012). From all these studies it can be inferred that tissue source plays an important role in MSCs proliferation and differentiation properties. Musculoskeletal tissue (synovium and bone marrow) derived MSCs may be preferable for musculoskeletal tissue regeneration.

Location of the particular tissue may also affect MSCs properties. Omentum and subcutaneous adipose tissue derived AD-MSCs had comparable trophic, vasculogenic and immuno-modulatory properties. Omentum, however, had significantly higher cell yield as compared to others (Bahamondes et al. 2017). dMSCs from subcutaneous fat had shown multipotent differentiation up to passage 8 while such a potential of MSCs from visceral (periovaric) fat had been only up to passage 4. The former cell type additionally had higher osteogenic expression (Yaneselli et al. 2018). Among various BM-MSCs sources ilium derived cells had been most proliferative (Volk et al. 2012). Mandible MSCs may carry higher cell proliferation in comparison to the femur (Bugueño et al. 2017). However, considering factors like collection technique and cell characteristics, humerus had been considered to be better source (Laura et al. 2008).

### 11.2.5 Effect of Donor Age on dMSCs

Ageing may reduce dMSCs proliferation and differentiation potential. MSCs from aged dog bone marrow had a significant (40%) reduction in proliferation potential, in addition to the osteogenic potential as compared to young ones (Volk et al. 2012). Other similar study had demonstrated young donor AD-MSCs carried higher population doubling potential as compared to those of aged dog MSCs. These cells had also expressed higher surface and pluripotency markers (Lee et al. 2017). Other study though had reported comparable osteogenic gene and fibroblast growth factor-10 expression in young and aged dMSCs but tumour necrosis factor and interferon gamma expression had increased in aged dAD-MSCs (Taguchi et al. 2019). Thus, available environment may affect dMSCs expression potential.

### 11.2.6 Effect of Drugs

dMSCs may be affected by steroids (Marycz et al. 2014; Guo et al. 2015; Jo et al. 2017) and vitamin C (Guo et al. 2015). Low dose steroids may improve dMSCs activity while higher doses might be cytotoxic (Marycz et al. 2014). Vitamin C may also improve dMSCs proliferation (Guo et al. 2015).

### 11.2.7 Breed Variability and MSCs

Dog breeds may vary in their genetics and are expected to show some cellular differences including of MSCs. Most of the breeds including German shepherd, Labrador, Golden retriever, Border collie and Malinois had comparable initial BM-MSCs population except one (Hovawart) that had much lower cell population. MSCs from these breeds (German shepherd, Labrador, Golden retriever and Border collie), however, had undergone early senescence. The cellular differentiation potential in these breeds may also vary (Bertolo et al. 2015).

## 11.3 In Vivo Potential Therapeutic Applications

MSCs are increasingly being utilized in dog clinical practice. dMSCs have been studied in various types of musculoskeletal and non-musculoskeletal tissues (Tables 11.2, 11.3, 11.4, 11.5, 11.6, 11.7, 11.8, 11.9, 11.10, 11.12, and 11.13) (Gugjoo et al. 2019a). The in vivo studies have largely supported dMSCs therapeutic potential (Hall et al. 2010; Sousa et al. 2011; Kim et al. 2011; Cuervo et al. 2014; Muir et al., 2016). Many of these studies have failed to report any major adverse effect (Cuervo et al. 2014; Bçrziòð et al. 2018; Villatoro et al. 2018) while some have reported adverse reactions like self-controllable localized inflammation and pulmonary parenchymal oedema and haemorrhage (Mokbel et al. 2011; Park et al. 2012a, b; Kang and Park 2014).

In vivo MSCs viability remains important for their effectiveness. MSCs in vivo implantation exposes them to noxious environment. As such the scaffold seeded cells are being implanted. dAD-MSCs encapsulated in alginate beads had improved retention and the cells were viable up to 7 days (Koh et al. 2017). The studies described in this chapter range from preclinical experimental models to the clinical trials. The clinical studies described are mostly uncontrolled and have been conducted as case report, case series and/or comparative studies.

### 11.3.1 Musculoskeletal Tissues

#### 11.3.1.1 Osteoarthritis

Osteoarthrosis (OA) affects larger dog population (20%) (Johnston, 1997). Currently, no successful clinical treatment is available (Gugjoo et al. 2016; Gugjoo et al. 2017; Li et al. 2018; Gugjoo et al. 2019b). As such stem cell applications in dog are being evaluated under experimental settings (Table 11.2) and in also clinical trials (Table 11.3).

**Table 11.2** In vivo chondrogenic preclinical experimental mesenchymal stem cell studies in dogs

| Model type | Number of animals | Model defect size/Study period | Biomaterial used | Cell dose | Evaluation criteria | Overall result | References |
|---|---|---|---|---|---|---|---|
| Partial thickness chondral defects of lateral femoral condyle | 32 (8 control; 12 group II; 12 group III) | 3 mm diameter and 1 mm depth/8 weeks | Autologous BM-MSCs | 1.4–1.6 × $10^6$ | Morphological, histological and fluorescence analysis | Recovery was significant both clinically and histologically in the two cell treated groups (group II and III) as compared to the control (group I). In the meantime, group II showed better results at 8 weeks than group III. Homing was confirmed by the incorporation of injected GFP-labelled MSCs within the newly formed cartilage | Mokbel et al. (2011) |
| Osteochondral defects of medial femoral condyles | Group I: Cell + scaffold; group II: Control | 4.2 mm diameter and 6 mm depth/3 months and 6 months | Chondrogenically-induced BMSCs and scaffold | – | Gross morphology and by histological, biochemical, biomechanical and micro-CT analyses | Statistically significant improvement in gross and histology and cartilage stiffness in cell/scaffold treated animals as compared to control. Comparable micro-CT analysis of the subchondral bone in two groups. Better results in later period than at early period | Yang et al. (2011) |

(continued)

**Table 11.2** (continued)

| Model type | Number of animals | Model defect size/Study period | Biomaterial used | Cell dose | Evaluation criteria | Overall result | References |
|---|---|---|---|---|---|---|---|
| Osteonecrosis of the femoral head | 24 (54 hip joints) group I: Transgenic BM-MSCs; group II: BM-MSCs group III: Control | 2 mm diameter and 2 mm width/12 weeks | VEGF 165 transgenic bone marrow mesenchymal stem cells or simple BM-MSCs | $2 \times 10^7$ | Radiography, single-photon emission, computed tomography, histopathology, histomorphometric analysis and immunofluorescent staining for von Willebrand factor | Better results in group I as compared to group II and group III. A regular arrangement of trabeculae and obvious bone regeneration with significantly increased capillaries in group I animals | Hang et al. (2012) |
| Osteochondral defects | 12 (24 (defects) group I: Cell + PRF treated; group II: Control | 6 mm diameter and depth of 5 mm/4–24 weeks | Autologous BM-MSCs + platelet rich fibrin (PRF) | $1 \times 10^6$ | Macroscopic and histopathology | Group I had statistically significant improvement in gross and histological parameters as compared to control at 16 weeks. Compared to cell treated group gross parameters were deteriorated with time in control group animals | Kazemi et al. (2017) |

| | | | | | | | |
|---|---|---|---|---|---|---|---|
| Chondral defects of stifle joint | 24 (48 defects) group I: Cell + HA treated; group II hyaluronic acid; group III: Control | 4 mm/28 weeks | BM-MSCs + hyaluronic acid (HA) | $1 \times 10^7$ | Macroscopy, magnetic resonance imaging (MRI), histopathology, immunohistochemistry for type II collagen | Group I had regenerated more cartilage-like tissue than did HA alone or saline. Macroscopic evaluation and histological assessment score were significantly improved in cartilage defects of group I as compared to those in the other groups. MRI demonstrated cartilage-like signal at defect sites, and the surface of cartilage was relatively smooth. No obvious defect was found and cartilage-like tissue with the same thickness of the surrounding normal tissue was formed in group I. for group II, cartilage-like signal was also observed, but the thicknesses of tissue at the defect sites were thinner than those at normal. For group III, no cartilage-like signal was observed | Li et al. (2018) |

**Table 11.3** In vivo chondrogenic clinical mesenchymal stem cell studies in dogs

| Clinical condition/ ailment | Number of animals included | Study period | Study type | Cell source | Cell dose | Evaluation criteria | Overall result | References |
|---|---|---|---|---|---|---|---|---|
| Chronic osteoarthritis of Coxo-femoral joint | 21 | 90 days | Case series (randomized, double blinded, placebo-controlled trial) | Autologous AD-MSCs | $5 \times 10^6$ (20 dogs) and $4.5 \times 10^6$ (1 dog) | Orthopaedic examination scores, lameness and composite scores and size effect | Statistically significant improvement in scores for lameness and the compiled scores for lameness, pain and range of motion as compared to control dogs | Black et al. (2007) |
| Chronic osteoarthritis of humeroradial joint | 14 | 180 days | Case series (randomized, double blinded, non-placebo-controlled trial) | Autologous AD-MSCs | $3–5 \times 10^6$ | Orthopaedic examination score and size effect | Statistically significant improvement in lameness, pain on manipulation, range of motion and functional disability in cell treated animals | Black et al. (2008) |
| Chronic osteoarthritis of humeroradial joint | 4 | 30 days | Case series (uncontrolled study) | Autologous AD-MSCs | $3–5 \times 10^6$ (laden in platelet rich plasma or hyaluronic acid) | Clinical tests like trot pain on palpation and functional improvement in disability | Improvement with time as per owner although without any statistical testing | Guercio et al. (2012) |
| Chronic arthritis of the hip joint | 8 | 180 | | Autologous AD-MSCs | $15 \times 10^6$ | Force platform analysis | Significant improvement in clinical parameters in cell treated cases | Vilar et al. (2013) |

| | | | | | | | | |
|---|---|---|---|---|---|---|---|---|
| Hip osteoarthritis | 39 (autologous AD-MSCs vs PRGF) | 180 days | Randomized comparative clinical trial | Autologous AD-MSCs and PRGF | $30 \times 10^6$ | Visual analog scale (VAS), bioearth scale assessment, range of motion (ROM), radiography | MSCs and PRGF implantation was safe and effective in the functional analysis at 1, 3 and 6 months; provide a significant improvement, reducing dog's pain, and improving physical function. With respect to basal levels for every parameter in patients with hip OA, MSCs showed better results at 6 months | Cuervo et al. (2014) |
| Hip dysplasia | 9 (group I: SVF, $n = 4$; group II: AD-MSCs; $n = 5$) | 30 days | Comparative study | Allogeneic AD-MSCs at acupoints or stromal vascular fraction (SVF) | $0.2–0.8 \times 10^6$ | Physical and orthopaedic examinations | Positive results were more clearly seen in the SVF-treated group. All dogs had shown improvement in range of motion, lameness at trot and pain on manipulation of the joints, except for one ASC-treated patient. These results had shown that autologous SVF or allogeneic ASCs can be safely used in acupoint injection for treating hip dysplasia in dogs and represent an important therapeutic alternative for this type of pathology | Marx et al. (2014) |

(continued)

**Table 11.3** (continued)

| Clinical condition/ ailment | Number of animals included | Study period | Study type | Cell source | Cell dose | Evaluation criteria | Overall result | References |
|---|---|---|---|---|---|---|---|---|
| Chronic arthritis of the hip joint | 15 (group I: MSCs treated, $n = 10$; sound dogs, $n = 5$) | 180 days | Case control (blinded control study) | Autologous AD-MSCs | $15 \times 10^6$ | X-ray and platform gait analysis: Mean values of peak vertical force (PVF) and vertical impulse (VI) | Mean values of PVF and VI were significantly improved within the first 3 months cell post-treatment in the OA group, increasing 9% and 2.5% body weight, respectively, at day 30. After this, the effect seems to decrease reaching initial values | Vilar et al. (2014) |
| Osteoarthritis of hip, elbow, stifle, and or shoulder joints | 82 (bilateral joints: Treated, $n = 35$; placebo control, n = 32) and (single joint: Treated: n = 8; placebo control, n = 7) | 60 days | Case series study (prospective, randomized, masked and placebo-controlled) | Allogeneic AD-MSCs | $12 \times 10^6$ (cryopreserved with 85.1% viability) | Owner client-specific outcome measurement (CSOM) and secondary measures included veterinary pain on manipulation, veterinary global score, and owner global score | Success in the primary outcome was variable. CSOM was statistically improved in the treated dogs compared to the placebo dogs. The veterinary pain on manipulation score and the veterinary global score were both statistically improved in treated dogs as compared to placebo | Harman et al. (2016) |

| | | | | | | | | |
|---|---|---|---|---|---|---|---|---|
| Elbow dysplasia and elbow osteoarthritis | 30 dogs (39 elbow joints) | 1 year | Case series (uncontrolled study) | Allogeneic AD-MSCs + hyaluronic acid (0.5%) | $12 \times 10^6 \pm 3.2 \times 10^6$cells | Owner and veterinarian examination, arthroscopy and histopathology | A highly significant improvement was achieved after cell implantation without any medication as demonstrated by the degree of lameness during the follow-up period. Control arthroscopy of 1 transplanted dog indicated that the cartilage had regenerated. Histological analysis of the cartilage biopsy confirmed that the regenerated cartilage was of hyaline type | Kriston-Pál et al. (2017) |
| Hip, knee, radiocarpal intercarpal, elbow and ulnar osteoarthritis | 10 patients ($n = 5$ evaluated for 90 days and $n = 5$ for 1–4 years) | 30–90 days/4 years (5 animals) | Case reports | Autologous AD-MSCs | $3 \times 10^7$ | Physical examination and assessment for lameness, pain on manipulation, range of motion of the joint and functional disability | Functional outcomes for all analysed characteristics improved significantly at the end of this evaluation compared with the baseline. Long-lasting positive effects on two out of five analysed characteristics | Dražilov et al. (2018) |

(continued)

**Table 11.3** (continued)

| Clinical condition/ ailment | Number of animals included | Study period | Study type | Cell source | Cell dose | Evaluation criteria | Overall result | References |
|---|---|---|---|---|---|---|---|---|
| Degenerative arthritis | 203 (divided into groups as per age, weight) | 10 weeks | Controlled clinical study | Allogeneic AD-MSCs | 128 received a single dose of intra-articular injection in the affected joint. 65 dogs received a single intravenous injection, and 10 dogs received both IA and IV injections | Quality of life score (QoL), lameness and pain score | Ninety percent of young dogs (<9 years) had shown excellent improvement in pain and mobility and were able to run and resume normal activity. Sixty percent of older dogs showed good improvement. However, 12% of dogs did not exhibit any change in symptoms; one dog showed worsening of the symptoms | Shah et al. (2018) |
| Osteoarthritis (hip, knee, elbow and tarsal joints) | 22 | 6 months and 2 years | Non-randomized, open and monocentric study | Allogeneic MSCs | $10 \times 10^6$ (repeated injection after 6 months in 8 dogs or 11 joints) | Clinical evaluation, flow Cytometric Crossmatch analysis of Humoral response against cellular product | Veterinary clinical evaluation showed a significant and durable clinical improvement (up to 6 months) following MSC administration. Eight dogs (11 joints) were re-injected 6 months apart, sustaining clinical benefits up to 1 year. Owner's global satisfaction reached 75% at 2 years post-treatment | Cabon et al. (2019) |

| | | | | | | | | |
|---|---|---|---|---|---|---|---|---|
| Elbow osteoarthritis | 6 | 6 weeks | Case control | Xenogeneic chondrogenic induced PB-MSCs (equine source) | $1 \times 10^6$ cells | Orthopaedic examination, synovial fluid examination, radiography | There was no significant difference in the orthopaedic examination parameters, the radiographic examination, synovial fluid sampling and pressure plate analysis between placebo and MSC treatment. A single intra-articular administration of MSCs proved to be a well-tolerated treatment, which reduced lameness and pain according to the owner's evaluations as compared to a placebo treatment | Daems et al. (2019) |

In these studies a single dose of dMSCs has been injected locally into joints except for a study that had implanted cells more than once (Cabon et al. 2019). One study had utilized acupoint as site of implantation (Marx et al. 2014). Either only cells have been utilized (Marx et al. 2014; Vilar et al. 2014) or along with platelet rich plasma/fibrin (Vilar et al. 2013; Kazemi et al. 2017), biphasic scaffold (Yang et al. 2011) or hyaluronic acid (Guercio et al. 2012; Kriston-Pál et al. 2017; Li et al. 2018). Even chondrogenically differentiated (Yang et al. 2011) or vascular endothelial growth factor transduced dMSCs have also been evaluated (Hang et al. 2012). In general, these studies have variably supported therapeutic applications of dMSCs.

Variable evaluation parameters have been undertaken in these studies. Variable follow-up period ranging from as low as 1 month (Guercio et al. 2012; Marx et al. 2014) to a maximum period of 5 years (Yoon et al. 2012) had been conducted. Therapeutic evaluation of dMSCs has been made on the basis of clinical evaluation parameters like Pain scales, and MRI, CT scan and histological examination. Quantitative Force Platform Gait Analysis tends to have greater accuracy and concordance for assessment of AO lameness as compared to pain assessment scales (Vilar et al. 2014). Improvement on the basis of clinical parameters, arthroscopic evaluation and histology has been demonstrated (Kriston-Pál et al. 2017). The histological scores of healed tissues in MSCs-treated groups were higher as compared to control but complete hyaline tissue was lacking. The healed tissues had been mixed fibrocartilage and/hyaline lacking complete integration to the native cartilage (Yang et al. 2011; Kazemi et al. 2017). Some of the studies, however, had failed to demonstrate significant improvement (Daems et al. 2019) and even some failed to demonstrate long-lasting improvement with MSCs applications (Vilar et al. 2014).

Several comparative studies have also been conducted. dAD-MSCs had been compared to platelet rich growth factors (PRGF) (Cuervo et al. 2014), stromal vascular fraction (SVF) (Marx et al. 2014), in addition to the VEGF transgenic dBM-MSCs against simple BM-MSCs. MSCs though had provided better outcome as compared to PRGF but not in comparison to SVF. SVF also contain growth factors and as such might have improved osteochondral defect healing (Kazemi et al. 2017). VEGF transgenic dBM-MSCs had better osteochondral healing potential as compared to simple MSCs (Hang et al. 2012). There is need for studies that compares dMSCs to the currently available surgical techniques.

#### 11.3.1.2 Osteogenic Defects

Bone usually heal fast but may sometimes be hampered, resulting in delayed union and/non-union. Varied conditions like infection, extensive damage, old age, tumours and/or osteoporosis impair bone healing. Dog fracture non-union rate remains at 3.4% of the total bone fractures (Millis and Jackson, 2003).

MSCs have been studied in various dog preclinical experimental models (Table 11.4) and in clinical cases (Table 11.5). MSCs have been implanted along with various scaffolds like β-TCP (Arinzeh et al. 2003), β-TCP/poly

**Table 11.4** In vivo Preclinical experimental osteogenic mesenchymal stem cell studies

| Model type | Number of animals | Model defect size/study period | Study type | Biomaterial used | Cell dose | Evaluation criteria | Overall result | References |
|---|---|---|---|---|---|---|---|---|
| Critical size defect of femora | 15 ($n = 6$ in cell treated and $n = 6$ ceramic treated and 3 in control) | 21 mm length/4–16 weeks | Comparative experimental study (treatment group Vs porous ceramic cylinder treated group Vs control) | Autologous BM-MSCs + porous ceramic cylinder containing hydroxyapatite (65%) and beta-tricalcium phosphate ceramic (35%) | – | Histology and morphometry, radiography | Atrophic non-union occurred in untreated defects. Only a small amount of trabecular bone formed at the cut ends of the cortex of the host bone in this group. In contrast, radiographic union was established rapidly at the interface between the host bone and the implants in cell loaded. Histological and morphometric analyses demonstrated that both woven and lamellar bone had filled the pores of the implants that had been loaded with mesenchymal stem cells. The amount of bone was significantly greater than the pores of the implants that had not been loaded with cells. The collar of bone remodelled during the sixteen-week period of study, resulting in a size and shape that were comparable with those of the segment of bone that had been resected. Callus did not develop around the cortex of the host bone or around the defect in any of the specimens in the other two groups. | Bruder et al. (1998) |

(continued)

**Table 11.4** (continued)

| Model type | Number of animals | Model defect size/study period | Study type | Biomaterial used | Cell dose | Evaluation criteria | Overall result | References |
|---|---|---|---|---|---|---|---|---|
| Critical size defect of femoral diaphysis | 12 | 21 mm length/16 weeks | Comparative experimental study (allogeneic MSCs + beta-tricalcium phosphate vs autologous MSCs vs control) | Allogeneic MSCs + beta-tricalcium phosphate | – | Histology and radiography | Defects treated with allogeneic mesenchymal stem cell implants showed no adverse host response. Histologically, no lymphocytic infiltration occurred and no antibodies against allogeneic cells were detected. Histologically, by eight weeks, a callus spanned the length of the defect, and lamellar bone filled the pores of the implant at the host bone-implant interface. Fluorescently labelled allogeneic cells were also detected. At sixteen weeks, new bone had formed throughout the implant. These results were consistent with those seen in implants loaded with autologous cells. Implants loaded with allogeneic or autologous stem cells had significantly greater amounts of bone within the available pore space than did cell-free implants at 16 weeks | Arinzeh et al. (2003) |

| | | | | | | | | |
|---|---|---|---|---|---|---|---|---|
| Radial critical size defect | 6 ($n = 3$ in treatment group and $n = 3$ in control) | 1.5 cm/12 weeks | Experimental study (allogeneic UCB-MSCs + beta-tricalcium phosphate Vs control) | Allogeneic UCB-MSCs + beta-tricalcium phosphate covered with β-TCP/poly L-lactide-co-ε-caprolactone composite supported with compression plate | $1 \times 10^6$ | Radiography, histology | Radiographically, continuity between implant and host bone was evident at only one of six interfaces in control by 12 weeks, but in three of six interfaces in cell treated group. Radiolucency was found only near the bone end in control group at 12 weeks after implantation, but in the entire graft in cell treated group. Histologically, bone formation was observed around beta-TCP in longitudinal sections of implant in both groups. Histomorphometric analysis revealed significantly increased new bone formation in cell treated group at 12 weeks after implantation. | Jang et al. (2008) |
| Critical size ulnar bone defect | 9 (18 forelimb defects) | 2.5 cm/12 weeks | Comparative study (groups of different BM-MSC concentration + rBMP concentration) | BM-MSCs + rBMP (E. coli) | 560 μg/$10^7$ cells, 560 μg/$10^5$ cells, 560 μg/0 cells, 140 μg/$10^7$. 140 μg/$10^5$, 140 μg/0 | Radiography, micro-CT, histological | Enhanced bone formation, with improved bone mineral density and reduced non-uniformity of the regenerated bone in cell and BMP treated group. Higher cell and growth factor concentration in general improved healing. Micro-CT confirmed bone mineral density was improved in the groups that received cells ($10^7$ or $10^5$) along with the 560 μg of rBMP. The non-uniformity of the bone mineral density was improved in animals that received 560 μg rBMP with $10^7$ and $10^5$ cells and 140 μg/$10^7$ cells. | Itoi et al. (2016) |

**Table 11.5** In vivo clinical osteogenic mesenchymal stem cell studies

| Clinical case | Number of animals included | Study period | Study type | Cell source | Cell dose | Route | Evaluation criteria | Overall result | References |
|---|---|---|---|---|---|---|---|---|---|
| Radial fracture with sclerotic ends | 1 | 90 days | Case report | Autologous AD-MSCs + hydroxyapatite and chitosan | $3.2 \times 10^7$ | Local implantation + compression plate | Radiography and lameness | The composition scaffold of HA containing ADSCs had induced new bone formation | Lee et al. (2009a) |
| Radial non-union fracture | 1 | 8 weeks | Case report | Umbilical cord MSCs + cortical allograft | $1 \times 10^7$ repeated after 1 month | Local implantation | Radiography | Fracture healing was identified by 6 weeks after cell injection. Increased opacity between radius and allograft bone, and a decreased radiolucent gap was checked. With time, the radiolucent site gradually decreased and the radiolucent fracture line disappeared | Jang et al. (2008) |
| Oligotrophic non-union left radial fracture | 1 dog | 9 weeks | Case report | Bone morphogenetic protein-7 AD-MSC sheet | $1 \times 10^4$ cells/cm$^2$ | Local implantation + locking plate | Lameness and radiography | Completely healed fracture | Song et al. (2017) |
| Tibial tuberosity osteotomy | 9 (control, $n = 5$; cell treated, $n = 4$) | 15, 30, 60 and 120 days | Double-blind experimental study | MSCs | $1.5 \times 10^6$ | Local application | Radiography, image analysis software | Cell treated group had significant ossification mean 36.45% greater than the control at 30 days | dos Santos et al. (2018) |

L-lactide-co-ε-caprolactone (Jang et al. 2008), coral scaffold (Cui et al. 2017) and porous ceramic cylinder (Bruder et al. 1998). All these studies have demonstrated improved bone density with uniform growth of regenerated tissue upon dMSCs implantation. Concentration of the cells and of the growth factor (rBMP) used too may have effect on bone healing (Itoi et al. 2016). Contrary to ex vivo reports, dMSCs from varied sources had provided comparable in vivo osteogenesis (Kang et al. 2012).

Apart from experimental studies, clinical application of MSCs too has supported their role in enhancement of healing. BMP-7 transgenic dAD-MSC sheets implanted in a clinical oligotrophic non-union radial fracture and autologous AD-MSCs in combination with hydroxyapatite and chitosan implanted in a radial fracture with sclerotic ends had complete healing in about 9 weeks (Song et al. 2017) and 12 weeks (Lee et al. 2009a), respectively. All these studies have utilized MSCs along with variable scaffolds and/growth factors. There is need to perform comparative uniform studies incorporating various scaffolds and growth factors. Besides, these studies have to be compared against the current surgical treatment options for their feasibility and effectiveness.

#### 11.3.1.3 Periodontal Defects

Periodontitis has limited outcome with currently available regenerative therapies (guided tissue regeneration and enamel matrix derivatives). These techniques are associated with poor clinical predictability (Esposito et al. 2009). To address the issue, stem cell application for periodontal regeneration is currently being evaluated (Table 11.6).

In experimental studies, dMSCs have improved periodontal defect repair, although variably. The in vivo implanted cells had migrated to these structures and showed differentiation into osteoblast- and fibroblast-like cells (Akizuki et al. 2005; Wei et al. 2010), and odontoblast-like cells (Shi and Gronthos, 2003). MSCs from various sources have been utilized (Table 11.6). dMSCs from alveolar periosteum, bone marrow and periodontal ligament (PDL) had variably improved healing. PDL-MSCs had led to the significant periodontal regeneration in comparison to the other cell types. The healed tissue with PDL-MSC implantation included well-oriented PDL fibres, regenerated nerve, cementum and alveolar bone (Tsumanuma et al. 2011). To improve cell survival and effectivity, MSCs along with scaffold matrices and osteo-inductive material had successfully regenerated the tissue in furcation defects (Takewaki et al. 2017). Further, cryopreserved MSCs had comparable periodontal healing potential to that of fresh MSCs (Li et al. 2009).

These model studies evaluated MSCs in variable classes of furcation defects. MSCs concentration, passage number, concentrations and sources has been variable in these studies. Addition of scaffolds and growth factors too makes these studies non-uniform.

**Table 11.6** In vivo preclinical experimental mesenchymal studies on periodontal regeneration

| Model type | Number of animals | Model defect size/study period | Study type | Biomaterial used | Cell dose | Evaluation criteria | Overall result | References |
|---|---|---|---|---|---|---|---|---|
| Periodontal fenestration defects | Five dogs (26 teeth) | 5 mm × 5 mm (apical to the cemento-enamel junction)/8 weeks | Comparative experimental study (fresh BM-MSCs Vs cryopreserved BM-MSCs) | Freshly isolated and cryopreserved BM-MSCs seeded on collagen scaffold | $5 \times 10^6$ | Histological and histomorphometric parameters | Both freshly isolated and cryopreserved BMSC transplantations induced significantly better periodontal regeneration with newly formed cementum, alveolar bone and periodontal ligament as compared to collagen scaffold applied cases. | Li et al. (2009) |
| Class II furcation defects | 7 (control group: Coronally positioned flap, bioabsorbable membranes: GTR, sponge group: Carrier + GTR, cell group: Carrier + PDL cells + GTR | 6 months | Experimental study (cell treated Vs control) | Autologous periodontal ligament derived MSCs | | Histometric parameters, scanning electron microscopy (SEM) | PDL cells had significantly promoted periodontal regeneration. Cell group presented a superior length of new cementum, a greater extension of periodontal regeneration and a larger area of new bone as compared to sponge, GTR and control group, respectively | Suaid et al. (2011) |

| | | | | | | | | |
|---|---|---|---|---|---|---|---|---|
| One-wall intrabony defect | 4 | 5 mm in depth, 5 mm mesio-distal width/8 weeks | Comparative study (PDL-MSCs Vs alveolar periosteum MSCs (AP) Vs BM-MSCs) | Autologous cells in concerned groups | $3–5 \times 10^4$ cell sheets | Histological and histometric parameters | Significant periodontal regeneration (Cementum and well-oriented PDL fibres) in the PDL-MSCs group than others. Nerve regeneration and alveolar bone formation are also better in PDL-MSC group. | Tsumanuma et al. (2011) |
| Class II furcation defects | 3 | 5 mm (apico-occlusal direction) and 2 mm (buccolingual direction)/8 weeks | Comparative experimental study (BM-MSCs + PRP Vs PRP Vs control) | Autologous BM-MSCs + PRP | $1 \times 10^7$ | Histomorphometric analysis and histological analysis | Periodontal regeneration with complete filling of class II furcation defects with cementum, alveolar bone and periodontal ligament all cases healed equally. | Simsek et al. (2012) |
| Class III furcation defects | 7 | 5.0 mm in the apico-occlusal direction/3 months | Comparative study (PDL-MSCs + sponge carrier vs sponge treated vs control) | Autologous periodontal ligament derived MSCs + carrier sponge | $3 \times 10^5$ | Histomorphometrically | PDL cells with sponge had significantly promoted periodontal regeneration | Suaid et al. (2012) |

(continued)

**Table 11.6** (continued)

| Model type | Number of animals | Model defect size/ study period | Study type | Biomaterial used | Cell dose | Evaluation criteria | Overall result | References |
|---|---|---|---|---|---|---|---|---|
| 3-walled periodontal defects in first lower premolar teeth | 10 (one side treated and other side untreated) | 3 mm buccolingually, 5 mm apico-coronally, and 8 mm mesiodistally created defect/8 weeks | Experimental study (cell treated Vs control) | Autologous dental pulp MSCs and bio-Oss granules after 4 weeks | $2 \times 10^7$ | Histologically and histomorpho-metrically | Cementum regeneration and periodontal ligament regeneration significantly higher in treated group than control | Khorsand et al. (2013) |
| Class III periodontal tissue defects | 8 | 5 mm (height from the top of the dental root bifurcation to the reduced alveolar crest in the class III defect) /2 months | Comparative experimental study (cell + PRP vs PRP Vs non treated groups) | Autologous AD-MSCs and PRP | $1.5 \times 10^7$ | Histologic, immuno-histologic and radiography | Periodontal tissue with the correct architecture including alveolar bone, cementum-like structures and periodontal ligament-like structures regenerated in cell and PRP treated cases | Tobita et al. (2013) |
| Premolar class III furcation defect | 7 (24 right and left premolar defects) | Height from the cement-enamel junction to the reduced alveolar crest was 4 mm/12 weeks | Experimental study (cell treated Vs control) | Autologous BM-MSCs and osteogenic induced BM-MSCs | $7 \times 10^4$ cell sheet | Micro-computed tomography, histological analysis and morphometric analysis | Successful periodontal tissue regeneration with BM-MSCs as compared to control. Rapid alveolar bone formation with osteogenic differentiated BM-MSCs | Takewaki et al. (2017) |

#### 11.3.1.4 Ligaments/Tendon/Muscle Injuries

Rupture of cranial cruciate ligament (CCL) causes hind limb lameness in dog (Johnson et al. 1994). Currently, no specific and effective treatment is available and as such MSCs are being evaluated (Table 11.7). Intra-articularly implanted dBM-MSCs tend to engraft CCL (Linon et al. 2014). An overall improvement, although variably has been demonstrated with MSCs (Canapp Jr. et al. 2016; Muir et al. 2016). A retrospective study had reported an improvement with AD-MSCs and BM-MSCs along with platelet rich plasma on partial cranial cruciate ligament rupture. Available arthroscopic findings had represented fully healed CCL (marked neo-vascularization and a normal fibre pattern) in the disrupted areas in most of the cases (Canapp Jr. et al. 2016). Further, MSCs implanted in a tibial plateau levelling osteotomy (TPLO) treated CCL had significantly improved bone healing at 1 month period in comparison to non-steroidal anti-inflammatory drugs (NSAIDs) treated ones. However, the improvement could not be maintained for long as non-significant differences in clinical parameters at later periods (up to 3 months) had been demonstrated (Taroni et al. 2017). Further extensive studies of longer follow-up periods are therefore desired.

MSCs may impart clinical healing either by fibroblastic differentiation or by the release of trophic factors (Yoon et al. 2012). MSCs like AD-MSCs may modulate tendon inflammation by increasing M2 macrophages, inhibiting inflammation (prostaglandin reductase 1). The cells may introduce progenitor cell, reduce mononuclear cell infiltration to the damaged site and increase pro-tenogenic proteins (periostin). Incorporation of recombinant bone morphogenetic protein-12 (rBMP-12) may further potentiate these effects possibly through induction of more activated M2 macrophages (Gelberman et al. 2017).

Athletic dogs undergo frequent muscle injuries (Lee et al. 2012). Muscle lack innate ability to regenerate. As MSCs tend to differentiate into myocyte like cells under ex vivo studies they have prospect to repair muscles under in vivo system. MSCs implanted into the incised infraspinatus tendon tend to align between its collagen fibres. These aligned cells may express higher tenomodulin, MMP13 and collagen type I in the composite (Omae et al. 2009). In dog clinical cases, these cells have promoted muscle repair and prevented fibrosis. The treated animals were able to work being characterized by improved functional gait and lack of lameness (Brown et al. 2012; Case et al. 2013; Gibson et al. 2017). Ultrasonographic examination, however, had demonstrated lack of proper arrangement of gastrocnemius muscle fibres (Case et al., 2013). Muscle derived MSCs may also improve muscular dystrophy. The implanted cells in immune-suppressed dogs had led to the gradual improvement of dystrophied muscles (Lorant et al. 2018).

#### 11.3.1.5 Myocardial Diseases

Dilated cardiomyopathy is one of the commonly seen cardiac ailments in dogs. AD-MSCs application in occult-dilated cardiomyopathy cases had been safe but without any observable clinical benefits (Pogue et al. 2013). Further studies are

**Table 11.7** In vivo clinical mesenchymal stem cell studies on tendon/ligament/muscle injuries

| Clinical case | Number of animals included | Study period | Study type | Cell source | Cell dose | Route | Evaluation criteria | Overall result | References |
|---|---|---|---|---|---|---|---|---|---|
| Chronic Chagas cardiomyopathy | 5 | 6 months | Case series | Autologous BM-MSC | $100 \times 10^6$ | Right and left coronary artery implantation | Electrocardiography and echocardiography | Significant improvement in cardiac function (peak velocity of aortic flow) after cell implantation | Sousa et al. (2011) |
| Semitendinosus myopathy | 2 | 19 weeks (case 1) and 22 weeks (case 2) | Case report | Autologous AD-MSCs | $3.7 \times 10^6$ | Local implantation | Ultrasonography, muscle biopsy, visual assessment of dog's improvement in affected tasks and surgeon assessments | AD-MSCs therapy enhanced muscle healing and prevented fibrosis in these clinical cases, similar to the reported literature in laboratory animals. The dogs returned to their previous training and occupations with a functional gait and no lameness. | Brown et al. (2012) |
| Grade IV patellar luxation, arthritis | 1 | 5 years | Case report | Autologous AD-MSC | $1 \times 10^6$ + hyaluronic acid | Local implantation | Lameness and radiography | Normal walking and reduced osteophytes and osteochondral lesions had been reported after cell implantation. | Yoon et al. (2012) |

| | | | | | | | | | | |
|---|---|---|---|---|---|---|---|---|---|---|
| Gastrocnemius tendon injury | 1 | | 631 days | Case report | Autologous BM-MSC | >20 × $10^6$ (cryopreservation) | Local implantation | Serial orthopaedic examinations, ultrasonography and long-term force-plate gait analysis | Lameness subjectively resolved and peak vertical force increased from 43% to 92% of the contralateral pelvic limb. Serial ultrasonographic examinations had revealed improved but incomplete restoration of normal linear fibre pattern of the gastrocnemius tendon. | Case et al. (2013) |

(continued)

**Table 11.7** (continued)

| Clinical case | Number of animals included | Study period | Study type | Cell source | Cell dose | Route | Evaluation criteria | Overall result | References |
|---|---|---|---|---|---|---|---|---|---|
| Partial cranial cruciate ligament tear treated with TPLO | 36 ($n = 19$ for BMC + PRP) and ($n = 17$ for AD-PCs + PRP) | 90 days | Retrospective comparative study | Bone marrow concentrate + PRP vs AD-MSCs + PRP | 2–4 mL (BMC + PRP) and 1–2 mL (AD-PCs + PRP) | Local joint implantation | Objective gait analysis, diagnostic arthroscopy, and validated functional questionnaire | Promising healing potential of these treatments was demonstrated. Arthroscopic findings of 13 cases out of 36 cases at 90 days interval were available. Nine dogs had fully healed CCL represented by marked neo-vascularization and a normal fibre pattern in the disrupted areas. Out of other four cases, one had improved healing and was subjected to another injection while rest three underwent tibial plateau levelling osteotomy | Canapp Jr. et al. (2016) |

| | | | | | | | | | |
|---|---|---|---|---|---|---|---|---|---|
| Cranial cruciate ligament rupture | 12 | 8 weeks | Case series (case controlled clinical study) | Autologous BM-MSCs | $2 \times 10^6$ and $5 \times 10^6$ | Intravenous and local intra-articular | Radiography, arthroscopy, serum and synovial cytokines and C-reactive protein | No adverse events. With lower circulating CD8+ T lymphocytes after cell treatment were detected. Serum C-reactive protein (CRP) was decreased at 4 weeks and serum CXCL8 was increased at 8 weeks. Synovial CRP in the complete CR stifle was decreased at 8 weeks. Synovial IFNγ was also lower in both stifles after cell injection. Systemic and intra-articular injection of autologous BM-MSCs in dogs with partial CR suppresses systemic and stifle joint inflammation, including CRP concentrations. | Muir et al. (2016) |

(continued)

**Table 11.7** (continued)

| Clinical case | Number of animals included | Study period | Study type | Cell source | Cell dose | Route | Evaluation criteria | Overall result | References |
|---|---|---|---|---|---|---|---|---|---|
| Semitendinosus myopathy | 9 dogs (11 cases) | Short-term (6 months) and long-term (1 year) follow-up | Uncontrolled retrospective comparative clinical trial | AD-MSCs | – | Local implantation | Ultrasonography, visual assessment score (VAS), | At short-term follow-up ultrasound and gait analysis showed a mean reduction in the overall intramuscular lesion size and reduction in the visual assessment score (VAS). At long-term follow-up, in 8 cases had a normal gait and in 3 cases the dogs had an improved gait as compared with initial examination. All 8 dogs had returned to active police work. | Gibson et al. (2017) |
| Cranial cruciate ligament rupture | 14 (9 MSCs and 5 NSAIDs) | 180 days | Comparative study (blinded) | Allogeneic foetal adnexa MSCs | $10 \times 10^6$ | Local implantation | Clinical score and gait and bone healing | Better healing in MSC treated animal at 1 month with comparable gait and clinical score at 6 months | Taroni et al. (2017) |

| | | | | | | | | | |
|---|---|---|---|---|---|---|---|---|---|
| Muscular dystrophy | 16 (group received cells with transient immunosuppression (GRMDMU/tr-IS), $n = 4$; group received cells without immunosuppression (GRMDMU/no-IS), $n = 4$; group received no cells without immunosuppression, $n = 3$; group received no cells with transient immunosuppression, $n = 5$) | 5 and 9 months | Case series | Allogeneic muscle derived MSCs + immunosuppression | $6.3 \times 10^7$ to $1.2 \times 10^8$ per kg | Intrafemoral injections | Immunohistochemistry, histomorphometry, mononuclear cell proliferation, western blotting | Persistent clinical improvement of the GRMDMU/tr-IS dogs was observed. GRMDMU/no-IS dogs exhibited no benefit. Histologically, only 9-month-old GRMDMU/tr-IS dogs showed an increased muscle regenerative activity. A mixed cell reaction with the host peripheral blood mononucleated cells (PBMCs) and corresponding donor cells revealed undetectable to weak lymphocyte proliferation in GRMDMU/tr-IS dogs compared with a significant proliferation in GRMDMU/no-IS dogs | Lorant et al. (2018) |

desired involving genetically modified cells to confirm their role in cardiac ailments, if any (Lu et al. 2013).

Dog has been used as myocardial ischemic model animal for human though the condition has little relevance in dog clinical practice. An increased vascularity and cardiac functional improvement (Memon et al. 2005; Silva et al. 2005; Bartunek et al. 2007; Perin et al. 2008) has been reported with MSCs implantation. As stem cells have self-renewal and multipotent differentiation properties, infarcted myocardium may have been repaired (Linke et al. 2005).

## 11.3.2 Non-Musculoskeletal Tissues

### 11.3.2.1 Wounds/Ulcers

Numerous MSCs studies are being conducted to evaluate the healing potential of MSCs in wound and ulcer. In dog MSCs have been employed in management of wounds (Tables 11.8 and 11.9), and ulcers (Table 11.8). Even dAD-MSCs may suffice as canine dermal papillae substitutes (Bae et al. 2015). Local implantation of the MSCs like BM-MSCs appears more effective in induced oral ulcers as compared to the systemic implantation (Aly et al. 2014).

An extensive study on full thickness wound model had shown that implantation of allogeneic BM-MSCs of various concentrations ($1 \times 10^4$, $1 \times 10^5$, $1 \times 10^6$ and $1 \times 10^7$) had led to wound epithelial regeneration by day 7. Although, collagen was higher in healed wounds implanted with highest cell concentration but was statistically insignificant. Further, pro-healing factor like bovine fibroblast growth factor (bFGF) was highly expressed while inflammatory mediators like IL-2 and INF-$\gamma$ were decreased at day 7 in comparison to control confirming their anti-inflammatory and anti-fibrotic role in skin wounds (Kim et al. 2013). MSCs extracellular vesicles (microvesicles) may also improve healing of wounds in dogs (El-Tookhy et al. 2017). In chronic clinical cases previously unresponsive to conventional therapy, MSCs had promoted healing (Madhu et al. 2014; Zubin et al. 2015). In our own experience allogeneic BM-MSCs had improved and led to faster healing of an extensive wound in a dog (Madhu et al. 2014).

MSCs tend to decrease inflammation (Yang et al. 2018), suppress collagenous matrix degradation (Jeon et al. 2010), favour angiogenesis (Kim et al. 2012) and accelerate epithelialization of wound (Rodriguez-Menocal et al. 2015).

### 11.3.2.2 Vocal Fold Injury

Vocal fold injury demands scar free healing for effective functioning (Rousseau et al. 2003). dMSCs and/conditioned medium had improved healing of vocal fold and laryngotracheal stenosis (Kanemaru et al. 2003; Lee et al. 2006; Ohno et al. 2011; Hu et al. 2014; Iravani et al. 2017) (Table 11.8). MSCs anti-inflammatory activity might prevent pseudomembrane formation and may also promote rational arrangement of the extracellular matrix proteins in the lamina propria (Kumai et al. 2010).

**Table 11.8** In vivo preclinical experimental mesenchymal stem cell studies on wound/oral ulcer healing

| Model type | Number of animals | Model defect size/study period | Study type | Biomaterial used | Cell dose | Evaluation criteria | Overall result | References |
|---|---|---|---|---|---|---|---|---|
| Oral ulcers (formosol induced) | 18 (3 groups with $n = 6$) group I; high dose; group II low dose and group III: Control | 0 day to 15 days | Experimental study (cell treated Vs control) | BM-MSCs | $1 \times 10^7$ (low dose) and $2 \times 10^7$ (high dose) | Clinical evaluation (wound assessment parameter scoring tool) and polymerase chain reaction for collagen gene | Groups I and II had shown clinical improvement in their wounds within 7 days following administration of MSCs, and the wounds showed a steady overall decrease in wound size. After 15 days, there was no statistically significant difference among the three groups. Groups II and III had shown better healing by histopathologic examination of oral tissue biopsies at 2 weeks as compared to the control group. An increased expression of both collagen and VEGF genes in MSCs-treated ulcers as compared to controls was demonstrated. Expression of VEGF was significantly higher in case of the group I as compared to group II | El-Menoufy et al. (2010) |

(continued)

**Table 11.8** (continued)

| Model type | Number of animals | Model defect size/study period | Study type | Biomaterial used | Cell dose | Evaluation criteria | Overall result | References |
|---|---|---|---|---|---|---|---|---|
| Full thickness skin wound | 10 (24 wounds) | 7 days | Experimental study | Allogeneic BM-MSCs | $1 \times 10^4$, $1 \times 10^5$, $1 \times 10^6$ and $1 \times 10^7$ cells /300 μL of PBS. 200 μL (wound bed and 100 μL in wound edges) | Histopathology, immunohisto-chemistry and PCR | Wound epithelial regeneration was seen with epithelial gap significantly smaller in cell treated groups as compared to untreated. Compared to control significantly higher proliferating cell nuclear antigen (PCNA), angiogenesis and collagen synthesis was seen in cell treated groups. Although, collagen was higher in $1 \times 10^7$ cell treated groups but was statistically non-significant. Bovine fibroblast growth factor (bFGF) was higher and IL-2 and INF-γ were lower at day 7 | Kim et al. (2013) |
| Induced Oral ulcers | 18 (6 in each group) | 15 days | Comparative experimental study (intravenous Vs local implantation Vs control) | BM-MSCs | $2 \times 10^7$ | Clinical and histopathological | The treatment resulted in dramatic wound edge activation and resurfacing of oral mucosa wound. Local application hastens such healing compared to systemic application | Aly et al. (2014) |
| Laryngotracheal stenosis | 7 | 6 weeks | Experimental study (cell/ conditioned media treated Vs control) | BM-MSCs | $2 \times 10^6$ and conditioned media | Histology | Prominent healing effect of the cells or conditioned media in wounds though statistically insignificant was reported | Iravani et al. (2017) |

**Table 11.9** In vivo clinical mesenchymal stem cell studies on wound healing

| Clinical case | Number of animals included | Study period | Study type | Cell source | Cell dose | Route | Evaluation criteria | Overall result | References |
|---|---|---|---|---|---|---|---|---|---|
| Large skin wound | 1 | 480 day | Case report | AD-MSCs + PRP | $2 \times 10^6$ | Local spraying | Clinical evaluation | Successful healing of complicated wound | Zubin et al. (2015) |
| Large skin wound | 1 | 50 days | Case report | Allogeneic BM-MSCs post caudal superficial epigastric axial pattern flap application | $5 \times 10^6$ | Local implantation on wound margins | Clinical evaluation | Uneventful wound healing | Madhu et al. (2014) |

#### 11.3.2.3 Immune-Mediated and/Anti-Inflammatory Effects

MSCs immuno-modulatory and anti-inflammatory properties makes them a suitable candidate to address immune-mediated disorders like inflammatory bowel disease (IBD), anal furunculosis or keratoconjunctivitis (Table 11.10) (Kang et al. 2008; Stevenson et al. 2012; Pérez-Merino et al. 2015a).

#### 11.3.2.4 Anal Furunculosis/Fistula

Anal furunculosis, an immune-mediated disorder tends to progress to perianal fistulas, commonly seen in middle aged German shepherd dogs (Day, 1993). Clinically, condition relapses frequently and thus, demands lifelong medication. The condition is akin to human anal condition known as Crohn's disease (Ferrer et al. 2016).

An open label trial had demonstrated significant improvement in the condition with local injection of single dose of human embryonic stem cell derived MSCs (hESC-MSCs) after 3 months. Animals had been free of fistulae by 3 months and required less than half of the dose of the cyclosporine. However, the condition had recurred in few cases after 6 months (Ferrer et al. 2016). Further studies are thus, desired that can incorporate multiple doses and be performed for longer periods to confirm utility of MSCs in anal fistula.

#### 11.3.2.5 Inflammatory Bowel Disease (IBS)

Allogeneic AD-MSCs had reduced the clinical symptoms in IBS (Pérez-Merino et al. 2015b). The intestinal tissues had better histology (Pérez-Merino et al. 2015a) (Table 11.10). Thus, further extensive studies are desired.

#### 11.3.2.6 Keratoconjunctivitis Sicca (KCS) and Corneal Ulcer

It is an immune-mediated disease of dog resembling to that of human (Barabino and Dana 2004; Abelson et al. 2009). Currently, KCS requires lifelong treatment (Lemp, 2008). Few clinical studies conducted on dog KCS had shown significantly improved results with the application of allogeneic AD-MSCs. Even immune-suppressants were not required in the treated cases (Villatoro et al. 2015; Bittencourt et al. 2016). These studies had shown improved results with variable dose of cells and had been conducted for a maximum of 12 months. Thus, uniform studies need to be conducted for a longer duration.

A single case of recurrent corneal ulcer in a diabetic poodle dog treated by topical application of two doses of allogeneic AD-MSCs had improved clinical parameters. The fluorescein test had been negative by day 69 (Arantes-Tsuzuki et al. 2019). The patho-physiological basis of the recurrent corneal ulcer, however, requires to be understood.

**Table 11.10** In vivo clinical mesenchymal stem cell studies on allergic/ auto-immune diseases

| Clinical case | Number of animals included | Study period | Study type | Cell source | Cell dose | Route | Evaluation criteria | Overall Result | References |
|---|---|---|---|---|---|---|---|---|---|
| Atopic dermatitis | 5 | 90 days | Case series (pilot study) | Autologous AD-MSCs | $1.3 \times 10^6$/kg b. wt. | Intravenous | Canine atopic dermatitis extent and severity score and visual pruritis score | Non-significant reduction in clinical signs | Hall et al. (2010) |
| Atopic dermatitis (for at least 12 months) | 26 (22 animals completed study) | 6 months | Controlled clinical case series | Allogeneic AD-MSCs | $1.5 \times 10^6$/kg b. wt | Intravenous | Clinical signs, haematological and biochemistry profiles, and AD severity (canine atopic dermatitis extent and severity index, version 4 (CADESI-04)), visual analogue scale | Pruritus and CADESI-04 scores decreased significantly after 1 week or month of treatment, respectively, and remained stable for 6 months. Owner's global assessment score was 2.15 ± 1.15 for all the animals in the study | Villatoro et al. (2018) |

(continued)

**Table 11.10** (continued)

| Clinical case | Number of animals included | Study period | Study type | Cell source | Cell dose | Route | Evaluation criteria | Overall Result | References |
|---|---|---|---|---|---|---|---|---|---|
| Inflammatory bowel disease | 11 | 42 | Case series (Unblinded, non-comparator study) | Allogeneic AD-MSCs | $2 \times 10^6$/kg b. wt | Intravascular | Clinical inflammatory bowel disease activity index (CIBDAI) and canine chronic Enteropathy clinical activity index (CCECAI), and C-reactive protein (CRP) | Significantly decreased CIBDAI and CCECAI but no effect on CRP | Pérez-Merino et al. (2015a, b) |
| Keratoconjunctivitis sicca | 15 (24 eyes) | 12 months | Case series | Allogeneic AD-MSCs | $1 \times 10^6$ | Lacrimal glands (dorsal and third eyelid) | Schirmer tear test, ocular surface evaluation | Statistically significant improvement in Schirmer tear test from 28 days | Bittencourt et al. (2016) |
| Keratoconjunctivitis sicca | 12 (24 eyes) | 9 months | Case series | Allogeneic AD-MSCs | $5 \times 10^6$ & $3 \times 10^6$ | Around lacrimal glands and third eyelid | Schirmer tear test, ocular surface evaluation | Statistically significant improvement in Schirmer tear test and ocular surface | Bittencourt et al. (2016) |

| | | | | | | | | | |
|---|---|---|---|---|---|---|---|---|---|
| Recurrent corneal ulcer | 1 | 69 days after first treatment | Case report | Allogeneic AD-MSCs | $3 \times 10^6$ (two doses at 48 days later) | Topical application | Clinical testing | Improved clinical parameters like blepharospasm, conjunctival hyperaemia, mucopurulent ocular discharge, photophobia, corneal opacity, chemosis, pigmentation, neo-vascularization and pain. The fluorescein test was negative by day 69 | Arantes-Tsuzuki et al. (2019) |
| Perianal fistula | 6 | 6 months | Case series | Human embryonic stem cell derived MSCs + cyclosporine | Commercial preparation | Local implantation | Clinical evaluation | Completely healed fistula at 3 months while 2 cases had relapse at 6 months | Ferrer et al. (2016) |

### 11.3.2.7 Atopic Dermatitis

One of the initial MSCs application study on dog allergic atopic dermatitis had failed to provide satisfactory results (Hall et al. 2010). However, a recent study had demonstrated improvement in clinical symptoms with MSCs as early as 1 week after treatment (Villatoro et al. 2018). Thus, further studies may be conducted to standardize the process.

### 11.3.2.8 Nervous System

Nervous tissue regeneration is currently not possible and as such MSCs are being evaluated for the purpose. To achieve therapeutic concentration of these cells in a brain, it is important to locate the suitable route of implantation. In case of meningo-encephalomyelitis of unknown origin (MUO), the stem cells implanted either by intrathecal and intra-arterial route had shorter time of reaction to therapy as compared to those receiving through intrathecal and intravenous route (Zeira et al. 2015). Another brain study had shown that intra-cerebro-ventricular implantation of xenogeneic UC-MSCs (human) in dogs had distributed to various brain structures. These cells had even enhanced sub-ventricular zone neural stem cell (Park et al. 2016).

In dogs spinal cord injury (SCI) is common and may occur due to spinal fractures and or disc degeneration (Fluehmann et al. 2006). The disease resembles to that of the human condition (Sarmento et al. 2014). As no successful therapy is available stem cells are being evaluated (Granger et al. 2012; Roszek et al. 2016).

Stem cells characteristic properties make them potential candidates to heal nervous tissues (Roszek et al. 2016). MSCs in various dog preclinical SCI models created mostly by balloon catheter have been evaluated (Table 11.11). In addition, MSCs regenerative potential has also been evaluated in peripheral nerve injury models (Ding et al. 2010) and disc disease models (Hiyama et al. 2008; Lee et al. 2009b). These cells have mostly been implanted after a week's period (Lee et al. 2011a; Park et al. 2012a, b). Apart from experimental studies, dMSCs have also been evaluated in clinical spinal cord injuries and meningo-encephalomyelitis of unknown origin (MUO) enlisted in Table 11.12.

MSCs from various sources like umbilical cord (UC), UCB (Lim et al. 2007; Lee et al. 2011b; Park et al. 2011; Besalti et al. 2015), WJ (Ryu et al. 2012), BM (Penha et al. 2014; Bhat et al. 2019) or AD (Park et al. 2012b) have been evaluated in SCI models. MSCs from autologous, allogeneic or xenogeneic sources have been implanted. Evaluation has been made on the basis of clinical scores, histology and/or magnetic resonance imaging (MRI). In general, MSCs have improved repair of injured spinal cord. In our own clinical study spinal cord injured dogs (n = 44), two doses of allogeneic BM-MSCs implanted in dogs at 2 weeks interval had significantly improved hind limb reflexes and Olby scores in comparison to those managed by the conventional treatment (Bhat et al. 2019).

dMSCs harvested from different tissues tend to show variable spinal cord repair potential. Umbilical cord blood MSCs (UCB-MSCs) had proven better results as

**Table 11.11** In vivo preclinical experimental mesenchymal stem cell studies on spinal cord and associated tissues

| Model type | Number of animals | Model defect size/study period | Study type | Biomaterial used | Cell dose | Evaluation criteria | Overall result | References |
|---|---|---|---|---|---|---|---|---|
| Spinal cord injury | 25 (5 per group) | Durotomy at L4/8 weeks | Comparative experimental study (recombinant methionyl human granulocyte colony-stimulating factor + UCB-MSCs Vs UCB-MSCs Vs saline treatment Vs control) | Allogeneic umbilical cord blood (UCB)-MSCs | $1 \times 10^6$ | The Olby score, magnetic resonance imaging, somatosensory-evoked potentials and histopathological examinations | Significant improvement in the nerve conduction velocity. Distinct structural consistency of the nerve cell bodies noted in the lesion of the spinal cord of the UCB-MSC and UCBG groups. | Lim et al. (2007) |
| Spinal cord injury | 10 (7 cell transplanted | 5 studied till end (between L2 and L3)/3 months to 1 year | Experimental study (cell treated vs no treatment) | Human umbilical cord derived MSCs | $1 \times 10^6$ or $2 \times 10^6$ | Behavioural assessment, MRI. Histology and immunohistochemistry | Remyelination of peripheral-type myelin sheaths | Lee et al. (2011a) |
| Spinal cord injury | 12 (4 per group) | Injury created at L1/8 weeks | Comparative experimental study (neural differentiated AD-MSC + Matrigel Vs Matrigel Vs control) | Neural differentiated AD-MSCs and Matrigel | $1 \times 10^7$ | Behavioural assessment, Histopathologic and immunofluorescence assessment, western blot | The combination of Matrigel NMSC produced beneficial effects in dogs with regard to functional recovery following SCI through enhancement of anti-inflammation, anti-astrogliosis, neuronal extension and neuronal regeneration effects | Park et al. 2012b) |

(continued)

**Table 11.11** (continued)

| Model type | Number of animals | Model defect size/study period | Study type | Biomaterial used | Cell dose | Evaluation criteria | Overall result | References |
|---|---|---|---|---|---|---|---|---|
| Spinal cord injury (hemilaminectomy, L4) (treated 7 days after injury) | 20 (control, Matrigel loaded, $n = 4$; Matrigel + BM-MSCs, $n = 4$; Matrigel + AD-MSCs, $n = 4$; Matrigel + WJ-MSCs, $n = 4$; Matrigel + UC-MSCs, $n = 4$) | 8 weeks | Experimental study | *BM-MSCs, AD-MSCs, WJ-MSCs, UC-MSCs, Matrigel* | $1 \times 10^5$ cells/μL (total 60 μL at cranial, epicentre and caudal lesions after 1 week of injury1 | Histopathological, immunofluorescence, quantitative analysis of inflammatory cytokines and surviving nerves | MSC groups showed significant improvements in locomotion. This recovery was accompanied by increased numbers of surviving neuron and neurofilament-positive fibres in the lesion site. Compared to the control, the lesion sizes were smaller, and fewer microglia and reactive astrocytes were found in the spinal cord epicentre of all MSC groups. Although there were no significant differences in functional recovery among the MSCs groups, UCB derived MSCs (UCSCs) induced more nerve regeneration and anti-inflammation activity ($P < 0.05$). Transplanted MSCs survived for 8 weeks and reduced IL-6 and COX-2 levels, which may have promoted neuronal regeneration in the spinal cord. | Ryu et al. (2012) |

| | | | | | | | | |
|---|---|---|---|---|---|---|---|---|
| Acute spinal cord injury | 16 | Hemilaminectomy (L1) | Comparative study (control, MPSS, AD-MSCs, and AD-MSCs + MPSS.) | Methylprednisolone sodium succinate (MPSS), AD-MSCs | $1 \times 10^7$ (3 doses intravenous) | Revised Tarlov scale, markers for oxidative metabolism (3-nitrotyrosine, 4-hydroxynonenal, and protein carbonyl) and inflammation (cyclooxygenase-2, interleukin-6, and tumour necrosis factor-α) | At 7 days post-treatment, hindlimb movement had improved in the AD-MSCs and AD-MSCs + MPSS groups. Hematoxylin and eosin staining revealed fewer haemorrhages and lesser microglial infiltration in the AD-MSCs group. Levels of 3-nitrotyrosine were decreased in the MPSS and AD-MSCs + MPSS groups; 4-hydroxynenonal and cyclooxygenase-2 levels were decreased in all treatment groups; and interleukin-6, tumour necrosis factor-α, and phosphorylated-signal transducer and activator transcription 3 levels were decreased in the AD-MSCs and AD-MSCs + MPSS groups | Kim et al. (2015) |

(continued)

**Table 11.11** (continued)

| Model type | Number of animals | Model defect size/study period | Study type | Biomaterial used | Cell dose | Evaluation criteria | Overall result | References |
|---|---|---|---|---|---|---|---|---|
| Acute spinal cord injury | 8 | 8 weeks | Comparative experimental study (AD-MSCs-Haemoxygenase-1 vs GFPAD-MSCs) | AD-MSCs (haemoxygenase-1 overexpression) or GFP AD-MSCs | $1 \times 10^7$ | Histopathology, immune-fluorescence, and western blot analyses | In AD- MSC-HO-1 group basso, Beattie, Bresnahan (BBB) score significantly improved at 7 weeks significantly higher motor recovery in the AD-MSC-HO-1 group and reduction of fibrotic changes as compared to GFPAD-MSC group | Lee et al. (2017c) |
| Intervertebral disc disease | 18 | Nucleotomy (L3/4, L4/5, and L5/6)/12 weeks | Experimental study (cell treated group Vs control) | Autologous BM-MSCs | $1 \times 10^6$ | Radiological, MRI, macroscopy, histology, immunohisto-chemistry, RT-PCR (FasL) | Regeneration of degenerated discs with expression of FasL gene in cell treated group | Hiyama et al. (2008) |
| Spontaneous degenerative lumbosacral IVD (L7-S1; decompressive surgery treated) | 20 (MSCs + microcarrier, $n = 11$; MSCs + TGF-β1 + microcarrier, $n = 6$; microcarrier only, n = 3) | 10 months | Prospective clinical study | Autologous BM-MSCs, collagen microcarrier | 3 × 106 cells/250 mg on collagen | MRI | In vivo injection was successful in all dogs, without any visible leakage, and clinical functioning was restored back to normality. Pfirrmann grade remained identical in all dogs, and formation of Schmorl's nodes was detected in 45% of dogs. This side effect was reduced by halving the injection volume, which was then observed only in 11% of dogs | Steffen et al. (2018) |

| | | | | | | | | |
|---|---|---|---|---|---|---|---|---|
| Sciatic nerve resection (a 35 mm long) | 20 (negative control, LOCC, hUC-MSC in combination with LOCC (LOCC/ hUC-MSC), and autograft (positive control) groups | 9 months | Human umbilical cord MSCs | $1 \times 10^6$, longitudinally oriented collagen conduit (LOCC) | Local implantation | Electrophysiological, electron microscopy, histological analysis | The LOCC and hUC-MSCs synergistically promoted regeneration and improved the functional recovery in a dog model of sciatic nerve injury | Cui et al. (2017) |

**Table 11.12** In vivo clinical mesenchymal stem cell studies on various neurological affections

| Clinical case | Number of animals included | Study period | Study type | Cell source | Cell dose | Route | Evaluation criteria | Overall result | References |
|---|---|---|---|---|---|---|---|---|---|
| Chronic spinal cord injury | 23 (10 BM-MSCs and 13 control) | 6 months to 35 months | Case series (uncontrolled, non-blinded clinical study) | Autologous BM-MSCs | $1.4 \times 10^6$–$5.6 \times 10^6$ | Subarachnoid space implantation between L5 and L6 | TSCIS system (involves evaluation of each pelvic limb separately and has 3 components: Gait, proprioceptive positioning and nociception) | Improvement in locomotor function without nociception in 6 cell treated dogs | Nishida et al. (2011) |
| Spinal cord injury | 1 | 180 days | Case report | Autologous bone marrow mononuclear cells (BMMNCs) | $20 \times 10^6$ & $4.16 \times 10^6$ BMMNCs + thermoreversible gelatin polymer | Local post at T12 left hemilaminectomy and intravenous implantation | Neurological examination, Olby scores | Functional recovery demonstrated with cell implantation | William et al. (2011) |
| Acute spinal cord injury | 85 (34 randomized subjects out of which 23 cell treated and 11 no cell treatment) | 6 months | Case series (controlled, randomized double-blinded trial) | Autologous olfactory mucosal stem cells | $5 \times 10^6$ cells | Intralesional | Gait analysis, somatosensory-evoked potential, Transcranial magnetic motor-evoked potentials, Urodynamics | Recipients of olfactory mucosal stem cell transplants gained significantly better forehand coordination than those dogs that did not received cells. Intraspinal olfactory mucosal cell transplantation improves communication across the damaged region of the injured spinal cord | Granger et al. (2012) |

| | | | | | | | | | |
|---|---|---|---|---|---|---|---|---|---|
| Spinal cord compression due to herniated intervertebral disks (T12 to L5 vertebrae) | 4 | 18 months | Case reports (uncontrolled clinical study) | Autologous BM-MSCs | $1 \times 10^6$ + collagen gelfoam scaffold | Local lesion implantation post-surgical decompression | Clinically, radiographically, and by nuclear magnetic resonance imaging (MRI) | Post MSCs implantation, a progressive recovery of the panniculus reflex and diminished superficial and deep pain response, although there were still low proprioceptive reflexes in addition to a hyperreflexia in the ataxic hind limb movement responses was demonstrated at day 10. Each dog demonstrated an improvement in these gains over time. Conscious reflex recovery occurred simultaneously with moderate improvement in intestine and urinary bladder functions in two of the four dogs. By the 18th month of clinical monitoring, a remarkable clinical amelioration accompanied by improved movement, in three of the four dogs was reported. However, no clinical gain was associated with alterations in magnetic resonance imaging. | Penha et al. (2014) |

(continued)

**Table 11.12** (continued)

| Clinical case | Number of animals included | Study period | Study type | Cell source | Cell dose | Route | Evaluation criteria | Overall result | References |
|---|---|---|---|---|---|---|---|---|---|
| Chronic spinal cord injury | 7 | 90 days | Case series (uncontrolled pilot study) | Allogeneic BM-MSCs | $1 \times 10^6$ | Intralesional and caudal to lesion implantation | MRI and Olby scale | All dogs improved locomotor and sensory function, tail tone was observed in seven dogs, pain reflexes and defecation return were observed in five dogs. | Sarmento et al. (2014) |
| Paraplegia | 13 | 12 months | Case series (uncontrolled clinical study) | Autologous neurogenically-induced BM-MSCs | $5.0 \times 10^6$ (2 doses after 21 days of interval post hemilaminectomy/ laminectomy | Percutaneous transplantation | Texas spinal cord injury scale (TSCIS), somatosensory-evoked potentials (SEP) and motor-evoked potentials (MEP) | Improvement in gait, nociception, proprioception, SEP and MEP results (2 cases), only gait score improvement (6 cases) and no improvement was recorded in (5 cases) | Besalti et al. (2015) |
| Chronic spinal injury/IVDD | 6 | 16 weeks | Case series (pilot clinical test) | Allogeneic AD-MSCs | $1 \times 10^7$ | Intralesional implantation | Functional evaluation of Locomotor activity | Only single dog regained ability to walk | Escalhão et al. (2017) |
| Meningo-encephalomyelitis of unknown origin (MUO) | 8 | 2 years | Clinical series | Autologous BM-MSC | $2 \times 10^6$, $4 \times 10^6$ (3 dogs) and $2 \times 10^6$, $0.5 \times 10^6$ (4 dogs) | Intrathecal (IT)/ right Intracarotid (IC) & intrathecal and intravenous | MRI, neurological examination and histopathology | All the dogs had presented early improvement in their general and neurological conditions, with particular effect on cervical pain. The group of dogs treated by IT+IA administration showed a shorter time of reaction to therapy compared to the group treated by IT+IC administration | Zeira et al. (2015) |

| | | | | | | | | | |
|---|---|---|---|---|---|---|---|---|---|
| Canine distemper | 8 (*n* = each group; group I: Cell + standard medicine; group II: Standard medicine) | 15 days | Case series (double-blind randomized trial) | Foetal olfactory epithelium stem cells | $1.0 \times 10^6$ /kg b. wt. | Intravenous | RT-PCR and laboratory investigations | MSC therapy resulted in no significant improvement when administered during the acute phase of canine distemper disease, and a prevalence of animals with high mortality rate was found in both groups due to the severity of symptoms | Pinheiro et al. (2016) |
| Canine distemper | 4 | 1 year | Case series | AD-MSCs | $1.0 \times 10^7$ (3 doses at monthly interval) | Intravenous | Neurological examination | During the first year after cellular therapy, the animals did not demonstrate significant changes in their locomotive abilities. However, the intense (grade V) myoclonus in three animals was reduced to a moderate (grade IV) level. At 1 year after the mesenchymal stem cell infusions, three animals regained functional ambulation (grade I), and all four dogs started to move independently (grades I and II). In two animals, the myoclonic severity had become mild (grade III) | Pinheiro et al. (2019) |

compared to BM-/AD-/WJ-MSCs. Among these cells BM-MSCs had lowest survival number (Ryu et al. 2012). Allogeneic or autologous BM-MSCs had improved the pelvic limb function scores in dogs. Olby scores, however, were higher in autologous cell implanted cases as compared to the allogeneic implanted ones (Jung et al. 2009). MSCs have mostly been implanted through local intrathecal route. Besides, MSCs percutaneous transplantation has also been made. Human UCB-MSCs applied through that route also had improved healing of the injured spinal cord (Lee et al. 2009b). dMSCs had been implanted at early (12 h), delayed (1 week) or late (2 weeks). The better results had been shown to occur by implanting the cells at a delay of 1 week. Peak inflammation at early injury period may hamper MSCs activity while more delayed cell implantation (2 weeks) may encounter already developed glial cell reaction (Park et al. 2011).

To further improve their effectiveness, various unorthodox options like MSCs genetic engineering, in vitro differentiated MSCs or their conditioned medium are being evaluated. In SCI, haemoxygenase-1 transfected dAD-MSCs had led to superior functional recovery as compared to the simple dAD-MSCs (Lee et al. 2017a). The haemoxygenase-1 has anti-oxidative and anti-inflammatory action, immune-modulation and might improve local microcirculation (Zhang et al. 2014). Combined local implantation and intravenous injection of brain derived nerve factor (BDNF) transfected MSCs had led to the significant functional recovery with improved histological features in SCI cases as compared to ones that received transfected dAD-MSCs locally (Lee et al. 2016). Contrarily, one of the studies had failed to demonstrate additive effects of BMP-2 and TGF-β1 on dBM-MSCs healing of nucleus pulposus (Bach et al. 2017). MSCs conditioned medium may also improve SCI repair (Al Delfi et al. 2016).

MSCs have also been demonstrated to improve peripheral nerve injury. MSCs loaded on chitosan/poly(lactic-co-glycolic acid) (PLGA)-based neural scaffold had led to the significant sciatic nerve regeneration and functional recovery in 50 mm sciatic nerve defect model. The healing had been comparable to that with nerve auto-grafting (Ding et al. 2010).

Disc regeneration is one of the important challenges. Degenerated disc repair may be improved with MSCs, although concentration has a role to play. One million cells had given better repair as compared to 0.1 million and ten million cells (Serigano et al. 2010). MSCs neural differentiation is currently questionable (Wu et al. 2003) but due to their trophic action nerve protection and repair may be enhanced (Ohta et al. 2004; Hokari et al. 2008; Saito et al. 2008; Park et al. 2011; Park et al. 2012a, b; Chung et al. 2013; Lim et al. 2017; Long et al. 2018).

### 11.3.3 Miscellaneous Studies

Miniscule number of MSCs studies related to various other ailments like diabetes (Zhu et al. 2011; Gautam et al. 2016), liver fibrosis (Matsuda et al. 2017) and acute kidney injury (Lee et al. 2017b) have also been undertaken as enlisted in Table 11.13. In a dog hepatocutaneous syndrome, multiple intrahepatic implantations of

**Table 11.13** Miscellaneous in vivo mesenchymal stem cell studies in dog

| Model type | Number of animals | Model defect size/study period | Study type | Biomaterial used | Cell dose | Evaluation criteria | Overall result | References |
|---|---|---|---|---|---|---|---|---|
| Diabetic model (alloxan and streptozotocin induced) | 30 dogs (6 in each group) | 16 weeks | Comparative experimental study (group 1: MSCs transfected with retrovirus carrying insulin and EGFP gene) Vs group 2: MSCs transfected with retrovirus carrying EGFP gene Vs group 3: MSCs without virus infection) Vs group 4: Cell-free group 5: Normal dogs) | Autologous BM-MSCs retrovirus carrying insulin and EGFP gene | 2.53106 / mL (total volume 20 mL injected | Analysis of C peptide, immunohistochemistry and immunocytochemistry, glucose tolerance test, measurement of blood glucose, insulin detection | Group 1 had insulin secretion through implanted cells; animals also responded to glucose challenge however β-cells were undetectable as no C peptide secretion was recorded. Rest of the groups had no control on glucose challenge. Glucose level was statistically controlled in group 1 compared to group 4 | Zhu et al. (2011) |

(continued)

**Table 11.13** (continued)

| Model type | Number of animals | Model defect size/study period | Study type | Biomaterial used | Cell dose | Evaluation criteria | Overall result | References |
|---|---|---|---|---|---|---|---|---|
| Dilated cardiomyopathy | 15 | 2 months | A pilot study | Stromal-derived factor-1 transduced AD-MSCs | Retrograde coronary venous delivery | Electrocardiograms (ECG), echocardiograms and Holter monitoring | Echocardiographic measurements of left ventricular size and function had shown continued progression of disease. Stem cell therapy in dogs with occult-dilated cardiomyopathy does not appear to offer advantage as compared to recently published survival data in similarly affected Doberman dogs | Pogue et al. (2013) |
| Acute kidney disease | 14 (3 in MSCs group, 8 in phosphate buffer saline treated group and 3 normal dogs | Bilateral renal corticomedullary/7 weeks junction induction of AKI | Experimental study (cell treated) Vs phosphate buffer saline Vs normal dogs | *UCB-MSCs* | $1 \times 10^6$ | Renal function and histology | Improved renal tubular function and moderate tubular necrosis in cell treated group compared to PBS group | Lee et al. (2017c) |

| | | | | | | | | |
|---|---|---|---|---|---|---|---|---|
| Canine liver fibrosis ($CCl_4$) | 16 (BM-MSCs injected, $n = 8$; control, $n = 8$) | 4 weeks | Experimental study (cell treated) Vs control | *Autologous BM-MSCs* | 4 x $10^5$/kg | CE-CT IMAGING, liver biopsy, histology, histomorphometry, DNA microarray analysis, PCR | BM-MSCs improved liver fibrosis without adverse effects. COL1A2, COL3, TIMP-1, EDNRA and EDNRB expression levels were significantly lower in the BMSC group compared with the control group. Alanine aminotransferase, albumin, and antithrombin 3 levels significantly improved in the BMSC group | Matsuda et al. (2017) |

allogeneic AD-MSCs had been made along with other managemental practices. An improvement in the clinical signs like hair regrowth and resolution of skin lesions was demonstrated. The dog, however, had died 32 months after diagnosis. The necropsy had revealed liver fibrosis and superficial necrolytic dermatitis (Nam et al. 2017). MSCs may also act drug (paclitaxel) carriers and can be used to target tumours like glioma and glioblastoma (Bonomi et al. 2017). Combined application of MSCs, recombinant human BMP-2 and cisplatin may prove useful in canine osteosarcoma (Rici et al. 2018). However, MSCs soluble factors may promote tumour cell proliferation and invasion (Teshima et al. 2018) and thus, studies are desired to confirm their safety, feasibility and actual utility.

## 11.4 Conclusion(S)

Numerous dog tissue sources harbour MSCs although concentration and properties may be variable. The cellular features also vary with age and health status of the donor. dMSCs have been evaluated for their therapeutic benefits on musculoskeletal and non-musculoskeletal tissue ailments. MSCs studies in dog tend to provide proof-of-therapeutic potential in human medicine. Due to the lack of understanding in their basic cellular properties, an initial step should be deliberated upon these properties. Further, case controlled uniform clinical trials need to be conducted to provide evidence based medicine for clinical conditions. Several issues like MSCs source, cell concentration, mode of delivery, cell survival on transplantation and frequency of stem cells need further deliberation and standardization.

## References

Abelson MB, Ousler GW, Maffei C (2009) Dry eye in 2008. Curr Opin Ophthalmol 20(4):282–286

Akizuki T, Odam S, Komaki M, Tsuchioka H, Kawakatsu N, Kikuchi A, Yamato M, Okano T, Ishikawa I (2005) Application of periodontal ligament cell sheet for periodontal regeneration: a pilot study in beagle dogs. J Periodontal Res 40:245–251

Al Delfi IR, Sheard JJ, Wood CR, Vernallis A, Innes JF, Myint P, Johnson WEB (2016) Canine mesenchymal stem cells are neurotrophic and angiogenic: an in vitro assessment of their paracrine activity. Vet J 217:10–17

Aly LA, El-Menoufy H, Sadeq HS, Ragae A, Sabry D (2014) Efficiency of systemic versus intralesional bone marrow derived stem cells in regeneration of oral mucosa after induction of formocresol induced ulcers in dogs. Dent Res J (Isfahan) 11(2):212–221

Arantes-Tsuzuki PDM, Mazzonetto PC, Lo Turco EG (2019) Treatment for canine corneal ulcer using adipose Tissuederived Mesenchymal stem cell therapy - Case report. Preprints 2019:120185

Arinzeh TL, Peter SJ, Archambault MP, van den Bos C, Gordon S, Kraus K, Smith A, Kadiyala S (2003) Allogeneic mesenchymal stem cells regenerate bone in a critical-sized canine segmental defect. J Bone Joint Surg Am 85-A:1927–1935

Ayala-Cuellar AP, Kim CW, Hwang KA, Kang JH, Lee G, Cho J, Choi KC (2019) Characterization of canine adipose tissue-derived mesenchymal stem cells immortalized by SV40-T retrovirus for therapeutic use. J Cell Physiol 234(9):16630–16642

Bach FC, Miranda-Bedate A, van Heel FW, Riemers FM, Müller MC, Creemers LB, Ito K, Benz K, Meij BP, Tryfonidou MA (2017) Bone morphogenetic Protein-2, but not Mesenchymal stromal cells, exert regenerative effects on canine and human nucleus Pulposus cells. Tissue Eng A 23(5–6):233–242

Bae S, Kim J, Li L, Lee A, Lim H, Jeong J, Lee SH, Kweon O-K, Kim WH (2015) Canine adipose-derived stem cell aggregates as A viable substitute to actual canine dermal papillae. Macedonian Vet Rev 38(2):167–173

Bahamondes F, Flores E, Cattaneo G, Bruna F, Conget P (2017) Omental adipose tissue is a more suitable source of canine Mesenchymal stem cells. BMC Vet Res 13:166

Baird AEG, Barsby T, Guest DJ (2015) Derivation of canine induced pluripotent stem cells. Reprod Domest Anim 50(4):669–676

Barabino S, Dana MR (2004) Animal models of dry eye: a critical assessment of opportunities and limitations. Invest Ophthalmol Vis Sci 45(6):1641–1646

Bartunek J, Croissant JD, Wijns W, Gofflot S, de Lavareille A, Vanderheyden M, Kaluzhny Y, Mazouz N, Willemsen P, Penicka M et al (2007) Pre-treatment of adult bone marrow mesenchymal stem cells with cardiomyogenic growth factors and repair of the chronically infarcted myocardium. Am J Physiol 292:H1095–H1104

Bçrziòð U, Matise-VanHoutana I, Pçtersone I, Dûrîtis I, Òikuïðins S, Bogdanova-Jâtniece A, Kâlis M, Svirskis Ð, Skrastiòa D, Ezerta A et al (2018) Characterisation and *In Vivo* safety of canine adipose-derived stem cells. Proc Latvian Acad Sci 72(3):160–171

Bearden RN, Huggins SS, Cummings KJ, Smith R, Gregory CA, Saunders WB (2017) In-vitro characterization of canine multipotent stromal cells isolated from synovium, bone marrow, and adipose tissue: a donor-matched comparative study. Stem Cell Res Ther 8:218

Bertolo A, Steffen F, Malonzo-Marty C, Stoyanov J (2015) Canine Mesenchymal stem cell potential and the importance of dog breed: implication for cell-based therapies. Cell Transplant 24:1969–1980

Besalti O, Can P, Akpinar E, Aktas Z, Elcin AE, Elcin YM (2015) Intraspinal transplantation of autologous neurogenically-induced bone marrow-derived mesenchymal stem cells in the treatment of paraplegic dogs without deep pain perception secondary to intervertebral disk disease. Turk Neurosurg 25:625–632

Bhat IA, Sivanarayanan TB, Somal A, Pandey S, Bharti MK, Panda BSK, Indu B, Verma M, Anand J, Sonwane A et al (2019) An allogenic therapeutic strategy for canine spinal cord injury using mesenchymal stem cells. J Cell Physiol 234(3):2705–2718

Bittencourt MKW, Barros MA, Martins JFP, Vasconcellos JPC, Morais BP, Pompeia C, Bittencourt MD, Evangelho KS, Kerkis I, Wenceslau CV (2016) Allogeneic Mesenchymal stem cell transplantation in dogs with Keratoconjunctivitis Sicca. Cell Med 8:63–77

Black LL, Gaynor J, Gahring D, Adams C, Aron D, Harman S, Gingerich DA, Harman R (2007) Effect of adipose-derived mesenchymal stem and regenerative cells on lameness in dogs with chronic osteoarthritis of the coxofemoral joints: a randomized, double-blinded, multicenter, controlled trial. Vet Ther 8:272–284

Black LL, Gaynor J, Adams C, Dhupa S, Sams AE, Taylor R, Harman S, Gingerich DA, Harman R (2008) Effect of intraarticular injection of autologous adipose-derived mesenchymal stem and regenerative cells on clinical signs of chronic osteoarthritis of the elbow joint in dogs. Vet Ther 9:192–200

Blecker D, Elashry MI, Heimann M, Wenisch S, Arnhold S (2017) New insights into the neural differentiation potential of canine adipose tissue-derived Mesenchymal stem cells. Anatomia Histologia Embryologia 46(3):304–315

Bonomi A, Ghezzi E, Pascucci L, Aralla M, Ceserani V, Pettinari L, Coccè V, Guercio A, Alessandri G, Parati E et al (2017) Effect of canine mesenchymal stromal cells loaded with paclitaxel on growth of canine glioma and human glioblastoma cell lines. Vet J 223:41–47

Brown SG, Harman RJ, Black LL (2012) Adipose-derived stem cell therapy for severe muscle tears in working German shepherds: two case reports. Stem Cell Discov 2(2):41–44

Bruder SP, Kraus KH, Goldberg VM, Kadiyala S (1998) The effect of implants loaded with autologous mesenchymal stem cells on the healing of canine segmental bone defects. J Bone Joint Surg Am 80:985–996

Bugueño J, Li W, Salat P, Qin L, Akintoye SO (2017) Canine mesenchymal stem cell bone regenerative capacity is regulated by sitespecific multi-lineage differentiation. Oral Surg Oral Med Oral Pathol Oral Radiol 123(2):163–172

Cabon Q, Febre M, Gomez N, Cachon T, Pillard P, Carozzo C, Saulnier N, Robert C, Livet V, Rakic R, Plantier N, Saas P, Maddens S, Viguier E (2019) Long-term safety and efficacy of single or repeated intra-articular injection of allogeneic neonatal Mesenchymal stromal cells for managing pain and lameness in moderate to severe canine osteoarthritis without anti-inflammatory pharmacological support: pilot clinical study. Front Vet Sci 6:10

Canapp SO Jr, Leasure CS, Cox C, Ibrahim V, Carr BJ (2016) Partial cranial cruciate ligament tears treated with stem cell and platelet-rich plasma combination therapy in 36 dogs: A retrospective study. Front Vet Sci 3:112

Case JB, Palmer R, Valdes-Martinez A, Egger EL, Haussler KK (2013) Gastrocnemius tendon strain in a dog treated with autologous mesenchymal stem cells and a custom orthosis. Vet Surg 42:355–360

Ceusters J, Lejeune J-P, Sandersen C, Niesten A, Lagneaux L, Serteyn D (2017) From skeletal muscle to stem cells: an innovative and minimally invasive process for multiple species. Sci Rep 7:696

Chow L, Johnson V, Regan D, Wheat W, Webb S, Koch P, Dow S (2017) Safety and immune regulatory properties of canine induced pluripotent stem cell derived mesenchymal stem cells. Stem Cell Res 25:221–232

Chung C-S, Fujita N, Kawahara N, Yui S, Nam E, Nishimura R (2013) A comparison of neurosphere differentiation potential of canine bone marrow-derived mesenchymal stem cells and adipose-derived mesenchymal stem cells. J Vet Med Sci 75:879–886

Csaki C, Matis U, Mobasheri A, Ye H, Shakibaei M (2007) Chondrogenesis, osteogenesis and adipogenesis of canine mesenchymal stem cells: a biochemical, morphological and ultrastructural study. Histochem Cell Biol 128(6):507–520

Cuervo B, Rubio M, Sopena J, Dominguez JM, Vilar J, Morales M, Cugat R, Carrillo JM (2014) Hip osteoarthritis in dogs: A randomized study using Mesenchymal stem cells from adipose tissue and plasma rich in growth factors. Int J Mol Sci 15:13437–13460

Cui L, Liu B, Liu G, Zhang W, Cen L, Sun J, Yin S, Liu W, Cao Y (2017) Repair of cranial bone defects with adipose derived stem cells and coral scaffold in a canine model. Biomaterials 28(36):5477–5486

Daems R, Van Hecke L, Schwarzkopf I, Depuydt E, Broeckx SY, David M, Beerts C, Vandekerckhove P, Spaas JH (2019) A feasibility study on the use of equine Chondrogenic induced Mesenchymal stem cells as a treatment for natural occurring osteoarthritis in dogs. Stem Cells Int 2019:4587594

Day M (1993) Immunopathology of anal furunculosis in the dog. J Small Anim Pract 34(8):381–389

de Bakker E, Van Ryssen B, De Schauwer C, Meyer E (2013) Canine mesenchymal stem cells: state of the art, perspectives as therapy for dogs and as a model for man. Vet Q 33(4):225–233

De Cesaris V, Grolli S, Bresciani C, Conti V, Basini G, Parmigiani E, Bigliardi E (2017) Isolation, proliferation and characterization of endometrial canine stem cells. Reprod Domest Anim 52(2):235–242

Ding F, Wu J, Yang Y, Hu W, Zhu Q, Tang X, Liu J, Gu X (2010) Use of tissue-engineered nerve grafts consisting of a chitosan/poly(lactic-co-glycolic acid)- based scaffold included with bone marrow mesenchymal cells for bridging 50-mm dog sciatic nerve gaps. Tissue Eng A 16:3779–3790

Dominici M, Le Blanc K, Mueller I, Slaper-Cortenbach I, Marini F, Krause D, Deans R, Keating A, Prockop D, Horwitz E (2006) Minimal criteria for defining multipotent mesenchymal stromal cells. The International Society for Cellular Therapy position statement. Cytotherapy 8(4):315–317 dos Santos CR, da Rocha Filgueiras R, Malard PF, da Cunha Barreto-Vianna AR, Nogueira K, da Silva Leite C, Mendes de Lima EM (2018) Mesenchymal stem cells in osteotomy repair after tibial tuberosity advancement in dogs with cranial cruciate ligament injury. J Exp Orthopaed 5:17

Dražilov SS, Mrkovački J, Spasovski V, Fazlagić A, Pavlović S, Nikčević G (2018) The use of canine Mesenchymal stem cells for the autologous treatment of osteoarthritis. Acta Vet Hung 66(3):376–389

El-Menoufy H, Aly LA, Aziz MT, Atta HM, Roshdy NK, Rashed LA, Sabry D (2010) The role of bone marrow-derived mesenchymal stem cells in treating formocresol induced oral ulcers in dogs. J Oral Pathol Med 39:281–289

El-Tookhy OS, Shamaa AA, Shehab GG, Abdallah AN, Azzam OM (2017) Histological evaluation of experimentally induced critical size defect skin wounds using Exosomal solution of Mesenchymal stem cells derived microvesicles. Inter J Stem cells 10(2):144–153

Enciso N, Ostronoff LLK, Mejías G, León LG, Fermín ML, Merino E, Fragio C, Avedillo L, Tejero C (2018) Stem cell factor supports migration in canine mesenchymal stem cells. Vet Res Commun 42(1):29–38

Escalhão CCM, Ramos IP, Hochman-Mendez C, Brunswick THK, Souza SAL, Gutfilen B, Goldenberg RCS, Coelho-Sampaio T (2017) Safety of allogeneic canine adipose tissue-derived Mesenchymal stem cell intraspinal transplantation in dogs with chronic spinal cord injury. Stem Cells Int 2017:3053759

Esposito M, Grusovin MG, Papanikolaou N, Coulthard P, Worthington HV (2009) Enamel matrix derivative (Emdogain(R)) for periodontal tissue regeneration in intrabony defects. Cochrane database. Syst Rev 2009:CD003875

Ferrer L, Kimbrel EA, Lam A, Falk EB, Zewe C, Juopperi T, Lanza R, Hoffman A (2016) Treatment of perianal fistulas with human embryonic stem cell-derived mesenchymal stem cells: a canine model of human fistulizing Crohn's disease. Regen Med 11(1):33–43

Fluehmann G, Doherr MG, Jaggy A (2006) Canine neurological diseases in a referral hospital population between 1989 and 2000 in Switzerland. J Small Anim Pract 47:582–587

Gautam P, Foale RD, Recino A, Zhao J, Gan SU, Wallberg M, Calne R, Lever AML (2016) Promoter optimisation of Lentiviral vectors for efficient insulin gene expression in canine Mesenchymal stromal cells: potential surrogate Beta cells. J Gene Med 18(10):312–321

Gelberman RH, Linderman SW, Jayaram R, Dikina AD, Sakiyama-Elbert H, Alsberg E, Thomopoulos S, Shen H (2017) Combined administration of ASCs and BMP-12 promotes an M2 macrophage phenotype and enhances tendon healing. Clin Orthopaed Relat Res 475(9):2318–2331

Gibson MA, Brown SG, Brown NO (2017) Semitendinosus myopathy and treatment with adipose-derived stem cells in working German shepherd police dogs. Can Vet J 58:241–246

Granger N, Blamires H, Franklin RJM, Jeffery ND (2012) Autologous olfactory mucosal cell transplants in clinical spinal cord injury: a randomized double-blinded trial in a canine translational model. Brain 135:3227–3237

Guercio A, Di Marco P, Casella S, Cannella V, Russotto L, Purpari G, Di Bella S, Piccione G (2012) Production of canine mesenchymal stem cells from adipose tissue and their application in dogs with chronic osteoarthritis of the humeroradial joints. Cell Biol Int 36:189–194

Gugjoo MB, Amarpal, Sharma GT, Kinjavdekar P, Aithal HP, Pawde AM (2016) Cartilage tissue engineering: role of mesenchymal stem cells along with growth factors and scaffolds. Indian J Med Res 144(3):339–347

Gugjoo MB, Amarpal, Abdelbaset-Ismail A, Aithal HP, Kinjavdekar P, Pawde AM, Kumar GS, Sharma GT (2017) Mesenchymal stem cells with IGF-1 and TGF- β1 in laminin gel for osteochondral defects in rabbits. Biomed Pharmacother 93:1165–1174

Gugjoo MB, Amarpal A, Sharma GT (2019a) Mesenchymal stem cell basic research and applications in dog medicine. J Cell Physiol 234(10):16779–16811

Gugjoo MB, Amarpal, Makhdoomi DM, Sharma GT (2019b) Equine mesenchyaml stem cells: properties, sources, characterization and potential therapeutic applications. J Equine Vet Sci 72:16–27

Guo P, Zeng J-J, Zhou N (2015) A novel experimental study on the fabrication and biological characteristics of canine bone marrow Mesenchymal stem cells sheet using vitamin C. Scanning 9999:1–7

Hall MN, Rosenkrantz WS, Hong JH, Griffin CE, Mendelsohn CM (2010) Evaluation of the potential use of adipose-derived mesenchymal stromal cells in the treatment of canine atopic dermatitis: a pilot study. Vet Ther 11:E1–E14

Hang D, Wang Q, Guo C, Chen Z, Yan Z (2012) Treatment of osteonecrosis of the femoral head with VEGF165 transgenic bone marrow mesenchymal stem cells in mongrel dogs. Cells Tissues Organs 195:495–506

Harman R, Carlson K, Gaynor J, Gustafson S, Dhupa S, Clement K, Hoelzler M, McCarthy T, Schwartz P, Adams C (2016) A prospective, randomized, masked, and placebo-controlled efficacy study of Intraarticular allogeneic adipose stem cells for the treatment of osteoarthritis in dogs. Front Vet Sci 3:81

Hiyama A, Mochida J, Iwashina T, Omi H, Watanabe T, Serigano K, Tamaru F, Sakai D (2008) Transplantation of mesenchymal stem cells in a canine disc degeneration model. J Orthop Res 26:589–600

Hodgkiss-Geere HM, Argyle DJ, Corcoran BM, Whitelaw B, Milne E, Bennett D, Argyle SA (2012) Characterisation and differentiation potential of bone marrow derived canine mesenchymal stem cells. Vet J 194:361–368

Hokari M, Kuroda S, Shichinohe H, Yano S, Hida K, Iwasaki Y (2008) Bone marrow stromal cells protect and repair damaged neurons through multiple mechanisms. J Neurosci Res 86:1024–1035

Hu R, Ling W, Xu W, Han D (2014) Fibroblast-like cells differentiated from adipose-derived mesenchymal stem cells for vocal fold wound healing. PLoS One 9(3):e92676

Huss R, Claudia L, Weissinger EM, Kolb HJ, Thalmeiler K (2000) Evidence of peripheral blood-derived, plastic-adherent $CD34^{-/low}$ hematopoietics stem cell clones with mesenchymal stem cell characteristics. Stem Cells 18:252–260

Iravani K, Sobhanmanesh A, Ashraf MJ, Hashemi SB, Mehrabani D, Zare S (2017) The healing effect of conditioned media and bone marrow-derived stem cells in laryngotracheal stenosis: a comparison in experimental dog model. World J Plast Surg 6(2):190–197

Itoi T, Harada Y, Irie H, Sakamoto M, Tamura K, Yogo T, Soeta S, Amasaki H, Hara Y, Tagawa M (2016) Escherichia coli-derived recombinant human bone morphogenetic protein-2 combined with bone marrow-derived mesenchymal stromal cells improves bone regeneration in canine segmental ulnar defects. BMC Vet Res 12:201

Jang BJ, Byeon YE, Lim JH, Ryu HH, Kim WH, Koyama Y, Kikuchi M, Kang KS, Kweon OK (2008) Implantation of canine umbilical cord blood-derived mesenchymal stem cells mixed with beta-tricalcium phosphate enhances osteogenesis in bone defect model dogs. J Vet Sci 9:387–393

Jeon YK, Jang YH, Yoo DR, Kim SN, Lee SK, Nam MJ (2010) Mesenchymal stem cells' interaction with skin: wound-healing effect on fibroblast cells and skin tissue. Wound Repair Regen 18:655–661

Jo K, Kim Y, Lee SH, Yoon YS, Kim WH, Kweon O-K (2017) Effect of canine cortical bone demineralization on osteogenic differentiation of adipose- derived mesenchymal stromal cells. Heliyon 3(8):e00383

Johnson JA, Austin C, Breur GJ (1994) Incidence of canine appendicular musculo-skeletal disorders in 16 veterinary teaching hospitals from 1980 through 1989. Vet Comp Orthop Traumatol 7:56–69

Johnston SA (1997) Osteoarthritis. Joint anatomy, physiology, and pathobiology. Vet Clin North Am Small Anim Pract 27:699–723

Jung DI, Ha J, Kang BT, Kim JW, Quan FS, Lee JH, Woo EJ, Park HM (2009) A comparison of autologous and allogenic bone marrow-derived mesenchymal stem cell transplantation in canine spinal cord injury. J Neurol Sci 285:67–77

Kafarnik C, McClellan A, Dziasko M, Daniels JT, Guest DJ (2020) Canine corneal stromal cells have multipotent mesenchymal stromal cell properties in vitro. Stem Cells Dev. https://doi.org/10.1089/scd.2019.0163

Kamishina H, Deng J, Oji T, Cheeseman JA, Clemmons RM (2006) Expression of neural markers on bone-marrow–derived canine mesenchymal stem cells. Am J Vet Res 67:1921–1928

Kanemaru S, Nakamura T, Omori K, Kojima H, Magrufov A, Hiratsuka Y, Hirano S, Ito J, Shimizu Y (2003) Regeneration of the vocal fold using autologous mesenchymal stem cells. Ann Oto Rhinol Laryngol 112:915–920

Kang MH, Park HM (2014) Evaluation of adverse reactions in dogs following intravenous mesenchymal stem cell transplantation. Acta Vet Scand 56:16

Kang JW, Kang K-S, Koo HC, Park JR, Choi EW, Park YH (2008) Soluble factors-mediated immunomodulatory effects of canine adipose tissue-derived mesenchymal stem cells. Stem Cells Develop 17(4):681–693

Kang BJ, Ryu HH, Park SS, Koyama Y, Kikuchi M, Woo HM, Kim WH, Kweon OK (2012) Comparing the osteogenic potential of canine mesenchymal stem cells derived from adipose tissues, bone marrow, umbilical cord blood, and Wharton's jelly for treating bone defects. J Vet Sci 13:299–310

Kazemi D, Asenjan S, Dehdilani K, Parsa H (2017) Canine articular cartilage regeneration using mesenchymal stem cells seeded on platelet rich fibrin. Bone Joint Res 6:98–107

Khorsand A, Eslaminejad MB, Arabsolghar M, Paknejad M, Ghaedi B, Rokn AR, Moslemi N, Nazarian H, Jahangir S (2013) Autologous dental pulp stem cells in regeneration of defect created in canine periodontal tissue. J of Oral Implantol 39(4):433–443

Kim YK, Lee SY, Park SJ, Lee SS, Kim JH, Lee HC (2011) Clinical application of mesenchymal stem cells in a dog with intervertebral disc disease. J Vet Clin 28:122–127

Kim S, Zhang H, Guo L, Kim J, Kim MH (2012) Amniotic mesenchymal stem cells enhance wound healing in diabetic NOD/SCID mice through high angiogenic and engraftment capabilities. PLoS One 7:1–11

Kim JW, Lee JH, Lyoo YS, Jung DI, Park HM (2013) The effects of topical mesenchymal stem cell transplantation in canine experimental cutaneous wounds. Vet Dermatol 24:242–e53

Kim Y, Jo S-H, Kim WH, Kweon O-K (2015) Antioxidant and anti-inflammatory effects of intravenously injected adipose derived mesenchymal stem cells in dogs with acute spinal cord injury. Stem Cell Res Ther 6:229

Kisiel AH, McDuffee LA, Masaoud E, Bailey TR, Esparza Gonzalez BP, Nino-Fong R (2012) Isolation, characterization, and in vitro proliferation of canine mesenchymal stem cells derived from bone marrow, adipose tissue, muscle, and periosteum. Am J Vet Res 73(8):1305–1317

Koh E, Jung YC, Woo H-M, Kang B-J (2017) Injectable alginate-microencapsulated canine adipose tissue-derived mesenchymal stem cells for enhanced viable cell retention. J Vet Med Sci 79(3):492–501

Kresic N, Simic I, Lojkic I, Bedekovic T (2017) Canine adipose derived mesenchymal stem cells transcriptome composition alterations: A step towards standardizing therapeutic. Stem Cells Int 2017:12. Art ID:4176292

Kriston-Pál É, Czibula Á, Gyuris Z, Balka G, Seregi A, Sükösd F, Süth M, Kiss-Tóth E, Haracska L, Uher F, Monostori É (2017) Characterization and therapeutic application of canine adipose mesenchymal stem cells to treat elbow osteoarthritis. Can J Vet Res 81:73–78

Kumai Y, Kobler JB, Park H, Galindo M, Herrera VL, Zeitels SM (2010) Modulation of vocal fold scar fibroblasts by adipose-derived stem/stromal cells. Laryngoscope 120:330–337

Laura C, Groza I, Oana L, Pall E, Pestean C, Cătană R, Cenariu M (2008) Canine Mesenchymal stem cells isolation from bone marrow aspirates. Bull UASVM Vet Med 65(2):96–101

Lee BJ, Wang SG, Lee JC, Jung JS, Bae YC, Jeong HJ, Kim HW, Lorenz RR (2006) The prevention of vocal fold scarring using autologous adipose tissue-derived stromal cells. Cells Tissue Organs 184:184–204

Lee HB, Chung YS, Heo SY, Kim NS (2009a) Augmentation of bone healing of nonunion fracture using stem cell based tissue engineering in a dog: a case report. Vet Med 54(4):198–203

Lee JH, Chang HS, Kang EH, Chung DJ, Choi CB, Lee JH, Hwang SH, Han H, Kim HY (2009b) Percutaneous transplantation of human umbilical cord blood-derived multipotent stem cells in a canine model of spinal cord injury. J Neurosurg Spine 11(6):749–757

Lee WS, Suzuki Y, Graves SS, Iwata M, Venkataraman GM, Mielcarek M, Peterson LJ, Ikehara S, Torok-Storb B, Storb R (2011a) Canine bone-marrow–derived mesenchymal stromal cells suppress alloreactive lymphocyte proliferation in vitro but fail to enhance engraftment in canine bone marrow transplantation. Biol Blood Marrow Transplant 17:465–475

Lee JH, Chung WH, Kang EH, Chung DJ, Choi CB, Chang HS, Lee JH, Hwang SH, Han H, Choe BY et al (2011b) Schwann cell-like remyelination following transplantation of human umbilical cord blood (hUCB)-derived mesenchymal stem cells in dogs with acute spinal cord injury. J Neurol Sci 300:86–96

Lee JC, Mitchell AWM, Healy JC (2012) Imaging of muscle injuries. Br J Radiol 85(1016):1173–1185

Lee KS, Cha SH, Kang HW, Song JY, Lee KW, Ko KB, Lee HT (2013) Effects of serial passage on the characteristics and chondrogenic differentiation of canine umbilical cord matrix derived mesenchymal stem cells. Asian Australas J Anim Sci 26(4):588595

Lee SH, Kim Y, Rhew D, Kim A, Jo KR, Yoon Y, Choi KU, Jung T, Kim WH, Kweon OK (2016) Impact of local injection of brain-derived neurotrophic factor-expressing mesenchymal stromal cells (MSCs) combined with intravenous MSC delivery in a canine model of chronic spinal cord injury. Cytotherapy S1465-3249(16):30540

Lee SH, Kim Y, Rhew D, Kim A, Jo KR, Yoon Y, Choi KU, Jung T, Kim WH, Kweon O-K (2017a) Effect of canine mesenchymal stromal cells overexpressing heme oxygenase-1 in spinal cord injury. J Vet Sci 18(3):377–386

Lee S-J, Ryu M-O, Seo M-S, Park S-B, Ahn J-O, Han S-M, Kang K-S, Bhang D-H, Youn H-Y (2017b) Mesenchymal stem cells contribute to improvement of renal function in a canine kidney injury model. In Vivo 31(6):1115–1124

Lee J, Lee KS, Kim C-L, Byeon JS, Gu N-Y, Cho I-S, Cha S-H (2017c) Effect of donor age on the proliferation and multipotency of canine adipose-derived mesenchymal stem cells. J Vet Sci 18(2):141–148

Lemp MA (2008) Management of dry eye disease. Am J Manag Care 14(3):S88–S101

Li H, Yan F, Lei L, Li Y, Xiao Y (2009) Application of autologous cryopreserved bone marrow mesenchymal stem cells for periodontal regeneration in dogs. Cells Tissues Organs 190:94–101

Li L, Duan X, Fan Z, Chen L, Xing F, Xu Z, Chen Q, Xiang Z (2018) Mesenchymal stem cells in combination with hyaluronic acid for articular cartilage defects. Sci Rep 8:9900

Lim JH, Byeon YE, Ryu HH, Jeong YH, Lee YW, Kim WH, Kang K-S, Kweon O-K (2007) Transplantation of canine umbilical cord blood-derived mesenchymal stem cells in experimentally induced spinal cord injured dogs. J Vet Sci 8:275–282

Lim J-H, Koh S, Thomas R, Breen M, Olby NJ (2017) Evaluation of gene expression and DNA copy number profiles of adipose tissue–derived stromal cells and consecutive neurosphere-like cells generated from dogs with naturally occurring spinal cord injury. Am J Vet Res 78:371–380

Linke A, Ller PM, Nurzynska D, Casarsa C, Torella D, Nascimbene A, Castaldo C, Cascapera S, Böhm M, Quaini F et al (2005) Stem cells in the dog heart are self-renewing, clonogenic, and multipotent and regenerate infracted myocardium, improving cardiac function. PNAS 102(25):8966–8971

Linon E, Spreng D, Rytz U, Forterre S (2014) Engraftment of autologous bone marrow cells into the injured cranial cruciate ligament in dogs. Vet J 202(3):448–454

Long C, Lankford L, Kumar P, Grahn R, Borjesson DL, Farmer D, Wang A (2018) Isolation and characterization of canine placenta-derived Mesenchymal stromal cells for the treatment of neurological disorders in dogs. Cytometry A 93(1):82–92

Lorant J, Larcher T, Jaulin N, Hedan B, Lardenois A et al (2018) Vascular delivery of allogeneic MuStem cells in dystrophic dogs requires only short-term immunosuppression to avoid host immunity and generate clinical/tissue benefits. Cell Transplant Cognizant Commun Corporat 27(7):1096–1110

Lu W, Yaoming N, Boli R, Jun C, Changhai Z, Yang Z, Zhiyuan S (2013) mHCN4 genetically modified canine Mesenchymal stem cells provide biological Pacemaking function in complete dogs with Atrioventricular block. PACE 36:1138–1149

Madhu DN, Ahmad RA, Amarpal, Gugjoo MB, Pawde AM, Kinjavedkar P, Aithal HP, Ansari M, Chandra V, Sharma GT (2014) Caudal superficial epigastric axial pattern flap and stem cell therapy for the management of large wound on medial aspect of thigh in a dog. Adv Anim Vet Sci 2(3):188–191

Malagola E, Teunissen M, van der Laan LJ, Verstegen MM, Schotanus BA, van Steenbeek FG, Penning LC, van Wolferen ME, Tryfonidou MA, Spee B (2016) Stem characterization and comparison of canine multipotent stromal cells derived from liver and bone marrow. Stem Cells Develop 25(2):139–150

Martinello T, Bronzini I, Maccatrozzo L, Mollo A, Sampaolesi M, Mascarello F, Decaminada M, Patruno M (2011) Canine adipose-derived-mesenchymal stem cells do not lose stem features after a long-term cryopreservation. Res Vet Sci 91(1):18–24

Marx C, Silveira MD, Selbach I, da Silva AS, Gomes de Macedo Braga LM, Camassola M, Nardi NB (2014) Acupoint injection of autologous stromal vascular fraction and allogeneic adipose-derived stem cells to treat hip dysplasia in dogs. Stem Cells Int 2014:391274

Marycz K, Smieszek A, Grzesiak J, Nicpoń JE (2014) Effects of steroids on the morphology and proliferation of canine and equine mesenchymal stem cells of adipose origin - in vitro research. Acta Vet Hung 62(3):317–333

Matsuda T, Takami T, Sasaki R, Nishimura T, Aibe Diaz Paredes B, Quintanilha LF, Matsumoto T, Ishikawa T, Yamamoto N, Tani K et al (2017) A canine liver fibrosis model to develop a therapy for liver cirrhosis using cultured bone marrow–derived cells. Hepatol Commun 1(7):691–703

Memon IA, Sawa Y, Miyagawa S, Taketani S, Matsuda H (2005) Combined autologous cellular cardiomyoplasty with skeletal myoblasts and bone marrow cells in canine hearts for ischemic cardiomyopathy. Journal of Thoracic & Cardiovascular Surgery 130:646–653

Millis DL, Jackson AM (2003) Delayed union, nonunion, and malunions. 1849–1861. In: Slatter D (ed) Textbook of small animal surgery, 3rd edn. W.R, Saunders, Philadelphia, p 2714

Mokbel A, El-Tookhy O, Shamaa AA, Sabry D, Rashed L, Mostafa A (2011) Homing and efficacy of intra-articular injection of autologous mesenchymal stem cells in experimental chondral defects in dogs. Clin Exp Rheumatol 29:275–284

Muir P, Hans EC, Racette M, Volstad N, Sample SJ, Heaton C, Holzman G, Schaefer SL, Bloom DD, Bleedorn JA et al (2016) Autologous bone marrow-derived Mesenchymal stem cells modulate molecular markers of inflammation in dogs with cruciate ligament rupture. PLoS One 11(8):e0159095

Nam A, Han S-M, Go D-M, Kim D-Y, Seo K-W, Youn H-Y (2017) Long-term management with adipose tissue-derived mesenchymal stem cells and conventional treatment in a dog with Hepatocutaneous syndrome. J Vet Intern Med 31:1514–1519

Neupane M, Chang C-C, Kiupel M, Yuzbasiyan-Gurkan V (2008) Isolation and characterization of canine adipose–derived mesenchymal stem cells. Tissue Eng A 14(6):1007–1015

Nishida H, Nakayama M, Tanaka H, Kitamura M, Hatoya S, Sugiura K, Suzuki Y, Ide C, Inaba T (2011) Evaluation of transplantation of autologous bone marrow stromal cells into the cerebrospinal fluid for treatment of chronic spinal cord injury in dogs. Am J Vet Res 72:1118–1123

Oh HJ, Park JE, Kim MJ, Hong SG, Ra JC, Jo JY, Kang SK, Jang G, Lee BC (2011) Recloned dogs derived from adipose stem cells of a transgenic cloned beagle. Theriogenology 75:1221–1231

Ohno S, Hirano S, Kanemaru S, Kitani Y, Kojima T, Tateya I, Nakamura T, Ito J (2011) Implantation of an atelocollagen sponge with autologous bone marrow- derived Mesenchymal stromal cells for treatment of vocal fold scarring in a canine model. Ann OtoRhinol Laryngol 120:401–408

Ohta M, Suzuki Y, Noda T, Ejiri Y, Dezawa M, Kataoka K, Chou H, Ishikawa N, Matsumoto N, Iwashita Y et al (2004) Bone marrow stromal cells infused into the cerebrospinal fluid promote functional recovery of the injured rat spinal cord with reduced cavity formation. Exp Neurol 187:266–278

Omae H, Zhao C, Sun YL (2009) Multilayer tendon slices seeded with bone marrow stromal cells: a novel composite for tendon engineering. J Orthop Res 27:937–942

Park SS, Byeon YE, Ryu HH, Kang BJ, Kim YS, Kim W-H, Kang K-S, Han H-J, Kweon O-K (2011) Comparison of canine umbilical cord blood derived mesenchymal stem cell transplantation times: involvement of astrogliosis, inflammation, intracellular actin cytoskeletal pathways, and neurotrophin-3. Cell Transplant 20:1867–1880

Park SS, Lee YJ, Lee SH, Lee D, Choi K, Kim WH, Kweon OK, Han HJ (2012a) Functional recovery after spinal cord injury in dogs treated with a combination of matrigel and neural-induced adipose-derived mesenchymal stem cells. Cytotherapy 14:584–597

Park S-B, Seo M-S, Kim H-S, Kang K-S (2012b) Isolation and characterization of canine amniotic membrane-derived multipotent stem cells. PLoS One 7(9):e44693

Park SE, Jung NY, Lee NK, Lee J, Hyung B, Myeong SH, Kim HS, Suh YL, Lee JI, Cho KR et al (2016) Distribution of human umbilical cord blood-derived mesenchymal stem cells (hUCB-MSCs) in canines after intracerebroventricular injection. Neurobiol Aging 47:192–200

Penha EM, Meira CS, Guimarães ET, Mendonça MVP, Gravely FA, Pinheiro CMB, Pinheiro TMB, Barrouin-Melo SM, Ribeiro-dos-Santos R, Soares MBP (2014) Use of autologous Mesenchymal stem cells derived from bone marrow for the treatment of naturally injured spinal cord in dogs. Stem Cells Int 2014:437521

Pérez-Merino EM, Uson-Casaus JM, Duque-Carrasco J, Zaragoza-Bayle C, Marinas-Pardo L, Hermida-Prieto M, Vilafrnca-Compte M, Barrera-Chacon R, Gualtieri M (2015a) Safety and efficacy of allogeneic adipose tissue-derived mesenchymal stem cells for treatment of dogs with inflammatory bowel disease: endoscopic and histological outcomes. Vet J 206:391–397

Pérez-Merino EM, Uson-Casaus JM, Zaragoza-Bayle C, Duque-Carrasco J, Marinas-Pardo L, Hermida-Prieto M, Barrera-Chacon R, Gualtieri M (2015b) Safety and efficacy of allogeneic adipose tissue-derived mesenchymal stem cells for treatment of dogs with inflammatory bowel disease: clinical and laboratory outcomes. Vet J 206:385–390

Perin EC, Silva GV, Assad JA, Vela D, Buja LM, Sousa AL, Litovsky S, Lin J, Vaughn WK, Coulter S, Fernandes MR, Willerson JT (2008) Comparison of intracoronary and transendocardial delivery of allogeneic mesenchymal cells in a canine model of acute myocardial infarction. J Mol Cel Cardiol 44:486–495

Pinheiro AO, Cardoso MT, Vidane AS, Casals JB, Passarelli D, Alencar ALF, Sousa RLM, Fantinato-Neto P, Oliveira VC, Lara VM et al (2016) Controversial results of therapy with mesenchymal stem cells in the acute phase of canine distemper disease. Genet Mol Res 15(2). https://doi.org/10.4238/gmr.15028310

Pinheiro LL, de Lima AR, Martins DM, de Oliveira EHC, Souza MPC, de Carvalho Miranda CMF, Beltrão-Braga PCB, Russo FB, Conceição Pignatari G, da Silva Filho E, Branco E (2019) Mesenchymal stem cells in dogs with demyelinating leukoencephalitis as an experimental model of multiple sclerosis. Heliyon 5(6):e01857

Pogue B, Estrada AH, Sosa-Samper I, Maisenbacher HW, Lamb KE, Mincey BD, Erger KE, Conlon TJ (2013) Stem-cell therapy for dilated cardiomyopathy: A pilot study evaluating retrograde coronary venous delivery. J Small Anim Pract 54:361–366

Rici REG, Will S, Luna ACL, Melo LF, Santos AC, Rodrigues RF, Leandro RM, Maria DA (2018) Combination therapy of canine osteosarcoma with canine bone marrow stem cells, bone morphogenetic protein and carboplatin in an in vivo model. Vet Comp Oncol 16(4):478–488

Rodriguez-Menocal L, Shareef S, Salgado M, Shabbir A, Van Badiavas E (2015) Role of whole bone marrow, whole bone marrow cultured cells, and mesenchymal stem cells in chronic wound healing. Stem Cell Res Ther 6:1–11

Roszek K, Makowska N, Czarnecka J, Porowinska D, Dabrowski M, Danielewska J, Nowak W (2016) Canine adipose-derived stem cells: purinergic characterization and neurogenic potential for therapeutic applications. J Cell Biochem 9999:1–8

Rousseau B, Hirano S, Scheidt TD, Welham NV, Thibeault SL, Ford CN, Bless DM (2003) Characterization of vocal fold scarring in a canine model. Laryngoscope 113:620–627

Ryu HH, Lim JH, Byeon YE, Park JR, Seo MS, Lee YW, Kim WH, Kang KS, Kweon OK (2009) Functional recovery and neural differentiation after transplantation of allogenic adipose-derived stem cells in a canine model of acute spinal cord injury. J Vet Sci 10:273–284

Ryu H-H, Kang B-J, Park S-S, Kim Y, Sung G-J, Woo H-M, Kim WH, Kweon O-K (2012) Comparison of Mesenchymal stem cells derived from fat, bone marrow, Wharton's jelly, and umbilical cord blood for treating spinal cord injuries in dogs. J Vet Med Sci 74(12):1617–1630

Saito F, Nakatani T, Iwase M, Maeda Y, Hirakawa A, Murao Y, Suzuki Y, Onodera R, Fukushima M, Ide C (2008) Spinal cord injury treatment with intrathecal autologous bone marrow stromal cell transplantation: the first clinical trial case report. J Trauma 64:53–59

Sancak IG, Özen A, Pinarli FA, Tiryaki M, Ceylan A, Acar U, Delibaşi T (2014) Limbal stem cells in dogs and cats their identification culture and differentiation into keratinocytes. Kafkas Üniversitesi Veteriner Fakültesi Dergisi 20(6):909–914

Sarmento CAP, Rodrigues MN, Bocabello RZ, Mess AM, Miglino MA (2014) Pilot study: bone marrow stem cells as a treatment for dogs with chronic spinal cord injury. Regenerat Med Res 2:9

Sasaki A, Mizuno M, Ozeki N, Katano H, Otabe K, Tsuji K, Koga H, Mochizuki M, Sekiya I (2018) Canine mesenchymal stem cells from synovium have a higher chondrogenic potential than those from infrapatellar fat pad, adipose tissue, and bone marrow. PLoS One 13(8):e0202922

Saulnier N, Loriau J, Febre M, Robert C, Rakic R, Bonte T, Buff S, Maddens S (2016) Canine placenta: A promising potential source of highly proliferativeand immunomodulatory mesenchymal stromal cells? Vet Immunol Immunopathol 171:47–55

Sedigh HS, Saffarpour A, Jamshidi S, Ashouri M, Nassiri SM, Sharifi D, Torkzaban P, Kajbafzadeh AM (2010) Canine periodontal stem cells: isolation, differentiation potential and electronic microscopic characterization. Iranian J Vet Surg 5(1,2):19–28

Seo MS, Jeong YH, Park JR, Park SB, Rho KH, Kim HS, Yu KR, Lee SH, Jung JW, Lee YS, Kang KS (2009) Isolation and characterization of canine umbilical cord blood-derived mesenchymal stem cells. J Vet Sci 10:181–187

Seo MS, Park SB, Kang KS (2012) Isolation and characterization of canine Wharton's jelly-derived Mesenchymal stem cells. Cell Transplant 21:1493–1502

Serigano K, Sakai D, Hiyama A, Tamura F, Tanaka M, Mochida J (2010) Effect of cell number on mesenchymal stem cell transplantation in a canine disc degeneration model. J Orthop Res 28:1267–1275

Shah K, Drury T, Roic I, Hansen P, Malin M, Boyd R, Sumer H, Ferguson R (2018) Outcome of allogeneic adult stem cell therapy in dogs suffering from osteoarthritis and other joint defects. Stem Cells Int 2018:7309201

Shi S, Gronthos S (2003) Perivascular niche of postnatal mesenchymal stem cells in human bone marrow and dental pulp. J Bone Miner Res 18:696–704

Silva GV, Litovsky S, Assad JA, Sousa AL, Martin BJ, Vela D, Coulter SC, Lin J, Ober J, Vaughn WK et al (2005) Mesenchymal stem cells differentiate into an endothelial phenotype, enhance vascular density, and improve heart function in a canine chronic ischemia model. Circulation 111:150–156

Simsek SB, Keles GC, Baris S, Cetinkaya BO (2012) Comparison of mesenchymal stem cells and autogenous cortical bone graft in the treatment of class II furcation defects in dogs. Clin Oral Investig 16:251–258

Song J, Kim Y, Kweon O-K, Kang B-J (2017) Use of stem-cell sheets expressing bone morphogenetic protein-7 in the management of a nonunion radial fracture in a toy poodle. J Vet Sci 18(4):555–558

Sousa MG, Paulino-Junior D, Pascon JPE, Pereira-Neto GB, Carareto R, Champion T, Camacho AA (2011) Cardiac function in dogs with chronic Chagas cardiomyopathy undergoing autologous stem cell transplantation into the coronary arteries. Can Vet J 52(8):869–874

Spencer ND, Chun R, Vidal MA, Gimble JM, Lopez MJ (2012) In vitro expansion and differentiation of fresh and revitalized adult canine bone marrow-derived and adipose tissue-derived stromal cells. Vet J 191(2):231–239

Starkey MP, Scase TJ, Mellersh CS, Murphy S (2005) Dogs really are man's best friend–canine genomics has applications in veterinary and human medicine! Brief Funct Genomic Proteomic 4:112–128

Steffen F, Bertolo A, Affentranger R, Ferguson SJ, Stoyanov J (2018) Treatment of naturally degenerated canine lumbosacral intervertebral discs with autologous Mesenchymal stromal cells and collagen microcarriers: A prospective clinical study. Cell Transplant. https://doi.org/10.1177/0963689718815459

Stevenson W, Chauhan SK, Dana R (2012) Dry eye disease: an immune-mediated ocular surface disorder. Arch Ophthalmol 130(1):90–100

Suaid FF, Ribeiro FV, Gomes TRLES, Silvério KG, Carvalho MD, Nociti FH Jr, Casati MZ, Sallum EA (2011) Autologous periodontal ligament cells in the treatment of class III furcation defects: a study in dogs. J Clin Periodontol 39:377–384

Suaid FF, Ribeiro FV, Silva Gomes TRLE, Gonzales Silvério K, Carvalho MD, Nociti FH Jr, Casati MZ, Antônio SE (2012) Autologous periodontal ligament cells in the treatment of class III furcation defects: a study in dogs. J Clin Periodontol 39(4):377–384

Sultana T, Lee S, Yoon H-Y, Lee JI (2018) Current status of canine umbilical cord blood-derived Mesenchymal stem cells in veterinary medicine. Stem Cells Int 2018:8329174

Taguchi T, Borjesson DL, Osmond C, Griffon DJ (2019) Influence of Donor's age on Immunomodulatory properties of canine adipose tissue-derived Mesenchymal stem cells. Stem Cells Dev 28(23):1562–1571

Takemitsu H, Zhao D, Yamamoto I, Harada Y, Michishita M, Arai T (2012) Comparison of bone marrow and adipose tissue-derived canine mesenchymal stem cells. BMC Vet Res 8:150

Takewaki M, Kajiya M, Takeda K, Sasaki S, Motoike S, Komatsu N, Matsuda S, Ouhara K, Mizuno N, Fujita T et al (2017) MSC/ECM cellular complexes induce periodontal tissue regeneration. J Dent Res 96(9):984–991

Taroni M, Cabon Q, Fèbre M, Cachon T, Saulnier N, Carozzo C, Maddens S, Labadie F, Robert C, Viguier E (2017) Evaluation of the effect of a single intra-articular injection of allogeneic neonatal Mesenchymal stromal cells compared to Oral non-steroidal anti-inflammatory treatment on the postoperative musculoskeletal status and gait of dogs over a 6-month period after Tibial plateau leveling osteotomy: A pilot study. Front Vet Sci 4:83

Teshima T, Matsumoto H, Koyama H (2018) Soluble factors from adipose tissue derived mesenchymal stem cells promote canine hepatocellular carcinoma cell proliferation and invasion. PLoS One 13(1):e0191539

Tharasanit T, Phutikanit N, Wangdee C, Soontornvipart K, Tantrajak S, Kaewamatawong T, Suwimonteerabutr J, Supaphol P, Techakumphu M (2011) Differentiation potentials of canine bone marrow Mesenchymal stem cells. Thailand J Vet Med 41(1):79–86

Tobita M, Uysal CA, Guo X, Hyakusoku H, Mizuno H (2013) Periodontal tissue regeneration by combined implantation of adipose tissue-derived stem cells and platelet-rich plasma in a canine model. Cytotherapy 15(12):1517–1526

Trindade AB, Therrien J, Garcia JM, Smith LC (2017) Mesenchymal-like stem cells in canine ovary show high differentiation potential. Cell Prolif 50(6):e12391

Tsumanuma Y, Iwata T, Washio K, Yoshida T, Yamada A, Takagi R, Ohno T, Lin K, Yamato M, Ishikawa I et al (2011) Comparison of different tissue-derived stem cell sheets for periodontal regeneration in a canine 1-wall defect model. Biomaterials 32:5819–5825

Uranio MF, Valentini L, Lange-Consiglio A, Caira M, Guaricci AC, L'abbate A, Catacchio CR, Ventura M, Cremonesi F, Dell'aquila ME (2011) Isolation, proliferation, cytogenetic, and molecular characterization and in vitro differentiation potency of canine stem cells from Foetal adnexa: A comparative study of amniotic fluid, amnion, and umbilical cord matrix. Mol Reprod Develop 78:361–373

Uranio MF, Dell'aquila ME, Caira M, Guaricci AC, Ventura M, Catacchio C, Martino NA, Valentini L (2014) Characterization and in vitro differentiation potency of early-passage canine amnion- and umbilical cord-derived mesenchymal stem cells as related to gestational age. Mol Reprod Develop 81:539–551

Vieira NM, Brandalise V, Zucconi E, Secco M, Strauss BE, Zatz M (2010) Isolation, characterization, and differentiation potential of canine adipose-derived stem cells. Cell Transplant 19:279–289

Vilar JM, Morales M, Santana A, Spinella G, Rubio M, Cuervo B, Cugat R, Carrillo JM (2013) Controlled, blinded force platform analysis of the effect of intraarticular injection of autologous adipose-derived Mesenchymal stem cells associated to PRGF-Endoret in osteoarthritic dogs. BMC Vet Res 9:131

Vilar JM, Batista M, Morales M, Santana A, Cuervo B, Rubio M, Cugat R, Sopena J, Carrillo JM (2014) Assessment of the effect of intraarticular injection of autologous adipose-derived mesenchymal stem cells in osteoarthritic dogs using a double blinded force platform analysis. BMC Vet Res 10:143

Villatoro AJ, Fernández V, Claros S, Rico-Llanos GA, Becerra J, Andrades JA (2015) Use of adipose-derived Mesenchymal stem cells in Keratoconjunctivitis Sicca in a canine model. Biomed Res Int 2015:527926

Villatoro AJ, Hermida-Prieto M, Fernández V, Fariñas F, Alcoholado C, Rodríguez-García MI, Mariñas-Pardo L, Becerra J (2018) Allogeneic adipose-derived mesenchymal stem cell therapy in dogs with refractory atopic dermatitis: clinical efficacy and safety. Vet Record 183(21):654. https://doi.org/10.1136/vr.104867

Volk SW, Wang Y, Hankenson KD (2012) Effects of donor characteristics and ex vivo expansion on canine Mesenchymal stem cell properties: implications for MSC-based therapies. Cell Transplant 21:2189–2200

Wei N, Gong P, Liao D, Yang X, Li X, Liu Y, Yuan Q, Tan Z (2010) Auto transplanted mesenchymal stromal cell fate in periodontal tissue of beagle dogs. Cytotherapy 12:514–521

Wei Y, Fang J, Cai S, Lv C, Zhang S, Hua J (2016) Primordial germ cell-like cells derived from canine adipose mesenchymal stem cells. Cell Prolif 49(4):503–511

William JB, Prabakaran R, Ayyappan S, Puskhinraj H, Rao D, Manjunath SR, Thamaraikannan P, Dedeepiya VD, Kuroda S, Yoshioka H et al (2011) Functional recovery of spinal cord injury following application of Intralesional bone marrow mononuclear cells embedded in polymer scaffold–two year follow-up in a canine. J Stem Cell Res Ther 1:110

Wu S, Suzuki Y, Ejiri Y, Noda T, Bai H, Kitada M, Kataoka K, Ohta M, Chou H, Ide C (2003) Bone marrow stromal cells enhance differentiation of cocultured neurosphere cells and promote regeneration of injured spinal cord. J Neurosci Res 72:343–351

Yaneselli KM, Kuhl CP, Terraciano PB, de Oliveira FS, Pizzato SB, Pazza K, Magrisso AB, Torman V, Rial A, Moreno M, Llambí S, Cirne-Lima E, Maisonnave J (2018) Comparison of the characteristics of canine adipose tissue-derived mesenchymal stem cells extracted from different sites and at different passage numbers. J Vet Sci 19(1):13–20

Yang Q, Peng J, Lu SB, Guo QY, Zhao B, Zhang L, Wang AY, Xu WJ, Xia Q, Ma XL, Hu YC, Xu BS (2011) Evaluation of an extracellular matrix-derived acellular biphasic scaffold/cell construct in the repair of a large articular high-load-bearing osteochondral defect in a canine model. Chin Med J (Engl) 124:3930–3938

Yang H-M, Song W-J, Qiang L, Kim S-Y, Kim H-J, Ryu M-O, Ahn J-O, Youn H-Y (2018) Canine mesenchymal stem cells treated with TNF-α and IFN-γ enhance anti-inflammatory effects through the COX-2/PGE2 pathway. Res Vet Sci 119:19–26

Yoon HY, Lee JH, Jeong SW (2012) Long-term follow-up after implantation of autologous adipose tissue derived mesenchymal stem cells to treat a dog with stifle joint osteoarthrosis. J Vet Clin 29:82–86

Zeira O, Asiag N, Aralla M, Ghezzi E, Pettinari L, Martinelli L, Zahirpour D, Dumas MP, Lupi D, Scaccia S et al (2015) Adult autologous mesenchymal stem cells for the treatment of suspected non-infectious inflammatory diseases of the canine central nervous system: safety, feasibility and preliminary clinical findings. J Neuroinflammation 12:181

Zhan X-S, El-Ashram S, Luo D-Z, Luo H-N, Wang B-Y, Chen S-F, Bai Y-S, Chen Z-S, Liu C-Y, Ji H-Q (2019) A comparative study of biological characteristics and transcriptome profiles of mesenchymal stem cells from different canine tissues. Int J Mol Sci 20:1485

Zhang LF, Qi J, Zuo G, Jia P, Shen X, Shao J, Kang H, Yang H, Deng L (2014) Osteoblast-secreted factors promote proliferation and osteogenic differentiation of bone marrow stromal cells via VEGF/heme-oxygenase-1 pathway. PLoS One 9(6):e99946

Zhu S, Lu Y, Zhu J, Xu J, Huang H, Zhu M, Chen Y, Zhou Y, Fan X, Wang Z (2011) Effects of intrahepatic bone-derived mesenchymal stem cells autotransplantation on the diabetic beagle dogs. J Surg Res 168:213–223

Zubin E, Conti V, Leonardi F, Zanichelli S, Ramoni R, Grolli S (2015) Regenerative therapy for the management of a large skin wound in a dog. Clin Case Rep 3(7):598–603

Zucconi E, Vieira NM, Bueno DF, Secco M, Jazedje T, Ambrosio CE, Passos-Bueno MR, Miglino MA, Zatz M (2010) Mesenchymal stem cells derived from canine umbilical cord vein—a novel source for cell therapy studies. Stem Cells Develop 19:395–402

# Equine Mesenchymal Stem Cell Basic Research and Potential Applications

# 12

M. B. Gugjoo, Amar Pal, D. M. Makhdoomi, and G. T. Sharma

**Abstract**

Regenerative medicine incorporating stem cells is being extensively studied to develop therapeutics for various incurable diseases. Mesenchymal stem cells (MSCs) due to their special features and easy isolation procedures are mostly being employed for therapeutic applications. These cells with or without other biomaterials have been studied under in vivo experimental and clinical trials of horse. MSCs therapeutic potential is seen to transgress their conventionally restricted healing potential for musculoskeletal system. The results of equine studies potentially can be construed for human medicine. The current chapter highlights horse MSCs sources, properties, and potential therapeutic applications.

M. B. Gugjoo (✉)
Division of Veterinary Clinical Complex, FVSc & AH, SKUAST-Kashmir, Jammu and Kashmir, India

A. Pal
Division of Surgery, Indian Veterinary Research Institute, Izatnagar, UP, India

D. M. Makhdoomi
Division of Surgery and Radiology, FVSc & AH, SKUAST-K, Srinagar, Jammu and Kashmir, India

G. T. Sharma
Division of Physiology and Climatology, Indian Veterinary Research Institute, Izatnagar, UP, India


M. B. Gugjoo, A. Pal (eds.), *Mesenchymal Stem Cell in Veterinary Sciences*, https://doi.org/10.1007/978-981-15-6037-8_12

## 12.1 Introduction

Equine industry is one of the biggest industries be it in the form of racing, polo, or for recreational and transportation purposes. In racehorses, musculoskeletal disorders reduce their performance and even destroy their carrier (Perkins et al. 2005). Among soft tissue ailments, skin wounds, eye affections, and reproductive disorders pose a significant problem to clinicians. Various horse and human developmental and autoimmune ailments have comparable pathophysiology (Gershwin 2007). Horse and human cartilage thickness and physiological features approximate to each other (Aigner et al. 2010). The cartilage defects associated with intense physical activity too resemble (Desancé et al. 2018). Thus, horse can make an excellent model animal to provide proof-of-therapeutic principle of mesenchymal stem cells (MSCs) in human medicine (McIlwraith et al. 2011; De Schauwer et al. 2013; Gugjoo et al. 2019b). Tissue engineering employing MSCs is being studied extensively. MSCs meager population in adult tissues/organs makes their culture expansion necessary for effective utilization. Innumerable in vitro studies have piled upon equine MSCs (eMSCs). These studies have been conducted on their culture characteristics, characterization, and their therapeutic properties. The in vivo studies have supported the therapeutic potentials of MSCs, although variably (Carrade et al. 2011a, b; Spaas et al. 2012; Burk et al. 2013) with mechanism(s) behind the healing largely being unrevealed. The current chapter deliberates on the eMSCs ex vivo properties and their potential therapeutic applications.

## 12.2 In Vitro Studies

### 12.2.1 Mesenchymal Stem Cell Sources, Culturing, and Characterization

Horse MSCs (eMSCs) have diverse sources ranging from adult tissues to fetal membranes as detailed in Table12.1. Donor tissues usually harbor small fraction of MSCs and are isolated along with the other cell types including progenitor cells (Chong et al. 2011). The characterization of eMSCs has been made as per the standard protocol of the International Society for Cell Therapy (ISCT) (Dominici et al. 2006). eMSCs and human MSCs tend to have comparable marker expression and differentiation potential (Hillmann et al. 2016). eMSCs invariably follow the plastic adherence and are multipotent (Guest et al. 2008; Ranera et al. 2011; De Schauwer et al. 2012). Their surface marker expression, however, remains controversial as these cells variably express different surface markers and lack expression of certain markers. This discrepancy in marker expression has mainly been observed in relation to the tissue source as shown in Table12.1. eMSCs also express pluripotency markers like Oct4, Sox-2, and Nanog and show extended differentiation like neural cell-like cells, myogenic cell-like cells, tenocyte-like cells, and hepatocyte-like cells (Violini et al. 2009; Coli et al. 2011; Gomiero et al. 2016; Pennington et al. 2016; Cabezas et al. 2018).

**Table 12.1** Horse mesenchymal stem cell sources and their characterization

| Positive markers | Negative markers | Differentiation | References |
|---|---|---|---|
| *Adipose tissue MSCs (AD-MSCs)* | | | |
| CD29, CD44, CD105, CD90 (limited expression) | CD45 | Osteogenic and adipogenic | Marycz et al. (2012) |
| CD29, CD44, and CD90 | MHC-II, CD34 | Adipogenic, chondrogenic, and osteogenic | Alipour et al. (2015) |
| CD29, CD44, CD105 (87%) | – | Adipogenic, chondrogenic, and osteogenic | Xie et al. (2013) |
| CD73, CD90, CD105 | | Adipogenic, chondrogenic, and osteogenic | González-Fernández et al. (2016) |
| CD105, CD90, CD49, CD44 | CD43, CD73 | Adipogenic, chondrogenic, and osteogenic | Pall et al. (2016) |
| CD90, CD105, CD44, CD73, CD13, CD133, CD29,OCT-4 | CD45,CD34, CD31, SDF-1 receptor CXCR4 | – | Angelone et al. (2017) |
| CD44, CD90, CD105 | CD45 | Adipogenic, chondrogenic, and osteogenic | Nawrocka et al. (2017) |
| CD29, CD44, CD90, CD105, CD166 | CD34, CD45, and CD79a | Chondrogenic and osteogenic | Bundgaard et al. (2018) |
| CD29, CD44, CD90, CD-105, MHC-I | MHC-II, CD45, OCT4, Sox2 | Adipogenic, chondrogenic, and osteogenic | Cabezas et al. (2018) |
| CD90, CD105, Oct4 | | Osteogenic | Elashry et al. (2018) |
| *Amniotic membrane MSCs (am-MSCs)* | | | |
| CD90, CD73, CD34, Oct4 (weak), IL-6, IL-β1, IL-8 (weak) | MHC-I, MHC-II, CD45, Sox2, Nanog, INF-γ, TNF-α, IL-4 | Adipogenic, chondrogenic, and osteogenic | Iacono et al. (2017) |
| CD29, CD44, CD106, CD105 | CD34, MHC-II | Adipogenic, chondrogenic, osteogenic, and neurogenic | Perrini et al. (2016) |
| *Bone marrow MSCs (BM-MSCs)* | | | |
| CD29, CD44, CD172a | CD34, CD45 | – | Fortier et al. (2010) |
| CD29, CD44, CD105 (64%) | – | Adipogenic, chondrogenic, and osteogenic | Xie et al. (2013) |
| CD11a/18, CD34, CD44, CD90 | | Adipogenic, chondrogenic, and osteogenic | Lombana et al. (2015) |

(continued)

**Table 12.1** (continued)

| Positive markers | Negative markers | Differentiation | References |
|---|---|---|---|
| CD90, CD29 | MHC-II, CD44, CD45RB | Adipogenic, chondrogenic, and osteogenic | Mitchell et al. (2015) |
| CD90, CD105, CD44, CD73 | CD34, CD45 | Adipogenic, chondrogenic, and osteogenic | Barrachina et al. (2016) |
| CD90, CD44, CD29 | MHC-II, CD86 | – | Clark et al. (2016) |
| CD73, CD90, CD105 | | Adipogenic, chondrogenic, and osteogenic | González-Fernández et al. (2016) |
| CD44, CD29, CD90, CD73 (at p3 and p4 only), CD105 (at P3 25% expression) | CD34, MHC-II, CD105 (P4) | Adipogenic, chondrogenic, and osteogenic | Branly et al. (2017) |
| CD29, CD44, CD90, MHC-I, Sox2 | CD34, CD105, MHC-II, Nanog, Oct4, SSEA-1, −3,-4 | Adipogenic, chondrogenic, and osteogenic | Zahedi et al. (2018) |
| CD29, CD44, CD90, CD105, CD166 | CD34, CD45, and CD79a | Chondrogenic and osteogenic | Bundgaard et al. (2018) |
| CD29, CD90 | CD34, CD45, CD14, CD73, and CD105, CD79a, DRB, NANOG, and SOX2 | Adipogenic, chondrogenic, osteogenic, and tenogenic | Korchunjit et al. (2019) |
| CD44, CD90, MHC-I, CD105 (33.9%) | CD11a/CD18, MHC-II | Chondrogenic and osteogenic | Schrock et al. (2018) |
| CD29, CD44, CD90, CD105, and MHCI, MHC-II | CD45 and CD79a | Chondrogenic | Gale et al. (2019) |
| *Dental pulp MSCs (DP-MSCs)* | | | |
| CD11a/CD18, CD44, CD90, CD105, MHC-I, MHC-II | CD34, CD45 | Adipogenic, chondrogenic, and osteogenic | Ishikawa et al. (2017) |
| *Dermal MSCs* | | | |
| CD90, CD105 (48%), CD29, CD44 | CD34, CD45, CD79α, CD14, MHC-II, CD73 | – | Michler et al. (2018) |
| *Endometrium MSCs (end-MSCs)* | | | |
| CD29, CD44, CD105, CD90, Pericyte (NG2, CD146) | CD34, CD45, MHC-II | Adipogenic, chondrogenic, osteogenic, and smooth muscle | Rink et al. (2017) |
| CD29, CD44, CD90, CD-105, MHC-I | MHC-II, CD45, OCT4, Sox2 | Adipogenic, chondrogenic, and osteogenic | Cabezas et al. (2018) |

(continued)

**Table 12.1** (continued)

| Positive markers | Negative markers | Differentiation | References |
|---|---|---|---|
| *Gingiva MSCs* | | | |
| CD90, CD105, Vimentin | CD31 | Adipogenic, chondrogenic, and osteogenic | Mensing et al. (2011) |
| *Hoof MSCs (H-MSCs)* | | | |
| CD44, CD29, CD105, Oct4, Sox2, keratin (K)14, K15, K19 | – | Adipogenic, osteogenic, and neurogenic | Yang and Lopez (2019) |
| *Peripheral blood MSCs (PB-MSCs)* | | | |
| CD90, CD105 (40.5%), CD44, MHC-1 (60.2%), CD11a/CD18 (42.1%) | CD34, CD45, MHC-II | Adipogenic, chondrogenic, and osteogenic | Ishikawa et al. (2017) |
| *Periodontal ligament (PL-MSCs)* | | | |
| CD90, CD105, Vimentin | CD31 | Adipogenic, chondrogenic, and osteogenic | Mensing et al. (2011) |
| CD44, CD29, MHC-II | | | Beerts et al. (2017) |
| *Synovial fluid MSCs (SF-MSCs)* | | | |
| CD45, CD105, CD90, CD34, CD117, CD133, TRAI-1-81, VEGF, LY6A, OCT3/4, Nanog | – | – | Prado et al. (2015) |
| CD90, CD44, PCNA, Vimentin, PGP-9.5, lysozyme | – | Adipogenic, chondrogenic (weak), and osteogenic | Fülber et al. (2016) |
| *Synovial membrane MSCs (Sy-MSCs)* | | | |
| CD90, CD44, PCNA, Vimentin, PGP-9.5, lysozyme | | Adipogenic, chondrogenic (weak), and osteogenic | Fülber et al. (2016) |
| CD44 and CD90, MHC-I, CD11a/18, CD105, MHC-II (weak), Sox2 | CD34, CD45. | Chondrogenic, osteogenic, and tenogenic | Yamasaki et al. (2018) |
| *Umbilical cord MSCs (UC-MSCs)* | | | |
| CD44, CD90 (weak) | MHC-II, CD34, Oct4 | Osteogenic and adipogenic | Maia et al. (2017) |
| *Wharton's jelly MSCs (WJ-MSCs)* | | | |
| CD90, CD73, CD34, Oct4, IL-6, IL-β1, IL-8 | MHC-I, MHC-II, CD45, Sox2, Nanog, INF-γ, TNF-α, IL-4 | Adipogenic, chondrogenic, and osteogenic | Iacono et al. (2016) |

*AD-MSCs* adipose tissue MSCs, *Am-MSCs* amniotic membrane MSCs, *BM-MSCs* bone marrow MSCs, *DP-MSCs* dental pulp MSCs, *End-MSCs endometrium MSCs*, *H-MSCs* hoof MSCs, *PB-MSCs* peripheral blood MSCs, *PL-MSCs* periodontal ligament, *SF-MSCs* synovial fluid MSCs, *Sy-MSCs* synovial membrane MSCs, *UC-MSCs* umbilical cord MSCs, *WJ-MSCs* Wharton's jelly MSCs

### 12.2.2 Tissue Source and MSCs Features

Peripheral blood (PB) contains the lowest concentration of eMSCs (ePB-MSCs), and as such the success rate to harvest these cells from such a tissue is lower. ePB-MSCs harvesting rate too varies and had been demonstrated to range from as low as 33% (Koerner et al. 2006), 44% (Martinello et al. 2010) and to the maximum of 66% (Carvalho et al. 2013b). Their differentiation potential too had been limited in comparison to horse bone marrow-derived MSCs (eBM-MSCs) (Koerner et al. 2006). The initial cell concentration of stromal vascular fraction (SVF) and/or mononuclear cell fraction (MNC) plays an important role to achieve better MSCs yield. Horse adipose tissue (eAD)-stromal vascular fraction (SVF) had ranged from $1.47 \times 10^6$ cells to $2.71 \times 10^6$ cells per gram of tissue (Nixon et al. 2008; Tuemmers et al. 2012) while adult horse bone marrow had yielded $6.4 \pm 3.4 \times 10^6$ nucleated cells (MNCs) per ml of aspirate (Hildebrand et al. 2002; Baghavan et al. 2009). MSCs yield from horse bone marrow (BM) remains lower than the adipose tissue (AD) (Toupadakis et al. 2010; Stewart and Stewart 2011).

Proliferation potential of eMSCs varies with tissue source. Bone marrow (BM)-derived and adipose tissue (AD)-derived MSCs tend to show higher proliferation *in* comparison to umbilical cord (UC)-derived MSCs (Barberini et al. 2014). MSCs harvested from a particular tissue system may be more tenable in relevant ailments. eBM-MSCs tend to show superior musculoskeletal tissue differentiation ability as compared to eAD-MSCs (Noel et al. 2008; Vidal et al. 2008; Frisbie et al. 2009; Toupadakis et al. 2010; Branly et al. 2017). Even among similar tissue/organ system, MSCs differentiation potential may vary. Synovial (Sy)-MSCs and BM-MSCs had comparable chondrogenic potential, but the latter cell type had better osteogenic potential (Gale et al. 2019). For musculoskeletal ailments, eBM-MSCs in comparison to eAD-MSCs may give better response. Altered sensitivity of eAD-MSCs to osteogenic growth factors like transforming growth factor (TGF-β) and bone morphogenetic protein (BMP) may predispose them to reduced chondrogenic differentiation. Such a reduction in chondrogenesis may be potentiated with BMP-6 (Estes et al. 2006; Hennig et al. 2007). eMSCs from a particular tissue but from varied locations (ilium vs. sternum) may not show any observable difference (Lombana et al. 2015).

### 12.2.3 Donor Health Status and MSCs Properties

Individual health status affects cell viability and as such may affect clinical results. AD-MSCs harvested from metabolic syndrome (insulin resistance disease) horses may be more senescent with an insubstantial differentiation potential. Oxidative stress that promotes accumulation of toxic compounds in the mitochondria might lead to the cell senescence. Autophagic switch to counter oxidative stress might decrease MSCs differentiation potential (Marycz et al. 2016a, b). In vitro application of *Spirulina platensis* had restored senescent eAD-MSCs morphology and functionality and could have been possible through the reduction in oxidative stress

(Nawrocka et al. 2017). eSy-MSCs harvested from osteoarthritic/osteochondrosis and healthy tissue tend to show comparable phenotype and differentiation potential except for chondrogenesis, being better in healthy tissue-derived MSCs (Fülber et al. 2016). In clinical settings, the problem can be countered with the implantation of higher cell doses (Frisbie and Smith 2010)

### 12.2.4 Effect of Aspiration and Implantation on MSCs

Cell aspiration/implantation technique affects the cell viability. The compromise in cell viability may be more during aspiration than during the injection. The aspiration creates more negative pressure leading to cell stress, and as such forcible aspiration of MSCs should be prevented (Williams et al. 2016; Lang et al. 2017). Injection of cells through wide bore needles (19-gauge) had not affected eMSCs viability, while those injected through small bore (23-gauge) had become apoptotic. The current recommended needle size of 21-gauge needle balances the cell viability and needle-induced damage at the site of injection (Garvican et al. 2016). However, the consensus on these findings is lacking (Williams et al. 2016). The material of transport/storage container, plastic, or glass had nonsignificant effect on the cell viability (Espina et al. 2016).

### 12.2.5 Effect of Drugs/Anesthetics on MSCs

Anesthetics and/drugs affect MSCs viability based on their class (Edmonds et al. 2016; Schrock et al. 2018). Romifidine and mepivacaine had nonsignificant effect, while detomidine and butorphanol had significantly reduced MSCs viability. Xylazine had also decreased MSCs viability but insignificantly. Commonly used steroids like methylprednisolone and triamcinolone may compromise eBM-MSCs viability (Edmonds et al. 2016). Therapeutic concentrations of antibiotics like aminoglycosides (gentamicin and amikacin) had deleteriously affected eBM-MSCs, while hyaluronic acid had proven beneficial for their proliferation (Bohannon et al. 2013).

## 12.3 Clinical Applications of eMSCs

Stem cell therapy in horse medicine mostly remains in experimental phase due to lack of the literature incorporating uniform case controlled studies (Imamura and Tanaka 2009; Clegg and Pinchbeck 2011; Volk and Theoret 2013; Dehghan et al. 2015; Ojeh et al. 2015). In horses, the therapeutic applications of stem cells had been mainly aimed to address the issues of musculoskeletal system and first of its kind had been conducted on superficial digital flexor tendonitis (Smith et al. 2003). As the potential goes beyond, MSCs potential applications are now being evaluated in nonmusculoskeletal tissues as well (Gugjoo et al. 2019a). Below are given the details of the in vivo studies conducted in horse.

### 12.3.1 Wound Repair

Horse wounds are predisposed to issues like proud flesh and/infections. The current standard therapeutic options heal only about 24% of wounds by primary healing (Wilmink et al. 2002). The limited availability of donor skin and more failures in "graft take" under grafting makes it a least opted therapeutic solution (Theoret 2009). Potential applications of stem cells in wound healing had come to the fore when a higher number of the cells exhibiting similar phenotype to that of BM-MSCs were observed in blood of burn patients (Mansilla et al. 2006; Fox et al. 2008). Since then stem cell therapies are being evaluated.

Ex vivo horse study had demonstrated that encapsulated MSCs and their conditioned media might promote the repair of skin wounds. The repair could occur through equine dermal fibroblast mobilization and increased pro-wound healing gene expression (Bussche et al. 2015). In vivo implanted eBM-MSCs tend to remain well oriented and integrated under the epithelium of repaired soft palate (Carstanjen et al. 2006). The locally injected allogeneic equine umbilical cord blood (eUCB)-MSCs had improved wound healing of hind limb wounds. The cell-injected wounds had exhibited better tissue regeneration with little inflammatory cells. The expression of healing factors (transforming growth factor-β1 and cyclooxygenase 2) was significantly higher in cell-treated wounds as compared to the control. MSCs healing potential was unaffected by their culture technique (normoxia vs. hypoxia) (Table12.2) (Textor et al. 2017).

In clinical studies (foal deep sore wounds and decubital ulcers), local implantation of allogeneic amnion-derived MSCs (Am-MSCs) along with carboxymethylcellulose (Iacono et al. 2016) or platelet-rich plasma (Table12.3) (Iacono et al. 2012) had led to earlier and improved wound healing as compared to wounds treated without the cells. The cells had been implanted multiple times and along with the other biomaterials (Iacono et al. 2016). The combined application (involving local and parenteral injection) of PB-MSCs had improved wound healing of 4 chronic dermal wounds of adult horse metatarsals. Granulation tissue had formed significantly much earlier, while scar formation was delayed in comparison to the control (Spaas et al. 2013). Thus, MSCs may be employed to repair equine wounds. Improved healing upon the MSCs transplantation might arise through regeneration of epidermal cells and/increased angiogenesis (Krause et al. 2001; Wu et al. 2007; Borena et al. 2009). However, all these studies have mostly been uncontrolled, and as such evidence-based studies involving large significant numbers of cases are desired.

**Table 12.2** In vivo preclinical experimental mesenchymal stem cell studies on cutaneous wounds in horse

| Model type | Number of animals | Model defect size/ study period | Biomaterial used | Cell dose/ assembly | Evaluation criteria | Overall result | References |
|---|---|---|---|---|---|---|---|
| Bilateral distal forelimb full thickness wounds (3 wounds; each $2.5 \times 2.5$ cm$^2$ spaced at 2.5 cm vertically aligned) | 6 horses (each wound in horse treated with hypoxic- and normoxia-preconditioned cells. Other wounds were treated with above cell types laden in autologous fibrin gel. Control with one saline and other fibrin gel-treated wounds | 6 weeks | Umbilical cord blood MSCs | 15–$20 \times 10^6$ per wound | Histology, thermography, gene expression, and wound surface area | Cell treated irrespective of prior preconditioning improved histological scoring and wound area. No significant advantage of hypoxic conditioning. An improved transforming growth factor beta and cyclooxygenase-2 gene expression at 1 week. Histology showed wounds treated with MSCs more of pro-healing than pro-inflammatory | Textor et al. (2017) |

**Table 12.3** In vivo clinical mesenchymal stem cell studies on cutaneous wounds/eye affection in horse

| Clinical case | Number of animals | Study type | Study period | Biomaterial used | Cell dose/ assembly | Evaluation criteria | Overall result | References |
|---|---|---|---|---|---|---|---|---|
| Soft palate defect | 1 | Case report | 14 days | BM-MSCs | $10 \times 10^6$ | Macroscopy, histochemistry | The cells were organized in a manner similar to that in native soft palate tissue, indicating successful engraftment | Carstanjen et al. (2006) |
| Decubital ulcers (septic neonatal foal) | 1 (multiple sores) | Case report | 7 months | Amniotic fluid-derived MSCs, platelet-rich plasma (PRP) | Ulcers were treated for 15 days with mesenchymal stem cells (MSCs) plus PRP, PRP gel alone, or aloe gel. $5 \times 10^6$ MSCs (applied twice a week for 4 consecutive times) | Wound area measurement | MSCs along with platelet-rich plasma (PRP) gel improve decubital ulcers earlier and effectively compared to those implanted with PRP or aloe vera gel only. At 7 months, an ulcer treated with aloe gel was still not completely healed | Iacono et al. (2012) |

| | | | | | | | | |
|---|---|---|---|---|---|---|---|---|
| Chronic dermal wounds of metatarsals (unresponsive to conventional therapies) | 4 | Case report | 4 weeks and 8 weeks Uncontrolled study | PB-MSCs | $5 \times 10^5$ (local), $1.25 \times 10^5$ (intravenous) | Wound-healing parameters (granulation tissue, crust formation, and scar formation) | MSCs injected locally as well as parenterally in 4 wounds showed positive visual and clinical outcome. Granulation tissue was formed within 4 weeks of the cell application, significantly earlier compared to the control. The scar formation, however, was delayed by more than 2 months as compared to the control | Spaas et al. (2013) |

(continued)

**Table 12.3** (continued)

| Clinical case | Number of animals | Study type | Study period | Biomaterial used | Cell dose/assembly | Evaluation criteria | Overall result | References |
|---|---|---|---|---|---|---|---|---|
| Deep sore wounds | 7 (9 spontaneous pressure sores on the carpus (4), fetlock (2), and hock (3) of 5 hospitalized foals | Case report | Case controlled (criteria for selection of cases: Assessing classification of skin wounds, classification of the National Pressure Ulcer Advisory Panel) | Carboxymethylcellulose-seeded heterologous amnion-derived MSCs | $5 \times 10^5$ cells/$cm^2$ (applied twice a week for 4 consecutive times) | Wound area measurement | Cell seeded scaffold implantation led to faster dermal wound healing as compared to control (applied commercial preparation of formosulfathiazole) | Iacono et al. (2016) |

| | | | | | | | | |
|---|---|---|---|---|---|---|---|---|
| Corneal ulcers and retinal detachment | 4 (corneal ulcers) and 1 (bilateral retinal detachment) | Case reports | 6 months | PB-MSCs | Ophthalmic artery and locally | Clinical examination, ultrasound | After cell treatment, animals were active and able to jump. PB-MSCs improved the optic nerve function and visual acuity in the left eye after 2 months. The animal could even jump and cross barriers after 6 months. However, no substantial improvement was seen in the right eye | Marfe et al. (2012) |

### 12.3.2 Keratitis and Other Ophthalmic Conditions

Laterally placed large eyes make horse conducive to their injuries (Wada et al. 2010; Clode and Matthews 2011). Corneal wounds, especially of ulcerative keratitis, account 68.5% (Andrew and Willis 2005) to 90.5% (Wada et al. 2010) of the ocular problems, among which infectious keratitis accounts for 35–39% of all the cases (Strubbe et al. 2000; Olivier et al. 2003). To address these issues, a study had demonstrated that a significant corneal epithelial restoration and reduction of inflammatory condition occur with conjunctival and/ophthalmic artery injection of PB-MSCs (Table12.3) (Spaas et al. 2011; Marfe et al. 2012).

Even in an old (2 months) case of equine bilateral retinal detachment, improvement in nerve function and visual acuity had been demonstrated in left eye with intravenous injection of PB-MSCs. However, right eye had inconsiderable improvement (Marfe et al. 2012). The variable pathophysiology involved and/or insufficient cellular concentration achieved in the right eye could have been the reasons for unfavorable results in right eye. Thus, more extensive studies are desired as per the pathophysiology with MSCs dose titration.

### 12.3.3 Tendon/Ligament Injuries

Tendonitis tends to lead to equine lameness (Barreira et al. 2008). This condition is managed by rest and rehabilitation, although recurrent tendonitis in 43% cases has been reported (Dyson 1997; Violini et al. 2009). Lack of basic understanding of tendon cell characteristics and their precursors in tendon healing currently challenges effective tendon healing (Bi et al. 2007). To gain insights and provide early healing of tendon, MSCs are being evaluated.

Numerous MSCs studies (experimental or clinical) on tendon or ligament injuries have been conducted as enlisted in Table 12.4 and 12.5. The models have been created either surgically (Geburek et al. 2017) or by enzymatic action (Nixon et al. 2008; Carvalho et al. 2013a). Apart from experimental models, MSCs therapeutic applications have also been evaluated in clinical tendinopathies and/desmitis (Renzi et al. 2013; Van Loon et al. 2014; Beerts et al. 2017). Most of the studies have evaluated BM-MSCs and AD-MSCs with two studies demonstrating umbilical cord (UC)-MSCs (Van Loon et al. 2014) and PB-MSCs (Beerts et al. 2017). The cells have been injected intralesionally after some delay. Some studies have additionally incorporated scaffolds like platelet-rich plasma (Renzi et al. 2013), serum (Pacini et al. 2007; Geburek et al. 2017), fibrinogen (Crovace et al. 2010), and plasma lysate (Del Bue et al. 2008). Growth factors like insulin-like growth factor (IGF-1) have also been added (Smith et al. 2003; Schnabel et al. 2009). In all these studies, a single dose of the variable cell concentration has been made barring two studies that had injected the cells twice (Van Loon et al. 2014; Vandenberghe et al. 2015).

Most of the studies have shown positive role of MSCs in tendon healing. One of the studies, however, had failed to demonstrate significant therapeutic effects of the MSCs (Caniglia et al. 2012). BM-MSCs suspended in fibrinogen (Crovace et al.

**Table 12.4** In vivo preclinical experimental stem cell studies on tendinitis in horse

| Model type | Number of animals | Model defect size/ study period | Biomaterial used | Cell dose/ assembly | Evaluation criteria | Overall result | References |
|---|---|---|---|---|---|---|---|
| Superficial digital flexor tendinitis (collagenase-induced) | 8 (4 cell treated and 4 control) | 6 weeks | AD-nucleated cell fraction | $13.83 \pm 3.41 \times 10^6$ cells (3 doses at 40–53 h range) | Biochemical and molecular and ultrasonography were performed | Ultrasonography revealed no difference in rate or quality of repair between groups. Histologic evaluation revealed a significant improvement in tendon fiber architecture; reductions in vascularity; inflammatory cell infiltrate; collagen type III formation; and improvements in tendon fiber density and alignment in ADNC-treated tendons. Repair sites did not differ in DNA, proteoglycan, or total collagen content. Gene expression of collagen type I and type III in treated and control tendons was similar. Gene expression of COMP was significantly increased in ADNC-injected tendons | Nixon et al. (2008) |

(continued)

**Table 12.4** (continued)

| Model type | Number of animals | Model defect size/ study period | Biomaterial used | Cell dose/ assembly | Evaluation criteria | Overall result | References |
|---|---|---|---|---|---|---|---|
| Superficial digital flexor tendinitis | 12 (6 BM-MSCs; 6 IGF-I gene enhanced BM-MSCs) | 8 weeks | BM-MSCs, IGF-I gene enhanced BM-MSCs | $10 \times 10^6$ (local implantation) | Histological and immunohistochemical scores, ultrasound | Expression of collagen types I and III, IGF-I, cartilage oligomeric matrix protein (COMP), matrix metalloproteinase-3 (MMP-3), matrix metalloproteinase-13 (MMP-13), and aggrecanase-1 (ADAMTS-4) was similar in MSC and control tendons. Both MSC and AdIGF-MSC injection resulted in significantly improved tendon histological scores | Schnabel et al. (2009) |
| Superficial digital flexor tendinitis | 6 (4 animals: All four limbs; 2 animals: 3 limbs) | 18 weeks after treatment | BM-MSCs; BM-MNCs | 5.5 × 106 cBMSCs or 1.2 × 108 BMMNCs (average) intralesionally | Histological and immunohistochemical scores | In cBMSC- and BMMNCS-treated tendons, a high expression of cartilage oligomeric matrix protein (COMP) and type I collagen, but low levels of type III collagen were revealed by immunohistochemistry, with a normal longitudinally oriented fiber pattern. Placebo-treated tendons expressed very low quantities of COMP and type I collagen but large numbers of randomly oriented type III collagen fibers. Both cBMSC and BMMNCS grafts resulted in a qualitatively similar healing improvement of tendon extracellular matrix, in terms of the type I/III collagen ratio, fiber orientation, and COMP expression | Crovace et al. (2010) |

| | | | | | | | |
|---|---|---|---|---|---|---|---|
| Superficial digital flexor tendinitis (collagenase-induced) | 8 (4 animals biopsy at 60 days; 4 animals biopsy at 150 days) | 60, 150 days | AD-MSCs | $10 \times 10^6$ (30 days after collagenase injection) | Histological and immunohistochemical scores, clinical and ultrasonographic features | No differences occurred between the groups regarding the clinical and ultrasonographic analyses; however, histopathological evaluation revealed a significant improvement in tendon fiber organization and diminished inflammatory infiltrate, whereas immunohistochemical analysis showed increased expression of type I collagen in the treated group as compared with controls. The cellular therapy model implanted in this experiment promoted increased perivascular inflammatory infiltrate, fibroblastic density, neovascularization, and qualitative healing improvement of tendon extracellular matrix, in terms of fiber orientation and type I/III collagen ratio; moreover, it was considered to be a safe and viable process | Carvalho et al. (2011) |
| Superficial digital flexor (SDFT) | 6 horses (2 wounds in each). One treated with MSCs and the other bone marrow supernatant | 12 weeks after cell implantation | Autologous BM-MSCs | $10 \times 10^6$ (ultrasound guidance) | Mass average diameter (MAD), collagen fibril index (CFI), area-dependent diameter (ADD) | MAD and CFI in addition to the bimodal distribution of collagen fibril size were significantly lower in treated and control injured animals as compared to the normal and control limbs | Caniglia et al. (2012) |

(continued)

**Table 12.4** (continued)

| Model type | Number of animals | Model defect size/ study period | Biomaterial used | Cell dose/ assembly | Evaluation criteria | Overall result | References |
|---|---|---|---|---|---|---|---|
| Bilateral forelimb superficial digital flexor tendons (SDFT) | 9 horses (2 in each horse) | 22 weeks | AD-MSCs | $1 \times 10^7$ injected intralesionally after 2 weeks | Ultrasonography and histology | Color Doppler signal at 2 weeks was significantly more extensive. Histology showed significantly more neovascularization after 22 weeks | Conze et al. (2014) |
| Bilateral forelimb superficial digital flexor tendons (SDFT) | 9 horses (2 in each horse) | 24 weeks | AD-MSCs in inactivated serum and inactivated serum only | Ultrasound-guided $10 \times 10^6$ cells were injected intrialesionally | Ultrasonography, clinical, biochemical, and biomechanical evaluation | Statistically nonsignificant differences in two treated groups. Mature collagen crosslink hydroxylysylpyridinoline was comparable to normal tendon collagen in AD-MSCs treated group, while it was lower in the other group. Stress at failure and modulus of elasticity in AD-MSCs group were significantly lower than the mature normal tendon | Geburek et al. (2017) |

**Table 12.5** In vivo clinical mesenchymal stem cell studies on tendinitis in horse

| Clinical case | Number of animals | Study period | Biomaterial used | Cell dose/ assembly | Evaluation criteria | Overall result | References |
|---|---|---|---|---|---|---|---|
| Superficial digital flexor tendinitis | 26 (11 cell treated; 15 control) | Uncontrolled study | Autologous BM-MSCs | $9.5 \times 10^6$ | Ultrasound and ability to return to race | Nine MSC-treated animals recovered from their clinical conditions had an excellent ultrasound image of tendons after a period ranging from 3 to 6 months and returned to racing with good or even optimal results in the previous category of competition in 9–12 months without any reinjuring event. All of them are still active more than 2 years from diagnosis. One of the 2 remaining horses received less than $1 \times 10^6$ of MSCs, and its tendon did not heal, relapsing after rehabilitation; the other was lost to follow-up. Small number of MSCs may be sufficient to repair damaged tendons without the use of scaffold support. Ultrasound scanning showed that fibers were correctly oriented in cell-treated tendons. Further, no ectopic bone deposition occurred | Pacini et al. (2007) |

(continued)

**Table 12.5** (continued)

| Clinical case | Number of animals | Study period | Biomaterial used | Cell dose/assembly | Evaluation criteria | Overall result | References |
|---|---|---|---|---|---|---|---|
| Tendinitis (case controlled study) | 16 | 15–240 days | Allogeneic AD-MSCs + platelet-rich plasma | Local intralesional injection (cell concentration not given) | Ultrasonography | 14 out of 16 treated horses recovered and went back to activity. Two did not recover, one among them was affected by chronic tendonitis previously treated with different therapies (infiltrations, blisters, etc.), while the other horse, after 8 weeks and a satisfactory recovery, went back to activity too early and the lesion recurred | Del Bue et al. (2008) |
| Tendinitis (among 23 animals, 14 had frontal right and 9 frontal left limb affected); desmitis (among 9, 5 frontal right and 4 frontal left limb cases. 1 case had bilateral SL lesion). Lesion of 15–30% of transversal tendon surface was included | 33 (18 BM-MSCs treated; 13 pin firing treated) | 8–15 months | Autologous BM-MSCs | $2 \times 10^6$ cells in PRP per $cm^2$ of lesion | Clinical evaluation and ultrasonography | Gradual healing was observed on ultrasonography, and at 24 weeks the tendon was anatomically restored. 13/18 in cell-treated group returned to race competitions, while none among pin fired horses | Renzi et al. (2013) |

| | | | | | | | |
|---|---|---|---|---|---|---|---|
| Superficial digital flexor tendon | 12 (6 animals cell treated; 6 control) | 6 months | BM-MSCs, growth factors, natural mechanical stimulus | $1 \times 10^7$ autologous BM-MSCs suspended in 2 mL of marrow supernatant | Morphology, biomechanical, ultrasonography, histology | Cell-treated tendons had significantly lower structural stiffness although there is no significant difference in calculated modulus of elasticity. Lower (improved) histological scoring of organization and crimp pattern, lower cellularity, DNA content, vascularity, water content, GAG content, and MMP-13 activity were seen in cell-treated tendons | Smith et al. (2013) |
| Tendinitis of superficial flexor tendons (SDFT) and deep digital flexor tendon (DDFT), and desmitis of superficial ligament (SL) and inferior check ligament (ICL) (first time injured) | 52 animals (young: 5–9 years; medium aged: 10–14 years; >15 years) | 6 months | Umbilical cord blood-derived MSCs | $2–10 \times 10^6$ (single dose) $20 \times 10^6$ (double dose in 5 horses at 4–36 weeks). 4/5 double-treated animals were from SL group | Clinical and ultrasonography | 40/52 (77%) animals returned to the normal activity. 20/23 SDFT cases had positive result. 15/22 cases of SL had positive result. 4/6 DDFT and ICL had positive result. Control cases (SDFT: 1/3; SL: 2/3) reached to normal or higher performance levels after 6 months or more following proper rehabilitation programme. No significant difference in results with respect to the age was recorded | Van Loon et al. (2014) |

(continued)

**Table 12.5** (continued)

| Clinical case | Number of animals | Study period | Biomaterial used | Cell dose/ assembly | Evaluation criteria | Overall result | References |
|---|---|---|---|---|---|---|---|
| Proximal suspensory ligament injury (treated for 4 months by NSAIDs and PRP; AAEP scale 2/5) | 01 (case report) | 1 year | Allogeneic tenogenically induced peripheral blood-derived MSCs | $0.5 \times 10^6$ cells tenogenically differentiated; second dose at 12 weeks after the first injection allogeneic tenogenically induced MSCs + PRP | American Association of Equine Practitioners (AAEP) scale and ultrasonography, radiographic examination, computed tomography | From 4 weeks after treatment, the lameness reduced to an AAEP grade 1/5 and a clear filling of the lesion could be noticed on ultrasound. At 4 weeks after the second injection, the horse trotted sound under all circumstances with a close to total fiber alignment. The horse went back to previous performance level at 32 weeks after the first regenerative therapy and is currently still doing so (i.e., 20 weeks later or 1 year after the first stem cell treatment) | Vandenberghe et al. (2015) |
| Superficial ligament and superficial digital flexor tendon lesions | SL ($n = 68$); SDFT ($n = 36$) | 24 months | Allogeneic tenogenically induced peripheral blood MSCs | Cryopreserved $2–3 \times 10^6$ (at least 50% viability) in 1 mL PRP | Clinical, lameness, and ultrasonographic evaluation | After 12th month, 83.8% of SDFT and 79.2 of SL returned to previous performance level. At 24th month, 82.4% and 85.7% of SL and SDFT cases returned to previous performance level, respectively. Cell therapy significantly lowered reinjury rate | Beerts et al. (2017) |

2010), autologous serum (Pacini et al. 2007), and plasma lysate (Del Bue et al. 2008; Renzi et al. 2013; Vandenberghe et al. 2015) had improved tendon repair especially in relation to the functional recovery.

The intralesional injected MSCs had remained viable, distributed, and integrated within the injured tissue. The cell numbers had decreased with the time, and small numbers had been identified in the peripheral blood up to 24 h and in the contralateral tendon lesions up to 24 weeks (Burk et al. 2016). Even up to 34 days, the cells had been located in the injected lesions (Guest et al. 2008). AD-MSCs tend to decrease inflammation markedly and improve alignment of collagen fibrils and tendon fibers (Nixon et al. 2008). The cell concentration, however, appears to have a role to play. One million cells had been insufficient for effective tendon healing (Smith 2008; Caniglia et al. 2012), while higher cell concentrations like ten million (Smith et al. 2009) and 20 million (Van Loon et al. 2014) had significantly improved tissue matrix approximating to the normal tendon tissue.

Delayed implantation (day 30) of AD-MSCs in superficial digital flexor tendonitis too had improved healing as better histological and immunohistochemical scores were demonstrated. However, an insignificant difference in sonographic and clinical signs had been demonstrated between the cell-treated or nontreated cases (Carvalho et al. 2011). Combined application of BM-MSCs and IGF-1 had improved healing of the superficial digital flexor tendonitis. Tendons treated with cells and IGF-1 were histologically better in comparison to those treated with BM-MSCs only. The cell-treated tendons had better histological scores as compared to the control (Schnabel et al. 2009).

Even clinical improvement with the application of MSCs together with the routine managemental procedures has been demonstrated. The clinical studies on ligament and tendon injuries had demonstrated that AD-MSCs and BM-MSCs had improved most of the cases (85%) and these animals had returned back to competition (Ferris et al. 2009; Leppanen et al. 2009). Animals of diverse age, gender, and breed had similar outcome (Ferris et al. 2009). Single case report of superficial digital flexor tendonitis locally treated with BM-MSCs and growth factors along with natural stimulus had improved tendon healing. The stimulus might have triggered MSCs differentiation into tendon fibroblasts (Smith et al. 2003). Apart from growth factors, platelet-rich plasma (PRP) scaffold laden allogeneic AD-MSCs (Del Bue et al. 2008) or PB-MSCs (Beerts et al. 2017) had improved tendonitis and tendon/ligament injuries, respectively. Most of these animals had returned back to the normal activity. Combined application of allogeneic PB-MSCs and platelet-rich plasma (PRP) had moderately improved suspensory ligament (SL) and superficial digital flexor tendon (SDFT) sonographic features at 6-week period that had further improved at 24 months. The reinjury rate has markedly reduced in these cell-treated cases (18%) in comparison to conventionally treated horses (44%) (Beerts et al. 2017). Similar biomaterial combination used at 4 weeks after diagnosis and repeated at 32-week period in a proximal suspensory ligament desmitis had completely restored its fiber alignment being observed under ultrasonography. These animals had improved performance almost similar to the initial levels (Vandenberghe et al. 2015). An equine tendonitis study had failed to demonstrate significant results with AD-MSCs in comparison to the autologous inactivated serum. However, better histopathological observations were observed in the AD-MSC-treated tendons than control (Geburek et al. 2017).

One of the studies had failed to observe improved effects of MSCs on tendonitis and might have been due to the delayed cell implantation (1 month) (Caniglia et al. 2012). Cell induced pro-healing effects may be affected by inflammation at early stage, while limited results at delayed implantation might occur due to the developed fibrous tissue (Dahlgren 2009; Carvalho et al. 2011; Carvalho et al. 2013a). Further, the evaluation criteria and period of evaluation may also affect the results (Caniglia et al. 2012).

These studies favor MSCs application in tendon/ligament injuries, although variably. The evaluation criteria followed must be streamlined and the studies that incorporate various biomaterials must be compared with standard to come to the conclusion. Cells of uniform batches with respect to source, concentration, and passage must be applied through standard routes for effective comparison. Advanced techniques may also be applied to demonstrate cellular fate (Geburek et al. 2016). The proper follow-up of the studies and for extended period may be made.

### 12.3.4 Cartilage Injury

Cartilage healing is a challenge as it has limited intrinsic regenerative potential (Kinner et al. 2005; Kang et al. 2008; Campos et al. 2012). The healing usually occurs with mechanically inferior fibrous tissue that shows nonintegration to the native tissue (Campos et al. 2012; Gugjoo et al. 2017).

The stem cell osteochondral defect healing has been studied for various joints but mostly in the stifle joint both in preclinical experimental (Table 12.6) and in clinical (Table 12.7) cases. The critical defect size has been variable. Various MSCs sources like bone marrow, adipose tissue, peripheral blood, and synovium have been evaluated for osteochondral healing though the former two sources have been subject of most of the studies. The cell concentrations implanted in these studies have been variable ranging from five million to 20 million. Apart from simple MSCs, the chondrogenically differentiated allogeneic MSCs have also been evaluated (Broeckx et al. 2019). As MSCs fail to adhere or stick to the osteochondral defects, scaffolds like collagen repair patch (González-Fernández et al. 2016), allogeneic plasma (Broeckx et al. 2019), allogeneic PRP (Broeckx et al. 2014), fibrin glue (Raheja et al. 2011), and hyaluronan (McIlwraith et al. 2011) have been incorporated. The healing or clinical improvement in these studies has been evaluated by variable parameters including clinical examination, radiography, computed tomography (CT) and magnetic resonance imaging (MRI), arthroscopy or biomechanical testing, and histochemical examination. Most of the osteoarthritis studies have failed to demonstrate effective cartilage repair (Wilke et al. 2007; Frisbie et al. 2009; McIlwraith et al. 2011), but beneficial effect of pain reduction has been observed (Ferris et al. 2009; González-Fernández et al. 2016).

Combined application of BM-MSCs and hyaluronic acid or autologous platelet enhanced fibrin scaffolds (APEF) had been made (McIlwraith et al. 2011; Goodrich et al. 2016). In both these studies, clinical improvement had failed to occur at 12 months. Slight improvement with respect to gross morphology and arthroscopy had been reported at 12-month period (McIlwraith et al. 2011). The cobblestone-like appearance of healed tissue bearing fair to good integration at 3 months was demonstrated. However with time the tissue had degraded (Goodrich et al. 2016). Thus, extended studies are desired to see the course of disease.

In one of the comparative study, an equine meniscal tear model treated with MSCs of two sources (BM-MSCs or AD-MSCs) laden in collagen repair patch had demonstrated that cells of different sources have comparable healing effect. The defects were repaired by fibrocartilage in cell-treated animals, while in control group repair was either partial or nil (González-Fernández et al. 2016). Another comparative study had demonstrated higher pro-healing effects of BM-MSCs on osteoarthritis as compared to uncultured stromal vascular fraction (SVF). Joints treated with SVF had unwanted increase in tumor necrosis factor-α (TNF-α) concentration in synovial fluid (Frisbie et al. 2009). The combined treatment involving bone marrow aspirate implantation and microfracture too had favored cartilage regeneration in comparison to the microfracture technique alone. Microfracture enhances bone marrow mononuclear cell fraction availability for cartilage healing but may not have been sufficient (Fortier et al. 2010). One of the studies had compared autologous BM-MSCs treatment to the arthroscopy in joint affections. Out of total cases (n = 33) treated with intra-articular cell (15,020 million) implantation, 43% reached to previous level of work, 33% reached to work, and 24% had failed to work. In three cases, transient joint flare had been observed within 24 h of cell treatment. Joints treated with arthroscopy had significantly lower recovery rate (60–63%) (Ferris et al. 2014).

Inflammatory environment tends to reduce cell therapeutic effects. MSCs express higher adhesion molecules and have decreased migration potential under inflammation (Barrachina et al. 2016; Reesink et al. 2017). Additionally, these cells show reduced potential to secrete cartilage-specific matrices in the presence of pro-inflammatory cytokines (Taylor et al. 2007; Zayed et al. 2016) although such properties of MSCs may also be affected by tissue source (Zayed et al. 2016). Thus, efforts to reduce inflammation in the joints before application of MSCs should be made (Ferris et al. 2014). Currently, the local MSCs implantation is made at 20 million along with the scaffold (hyaluronan; González-Fernández et al. 2016).

### 12.3.5 Bone Injury

Bone heals naturally with healed tissue approximating to natural native tissue (Taylor et al. 2007). The healing of a fractured bone in horse is challenging due to its uncontrolled movement. In case of extensive fractures or tumors, the ability to heal becomes even more challenging. To improve healing in such conditions, tissue

**Table 12.6** In vivo chondrogenic preclinical experimental mesenchymal stem cell studies in horse

| Model type | Number of animals | Model defect size/study period | Biomaterial used | Cell dose | Evaluation criteria | Overall result | References |
|---|---|---|---|---|---|---|---|
| Chronic full thickness cartilage defects in femoro-patellar articulations | 6 (12 defects) group I: Autogenous fibrin vehicle containing MSCs; group II: Autogenous fibrin alone in control joints | 15 mm/1 month and 8 months | MSCs loaded into self-polymerizing autogenous fibrin vehicle | MSCs/fibrinogen mixture containing $12 \times 10^6$ MSC/mL | Histology, histochemistry, collagen type I and type II, immunohistochemistry, collagen type II in situ hybridization, and matrix biochemical assays | Arthroscopic scores in group I were significantly improved at the day 30. Biopsy showed MSC-implanted defects contained increased fibrous tissue with several defects containing predominantly type II collagen. At 8 months, there was no significant difference between stem cell-treated and control defects | Wilke et al. (2007) |
| Middle carpal joint osteochondral defect | 24 (8 each group) group I: AD-MSCs; group II: BM-MSC; group III: Control | 15 mm/70 days | AD-MSCs and BM-MSCs | $16.3 \times 10^6$ (AD-MSCs) and $10.5 \times 10^6$ (BM-MSCs) | Clinical outcome, macroscopy, histopathology, articular cartilage matrix evaluation | Nonsignificant differences in healing improvement with cell treatment were seen compared to control except improvement in PGE2 levels | Frisbie et al. (2009) |

| | | | | | | | |
|---|---|---|---|---|---|---|---|
| Full thickness femoral condyle defects followed by microfracture | 10 (20 defects in stifle joint) group I: BM-MSCs and microfracture; group II: Microfracture | 10 mm/6 months and 12 months | MSCs and hyaluronan | $20 \times 10^6$MSCs + 22 mg hyaluronan | Radiography, arthroscopy, magnetic resonance imaging and gross, histologic, histomorphometric, immunohistochemical, and biochemical examinations | Nonsignificant evidence of any clinically significant improvement in the joints with BM-MSCs. Arthroscopic and gross evaluation confirmed a significant increase in repair tissue firmness and a trend for better overall repair tissue quality (cumulative score of all arthroscopic and gross grading criteria) in BM-MSC-treated joints. Immunohistochemical analysis showed significantly greater levels of aggrecan in repair tissue treated with BM-MSC injection. There were no other significant treatment effects | McIlwraith et al. (2011) |

(continued)

**Table 12.6** (continued)

| Model type | Number of animals | Model defect size/study period | Biomaterial used | Cell dose | Evaluation criteria | Overall result | References |
|---|---|---|---|---|---|---|---|
| Chronic chondral defects in the medial femoral trochlea | 4 (8 defects) group I; MSCs; group II: Control | 10 mm/ months | AD-MSCs | $1.35 \times 10^7$ | Macroscopic, histopathological, and histochemical evaluations | The use of MSC in the treatment of chondral defects minimized joint inflammation, as confirmed by synovial fluid analysis. The treatment resulted in an improved repair tissue, verified by macroscopic examination and histochemical and histopathological analysis | Yamada et al. (2013) |
| Chondral defect model | 12 (24 defects) group I: BM-MSCs + APEF; group II: APEF | 15-mm defect in lateral trochlear ridge/3 months and 12 months | BM-MSCs along with autologous platelet enhanced fibrin scaffolds (APEF) | $12 \times 10^6$ | Arthroscopy, histological examination, magnetic resonance imaging (MRI), micro-computed tomography (micro-CT), and biomechanical testing | Addition of BM-MSCs to APEF did not enhance cartilage repair and stimulated bone formation in some cartilage defects | Goodrich et al. (2016) |

| | | | | | | | |
|---|---|---|---|---|---|---|---|
| Osteochondral defect (bilateral distal radius, 6 mm × 5 mm) | 3 (right-side implanted with cells; left without cells) | 9 weeks | Synovium-derived MSCs | $5 \times 10^6$ (each in right defect and radiocarpal joint) | CT and histology | Radiolucent volumes of defects calculated by analysis of postmortem CT images were decreased in implanted sites compared with control sites in all horses. The average scores for ICRS gross and histopathological grading scales in implanted sites were equal to or higher than those of the controls. These results suggest that allogeneic implantation and subsequent autologous injection of SM-MSCs might not obstruct subchondral bone formation in defects | Yamasaki et al. (2018) |

(continued)

**Table 12.6** (continued)

| Model type | Number of animals | Model defect size/study period | Biomaterial used | Cell dose | Evaluation criteria | Overall result | References |
|---|---|---|---|---|---|---|---|
| Stifle injury/femorotibial lesions (meniscal, cartilage, or ligamentous; case controlled) | 39 (33 had reports available) | 24 months | Autologous BM-MSCs + 20 mg of HA | 15–20 × $10^6$ | American Association of Equine Practitioners (AAEP) lameness score, graded severity of lesion | 43% of horses returned to previous level of work, 33% returned to work, and 24% failed to return to work. In horses with meniscal damage (n¼24), a higher percentage in the current study (75%) returned to some level of work compared to those in previous reports (60–63%) that were treated with arthroscopy alone, which resulted in a statistically significant difference between studies (P¼0.038). Joint flare post-injection was reported in 3 horses (9.0%); however, no long-term effects were noted | Ferris et al. (2014) |

**Table 12.7** In vivo chondrogenic clinical mesenchymal stem cell studies in horse

| Clinical condition/ailment | Number of animals | Study period | Biomaterial used | Cell dose | Evaluation criteria | Overall result | References |
|---|---|---|---|---|---|---|---|
| Bilateral articular cartilage fissure defects of the medial femoral condyles and concurrent cranial cruciate ligament injury | Single case | 4 months and 15 months | Autologous BM-MSCs + fibrin glue | $1 \times 10^7$ implanted at 90 days, 3 months, and 13 months period | Arthroscopic examination and performance records | At 4 months marked cartilage surface smoothing and reduction in the cartilage defect depth. Further, moderate improvement in the cranial cruciate ligament was observed. After 15 months of the initial MSC treatment, the horse returned to racing and had comparable race earning to that of preinjury records | Raheja et al. (2011) |
| Chronic degenerative joint disease of pastern joint | Single case | 4, 8, and 12 weeks and 1 year period | Peripheral blood-derived MSCs | $2.5 \times 10^6$ | American Association of Equine Practitioners (AAEP), radiography and pressure plate analysis | Clinical improvement observed and pressure plate analysis showed load rate (LR) symmetry ratio increased considerably in both gaits, indicating an increased speed of loading at the walk as well as at the trot. A clear improvement in peak vertical force (PVF) and vertical impulse (VI) symmetry ratios was evident at the trot, indicating an increased symmetry of the weight-bearing function of the forelimbs | Spaas et al. (2012) |

(continued)

**Table 12.7** (continued)

| Clinical condition/ailment | Number of animals | Study period | Biomaterial used | Cell dose | Evaluation criteria | Overall result | References |
|---|---|---|---|---|---|---|---|
| Degenerative joint disease of fetlock | 20 (group I: PRP; group II: MSCs; group III: MSCs and PRP; or group IV: Chondrogenic-induced MSCs and PRP) | 6 and 12 weeks and 6 and 12 months | Peripheral blood MSCs, platelet-rich plasma (PRP) | $6.7 \times 10^3$ MSCs/$cm^2$ cells chondrogenically induced + $200 \times 10^6$ platelets | American Association of Equine Practitioners (AAEP) | Group 4 animals generated the highest evolution scores, although it had statistically insignificant difference against group 4 in the early or late evolution score | Broeckx et al. (2014) |
| Bilateral meniscus defect (10 mm in length and 6 mm in depth) | 6 (3 treated with BM-MSCs; 3 treated with AD-MSCs) | 12 months | BM-MSCs and AD-MSCs + collagen repair patch | $0.5 \times 10^6$ cells | Macroscopic assessment of meniscal repair, histochemical analysis | At 12 months after treatment, cell-treated defects were regenerated with fibrocartilaginous tissue, whereas untreated defects were partially repaired or not repaired | González-Fernández et al. (2016) |
| Osteochondral fragment-induced osteoarthritis | 10 (cell and surgery treated, $n = 5$; surgery treated only, $n = 5$) | 30, 60, and 90 days | Allogeneic BM-MSCs | $1 \times 10^7$ | Physical examination, osteophyte formation, racing performance, and synovial fluid analysis | MSC treatment group had an improved physical examination score, better racing performance, lesser osteophyte formation, and mononuclear cells in synovial fluid than the conventional treatment group | Korchunjit et al. (2019) |
| Early stage fetlock degenerative joint disease | 75 (50 in treated and 25 in control) | 3, 6, and 18 weeks and 1 year | Chondrogenically induced allogeneic MSCs + allogeneic plasma | $2 \times 10^6$ cells + $\sim 85 \times 10^6$ platelets | AAEP score system | Lameness scores ($P < 0.001$), flexion test responses ($P < 0.034$), and joint effusion scores ($P < 0.001$) | Broeckx et al. (2019) |

engineering employing MSCs may be evaluated. In case of horse, very limited MSCs osteogenic studies, however, could be located.

An experimental study had been aimed to enhance the arthrodesis of pastern joint by utilizing tissue engineered construct (stabilized by locking compression plate). The engineered construct composed of BM-MSCs and scaffold and growth factor had enhanced bone formation and resulted in arthrodesis (Table12.8) (Seo et al. 2014). In metacarpal ostectomy model, periosteum MSCs committed to osteoprogenitor cells along with fibrin glue, however, had failed to show improved radiographic and histological features in comparison to the fibrin glue alone (McDuffee et al. 2012). This questions the role of osteoprogenitor cells in bone healing, although other reasons including limited osteoprogenitor cell concentration and scaffold type involved may also affect the outcome. A single horse clinical case report had demonstrated improved hoof regeneration with AD-MSCs after 3 weeks and complete fracture healing at 3 months (Table12.9) (Marycz et al. 2012). However, detailed in vivo osteogenic studies of eMSCs are currently unavailable.

### 12.3.6 Immune-Mediated, Inflammatory, and Ischemic Diseases

A lot of literature focusing on MSCs immunomodulation is available. These studies have demonstrated cellular ability to control immune reactions. MSCs have prevented such reactions under in vitro and/or in vivo experimental studies (Augello et al. 2005; Einstein et al. 2007; Sordi and Piemonti 2011). MSCs applications in equine recurrent uveitis, systemic lupus erythematosus (SLE), and endotoxemia that are immune-mediated and/inflammatory diseases may also be evaluated. Such studies can additionally form basis of therapeutics in human medicine as the condition in two species shares pathophysiology (Deeg et al. 2002; Liang et al. 2010; Sun et al. 2009; Shi et al. 2012). MSCs inhibit proliferation of B cell and memory $CD8^+$ T cell while T regulatory cell proliferation is promoted (Peroni and Borjesson 2011; Im et al. 2014; Cortinovis et al. 2015).

### 12.3.7 Laminitis

Laminitis is an ischemic clinical condition in horse usually exhibiting loss of p63 positive epidermal stem cells (Carter et al. 2011). Application of MSCs in clinical laminitis has been demonstrated to provide add-on benefits to the traditional managemental care (Table12.10). Route of MSCs application seems to play an important role. Intravenous injection of allogeneic BM-MSCs had shown better results as compared to their epidermal injection (Sole et al. 2012). The intra-arterial infusion had resulted in retention of MSCs in the distal limb, with a symmetrical and homogeneous distribution. Contrarily, intravenous regional limb perfusion (IVRLP) had resulted in an asymmetrical and heterogeneous distribution within the hoof. The IVRLP performed on standing, sedated horses had resulted in lower and more

**Table 12.8** In vivo preclinical experimental mesenchymal stem cell studies on osteogenesis in horse

| Model type | Number of animals | Model defect size/study period | Biomaterial used | Cell dose/ assembly | Evaluation criteria | Overall result | References |
|---|---|---|---|---|---|---|---|
| 4th metacarpal ostectomy | 10 (20 metacarpal, 10 control, and 10 experimental) | 1 cm/12 weeks | Periosteum-derived osteoprogenitor cells (osteogenic lineage differentiated) | $20 \times 10^6$ | Radiography and histology | MSCs along with fibrin glue-treated animals failed to show improved radiographic and histological features in comparison to the fibrin glue alone | McDuffee et al. (2012) |
| Pastern joint arthrodesis | 6 (12; 6 control and 6 experimental) | Locking compression plate stabilization | BM-MSC, gelatin/β--tricalcium phosphate and BMP-2 | $1 \times 10^7$ | Radiography, histology, computed tomography | An improved bone formation and subsequent arthrodesis in comparison to the use of locking compression plate application had occurred | Seo et al. (2014) |
| Splint bones defect (4 defects each in second metacarpal and metatarsal bone of right limbs and 4thmetacarpal and metatarsal in left limbs) | 6 animals (group A: BM-MSCs + BMP-2 in sponge; group B: BM-MSCs in sponge; group C: BMP-2 in sponge; group D: Saline) | 1 cm/16 weeks | BM-MSCs, BMP-2, and gelatin β-tricalcium phosphate sponges | Autologous BM-MSCs and BMP-2 in gelatine sponge | Radiography, CT, and histology | Group A had statistically significant radiographic and histological scores compared to control, while CT scores were statistically significant than other groups | Tsuzuki et al. (2014) |

**Table 12.9** In vivo clinical mesenchymal stem cell studies on osteogenesis in horse

| Clinical condition/ailment | Number of animals | Study period | Biomaterial used | Cell dose | Evaluation criteria | Overall result | References |
|---|---|---|---|---|---|---|---|
| Phalanx digitalis distalis fracture | 1 | 12 weeks after surgery | Orthopedic screw wrapped with plasma protein adhesive/ MSCs biofilm | $6 \times 10^6$ in plasma protein adhesive biofilm | Clinical evaluation and radiography | After 3 months, hoof was significantly regenerated with bone completely regenerated | Marycz et al. (2012) |

**Table 12.10** In vivo mesenchymal stem cell applications in horse laminitis

| Clinical condition/ailment | Number of animals | Study period | Biomaterial used | Cell dose | Evaluation criteria | Overall result | References |
|---|---|---|---|---|---|---|---|
| Chronic laminitis | 12 | 14 months | Allogeneic UC-MSCs | $20–25 \times 10^6$ (3–4 treatments) (retrograde venous digital perfusion) | – | 10/12 horses (83%) responded well. However, long-term success rates still need to be determined. The first "successful" case is only 14 months out from initial treatment | Morrison et al. (2014) |
| Chronic laminitis | 30 (onset of laminitis < 30 days [10 cases], 30–60 days [4 cases], 60–90 days [2 cases], and > 90 days [14 cases]) | 4 months | Allogeneic UC-MSCs or BM-MSCs | $20–30 \times 10^6$ (4 treatments at monthly interval) (retrograde venous digital perfusion, using the palmar/plantar digital vein) | Radiographically | Of the 30 cases, 21 patients (70%) had a successful outcome. In the cases receiving stem cells <30 days after the onset of laminitis, the success rate was 100%. The success rate in the cases receiving stem cells >90 days after the onset of laminitis was only 50%. Statistical analysis is pending | Dryden et al. (2013) |
| Chronic laminitis (III, IV grade) | 9 | 1 year | Allogeneic AD-MSCs (first dose) and autologous AD-MSCs (two doses) | $15 \times 10^6$ in 15 mL PRP (implanted thrice at 1 month interval) | Clinical observations and venography | Progressive amelioration of vascularization of foot. An improvement in structure and function of hoof. At 1 year, 7/9 horses were in activity with 2 developing recurrent laminitis. After 1 year, only two animals could be seen in activity and rest either deceased due to colic or were euthanized (recurrent laminitis cases). Two cases were not observed after 1 year | Angelone et al. (2017) |

variable retention of MSCs and poor distribution within the hoof. Coronary band injection had failed to show any migration of MSCs within the hoof (Spriet et al. 2013).

MSCs may prove beneficial within 30 days of third phalanx displacement after which lamellar wedge develops (Dryden et al. 2013; Morrison et al. 2014). Post lamellar wedge development, little observable improvement with MSCs has been reported. Autologous or allogeneic MSCs from adipose tissue and umbilical cord have been used. These cells have been injected more than once along with other biomaterials like platelet-rich plasma. An overall improvement in the hoof anatomy (based on quality and shape) and laminitic foot weight bearing and mobility had been demonstrated (Angelone et al. 2017).

### 12.3.8 Neurological Problems

Neuron is a highly differentiated cell with almost no regenerative power. Stem cells though are present in the nervous tissue, but their role is yet to be understood. Various studies including that of horse have variably supported neural or mesenchymal stem cell differentiation into the neuron-like cell (Hoynowski et al. 2007; Jamnig and Lepperdinger 2012; Villagrán et al. 2016). Endogenous neural stem cells (NSCs) tend to generate scar tissue that limits further injury but may hamper neurological activity (Stenudd et al. 2015; Watanabe et al. 2015). Currently, stem cell pro-healing effects may be due to their ability to secrete trophic factors (Sadan et al. 2009). In vivo preliminary MSCs studies in other species appear promising as improved clinical parameters have been demonstrated post-spinal cord injury (Ra et al. 2011; Bhat et al. 2019). In case of horse, commonly encountered affections are of spinal cord, peripheral nerves, and laryngeal hemiplegia. However, the relevant studies remain to be conducted. One study utilizing BM-MSCs as therapeutic for transected portion of peripheral nerve had failed to demonstrate any observable improvement (Table 12.11) (Villagrán et al. 2016). An extensive research in the field is desired to conclude MSCs role, if any in the nerve regeneration. Apart from peripheral nerve injuries, MSCs role in neurological disorders like equine myeloencephalopathy and equine motor neuron diseases may also be evaluated.

### 12.3.9 Reproductive Disorders

MSCs role has been evaluated in horse endometriosis (Mambelli et al. 2013; Corradetti et al. 2014; Falomo et al. 2015; Rink et al. 2018). MSCs in utero implantation has been made like that of artificial insemination. MSCs derived from amnion (AM-MSCs) or endometrium (End-MSCs) may potentially replenish endometrium in pregnancy failure (Corradetti et al. 2014; Falomo et al. 2015). Intrauterine-implanted cells had failed to migrate to healthy endometrium while these cells had migrated endometriosed glandular and periglandular tissues (Mambelli et al. 2013;

**Table 12.11** In vivo preclinical mesenchymal stem cell study on nerve regeneration in horse

| Model type | Number of animals | Model defect size/study period | Biomaterial used | Cell dose/assembly | Evaluation criteria | Overall result | References |
|---|---|---|---|---|---|---|---|
| Excision of lateral and medial stumps of ramus communicans | 3 horses (bilateral incision) | 45 days | Allogeneic BM-MSCs | $10 \times 10^6$ | Clinical and histological examination | No evidence of nerve regeneration at 45 days. No difference in histology of treated and control animals | Villagrán et al. (2016) |

Rink et al. 2018). These cells had shown immunomodulation (Falomo et al. 2015) and decreased inflammation (Rink et al. 2018) in endometritis.

MSCs may also play role to improve fertility by enhancing the follicular growth. MSCs tend to increase follicular number in ovary, although with an insignificant effect on dynamism (Grady et al. 2016).

### 12.3.10 Miscellaneous Conditions

MSCs may also be evaluated in inflammatory lung disease like recurrent airway obstruction (heaves). Stem cells tend to proliferate pulmonary vessel cells, decrease apoptotic cell number, and inhibit pulmonary vessel muscularization (Lee et al. 2015). MSCs have been trans-differentiated to other cell types like hepatocyte-like cells and myocyte-like cells (Reed and Johnson 2008; Lin et al. 2015; Pennington et al. 2016), but their in vivo therapeutic applications remain to be evaluated.

## 12.4 Conclusion(S)

Stem cells hold a great potential to act as an all-in-one therapeutic option for diverse ailments. MSCs generally follow principles of ISCT but may show differences in cell surface marker expression with respect to their sources. eMSCs from bone marrow, adipose tissue, or fetal tissues have mainly been evaluated for therapeutics. Numerous horse stem cell studies have demonstrated favorable results though without any definitive conclusion(s). MSCs from various tissue sources, together with variable routes, concentration, and biomaterials have been utilized in these studies. The conducted studies are uncontrolled, lacking sufficient sample size. Thus, extensive case controlled blinded studies involving uniform cell source, concentration, and biomaterials are desired in order to arrive at any conclusion.

## References

Aigner T, Cook J, Gerwin N, Glasson S, Laverty S, Little C, McIlwraith W, Kraus V (2010) Histopathology atlas of animal model systems—overview of guiding principles. Osteoarthr Cartil 18:S2–S6

Alipour F, Parham A, Kazemi Mehrjerdi H, Dehghani H (2015) Equine adipose-derived mesenchymal stem cells: phenotype and growth characteristics, gene expression profile and differentiation potentials. Cell J 16(4):456–465

Andrew SA, Willis AM (2005) Diseases of the cornea and sclera. In: Gilger BC (ed) Equine ophthalmology. Elsevier Saunders, St. Louis, pp 157–251

Angelone M, Conti V, Biacca C, Battaglia B, Pecorari L, Piana F, Gnudi G, Leonardi F, Ramoni R, Basini G, Dotti S, Renzi S, Ferrari M, Grolli S (2017) The contribution of adipose tissue-derived mesenchymal stem cells and platelet-rich plasma to the treatment of chronic equine laminitis: a proof of concept. Int J Mol Sci 18:2122

Augello A, Tasso R, Negrini SM, Amateis A, Indiveri F, Cancedda R, Pennesi G (2005) Bone marrow mesenchymal progenitor cells inhibit lymphocyte proliferation by activation of the programmed death 1 pathway. Eur J Immunol 35:1482–1490

Baghavan M, Taghiyar L, Dehghan M, Fallahi F, Kazemi H (2009) Equine marrow-derived mesenchymal stem cells: isolation, differentiation and culture optimization. Iran J Vet Res 10:1–26
Barberini DJ, Paiva Freitas NP, Magnoni MS, Maia L, Listoni AJ, Heckler MC, Sudano MJ, Golim MA, da Cruz Landim-Alvarenga F, Amorim RM (2014) Equine mesenchymal stem cells from bone marrow, adipose tissue and umbilical cord: immuno-phenotypic characterization and differentiation potential. Stem Cell Res Ther 5(1):25
Barrachina AR, Remachaa A, Romeroa FJ, Fortier LA (2016) Effect of inflammatory environment on equine bone marrow derived mesenchymal stem cells immunogenicity and immunomodulatory properties. Vet Immunol Immunopathol 171:57–65
Barreira APB, Alves ALG, Saito ME, Gómeza J, Jarrína CD, Ramírezb A, Galecioa JS (2008) Autologous implant of bone marrow mononuclear cells as treatment of induced equine tendinitis. Int J Appl Res Vet Med 6:46–54
Beerts C, Suls M, Broeckx SY, Seys B, Vandenberghe A, Declercq J, Duchateau L, Vidal MA, Spaas JH (2017) Tenogenically induced allogeneic peripheral blood mesenchymal stem cells in allogeneic platelet-rich plasma: 2-year follow-up after tendon or ligament treatment in horses. Front Vet Sci 4:158
Bhat IA, Sivanarayanan TB, Somal A, Pandey S, Bharti MK, Panda BSK, Indu B, Verma M, Anand J, Sonwane A et al (2019) An allogenic therapeutic strategy for canine spinal cord injury using mesenchymal stem cells. J Cell Physiol 234(3):2705–2718
Bi Y, Ehirchiou D, Kilts T et al (2007) Identification of tendon stem/progenitor cells and the role of the extracellular matrix in their niche. Nature Med 13:1219–1227
Bohannon LK, Owens SD, Walker NJ, Carrade DD, Galuppo LD, Borjesson DL (2013) The effects of therapeutic concentrations of gentamicin, amikacin and hyaluronic acid on cultured bone marrow-derived equine mesenchymal stem cells. Equine Vet J 45:732–736
Borena BM, Pawde AM, Pal A, Aithal HP, Kinjavdekar P, Singh R, Kumar D (2009) Autologous bone marrow-derived cells for healing excisional dermal wounds of rabbits. Vet Rec 165(19):563–568
Branly T, Bertoni L, Contentin R, Rakic R, Gomez-Leduc T, Desancé M, Hervieu M, Legendre F, Jacquet S, Audigié F, Denoix JM, Demoor M, Galéra P (2017) Characterization and use of equine bone marrow mesenchymal stem cells in equine cartilage engineering. Study of their hyaline cartilage forming potential when cultured under hypoxia within a biomaterial in the presence of BMP-2 and TGF-ß1. Stem Cell Rev Rep 13(5):611–630
Broeckx S, Zimmerman M, Crocetti S, Suls M, Marien T, Ferguson SJ, Chiers K, Duchateau L, Franco-Obregón A, Wuertz K et al (2014) Regenerative therapies for equine degenerative joint disease: a preliminary study. PLoS One 9(1):e85917
Broeckx SY, Seys B, Suls M, Vandenberghe A, Marien T, Adriaensen E, Declercq J, Van Hecke L, Braun G, Hellmann K et al (2019) Equine allogeneic chondrogenic induced mesenchymal stem cells are an effective treatment for degenerative joint disease in horses. Stem Cells Dev 28(6):410–422
Bundgaard L, Stensballe A, Elbæk KJ, Berg LC (2018) Mapping of equine mesenchymal stromal cell surface proteomes for identification of specific markers using proteomics and gene expression analysis: an in vitro cross-sectional study. Stem Cell Res Ther 9:288
Burk J, Badylak SF, Kelly J, Brehm W (2013) Equine cellular therapy-from stall to bench to bedside? Cytometry A 83(1):103–113
Burk J, Berner D, Brehm W, Hillmann A, Horstmeier C, Josten C, Paebst F, Rossi G, Schubert S, Ahrberg AB (2016) Long-term cell tracking following local injection of mesenchymal stromal cells in the equine model of induced tendon disease. Cell Transplant 25(12):2199–2211
Bussche L, Harman RM, Syracuse BA, Plante EL, Lu Y-C, Curtis TM, Ma M, Van de Walle GR (2015) Microencapsulated equine mesenchymal stromal cells promote cutaneous wound healing in vitro. Stem Cell Res Ther 6:66
Cabezas J, Rojas D, Navarrete F, Ortiz R, Rivera G, Saravia F, Rodriguez-Alvarez L, Castro FO (2018) Equine mesenchymal stem cells derived from endometrial or adipose tissue share significant biological properties, but have distinctive pattern of surface markers and migration. Theriogenology 106:93–102

Campos DFD, Drescher W, Rath B, Tingart M, Fischer H (2012) Supporting biomaterials for articular cartilage repair. Cartilage 3(3):205–221
Caniglia CJ, Schramme MC, Smith RK (2012) The effect of intralesional injection of bone marrow derived mesenchymal stem cells and bone marrow supernatant on collagen fibril size in a surgical model of equine superficial digital flexor tendonitis. Equine Vet J 44:587–593
Carrade DD, Affolter VK, Outerbridge CA, Watson JL, Galuppo LD, Buerchler S, Kumar V, Walker NJ, Borjesson DL (2011a) Intradermal injections of equine allogeneic umbilical cord-derived mesenchymal stem cells are well tolerated and do not elicit immediate or delayed hypersensitivity reactions. Cytotherapy 13(10):1180–1192
Carrade DD, Owens DS, Galuppo LD, Vidal MA, Ferraro GL, Librach F, Buerchler S, Friedman MS, Walker NJ, Borjesson DL (2011b) Clinicopathologic findings following intra-articular injection of auto-logous and allogeneic placentally derived equine mesen-chymal stem cells in horses. Cytotherapy 13:419–430
Carstanjen B, Desbois C, Hekmati M, Behr L (2006) Successful engraftment of cultured autologous mesenchymal stem cells in a surgically repaired soft palate defect in an adult horse. Can J Vet Res 70:143–147
Carter RA, Engiles JB, Megee SO, Senoo M, Galantino-Homer HL (2011) Decreased expression of p63, a regulator of epidermal stem cells, in the chronic laminitic equine hoof. Equine Vet J 43(5):543–551
Carvalho AM, Alves ALG, Gomes de Oliveira PG, Álvarez LEC, Renée CAH, Elenice D (2011) Use of adipose tissue-derived mesenchymal stem cells for experimental tendinitis therapy in equines. J Equine Vet Sci 31:26–34
Carvalho AD, Badial PR, Alvarez LEC, Yamada AL, Borges AS, Deffune E, Hussni CA, Garcia Alves AL (2013a) Equine tendonitis therapy using mesenchymal stem cells and platelet concentrates: a randomized controlled trial. Stem Cell Res Ther 4(85):1–13
Carvalho AM, Yamada ALM, Martins JRB, Maia L, Golim MA, Deffune E, Hussni CA, Alves ALG (2013b) Isolation and characterization of equine peripheral blood-derived multipotent mesenchymal stromal cells. Pesqui Vet Bras 33(9):1151–1154
Chong PP, Selvaratnam L, Abbas AA, Kamarul T (2011) Human peripheral blood derived mesenchymal stem cells demonstrate similar characteristics and chondrogenic differentiation potential to bone marrow derived mesenchymal stem cells. J Orthop Res 30:634–642
Clark KC, Kol A, Shahbenderian S, Granick JL, Walker NJ, Borjesson DL (2016) Canine and equine mesenchymal stem cells grown in serum free media have altered immunophenotype. Stem Cell Rev Rep 12:245–256
Clegg P, Pinchbeck GL (2011) Evidence-based medicine and stem cell therapy: how do we know such technologies are safe and efficacious? Vet Clin North Am Equine Pract 27:373–382
Clode AB, Matthews A (2011) Diseases and surgery of the cornea. In: Gilger BC (ed) Equine ophthalmology, 2nd edn. Elsevier Saunders, St. Louis, pp 181–266
Coli A, Nocchi F, Lamanna R, Iorio M, Lapi S, Urciuoli P, Scatena F, Giannessi E, Stornelli MR, Passeri S (2011) Isolation and characterization of equine amnion mesenchymal stem cells. Cell Biol Int Rep 18(1):23
Conze P, van Schie HTM, van Weeren R, Staszyk C, Conrad S, Skutella T, Hopster K, Rohn K, Stadler P, Geburek F (2014) Effect of autologous adipose tissue-derived mesenchymal stem cells on neovascularization of artificial equine tendon lesions. Regen Med 9(6):743–757
Corradetti B, Correani A, Romaldini A, Marini MG, Bizzaro D, Perrini C, Cremonesi F, Lange-Consiglio A (2014) Amniotic membrane-derived mesenchymal cells and their conditioned media: potential candidates for uterine regenerative therapy in the horse. PLoS One 9(10):e111324
Cortinovis M, Casiraghi F, Remuzzi G, Perico N (2015) Mesenchymal stromal cells to control donor-specific memory T cells in solid organ transplantation. Curr Opin Organ Transplant 20(1):79–85
Crovace A, Lacitignola L, Rossi G, Francioso E (2010) Histological and immunohistochemical evaluation of autologous cultured bone marrow mesenchymal stem cells and bone marrow

mononucleated cells in collagenase-induced tendinitis of equine superficial digital flexor tendon. Vet Med Int 2010:1

Dahlgren LA (2009) Fat-derived mesenchymal stem cells for equine tendon repair. Regen Med 4(Suppl 6):2

De Schauwer C, Piepers S, Van de Walle GR, Demeyere K, Hoogewijs MK, Govaere JL, Braeckmans K, Van Soom A, Meyer E (2012) In search for cross-reactivity to immunophenotype equine mesenchymal stromal cells by multicolor flow cytometry. Cytometry A 81(4):312–323

De Schauwer C, Van de Walle GR, Soom AV, Meyer E (2013) Mesenchymal stem cell therapy in horses: useful beyond orthopedic injuries? Vet Q 33(4):234–241

Deeg CA, Ehrenhofer M, Thurau SR, Reese S, Wildner G, Kaspers B (2002) Immunopathology of recurrent uveitis in spontaneously diseased horses. Exp Eye Res 5:127–133

Dehghan MM, Eslaminejad MB, Motallebizadeh N, Ashrafi Halan J, Tagiyar L, Soroori S, Nikmahzar A, Pedram M, Shahverdi A, Kazemi Mehrjerdi H, Izadi S (2015) Transplantation of autologous bone marrow mesenchymal stem cells with platelet-rich plasma accelerates distraction osteogenesis in a canine model. Cell J 17(2):243–252

Del Bue M, Ricco S, Ramoni R, Conti V, Gnudi G, Grolli S (2008) Equine adipose-tissue derived mesenchymal stem cells and platelet concentrates: their association in vitro and in vivo. Vet Res Commun 32:S51–S55

Desancé M, Contentin R, Bertoni L, Gomez-Leduc T, Branly T, Jacquet S, Betsch J-M, Batho A, Legendre F, Audigié F, Galéra P, Demoor M (2018) Chondrogenic differentiation of defined equine mesenchymal stem cells derived from umbilical cord blood for use in cartilage repair therapy. Int J Mol Sci 19:537

Dominici M, Le Blanc K, Mueller I, Slaper-Cortenbach I, Marini F, Krause D, Deans R, Keating A, Prockop D, Horwitz E (2006) Minimal criteria for defining multipotent mesenchymal stromal cells. The International Society for Cellular Therapy position statement. Cytotherapy 8(4):315–317

Dryden VC, Morrison S, Bras R, Morrell SA (2013) Using stem cells in clinical cases. J Equine Vet Sci 33(10):872–873

DysonS (1997) Treatment of superficial digital flexor tendinitis: a comparison of conservative management, sodium hyaluronate, and glycosaminoglycan polysulfate. In: Proceedings of the 43rd Annual Convention of the American Association of Equine Practitioners, pp 297–300

Edmonds RE, Garvican ER, Smith RKW, Dudhia J (2016) Influence of commonly used pharmaceutical agents on equine bone marrow-derived mesenchymal stem cell viability. Equine Vet J 49(3):352–357

Einstein O, Fainstein N, Vaknin I, Mizrachi-Kol R, Reihartz E, Grigoriadis N, Lavon I, Baniyash M, Lassmann H, Ben-Hur T (2007) Neural precursors attenuate autoimmune encephalomyelitis by peripheral immunosuppression. AnnNeurol 61:209–218

Elashry MI, Baulig N, Heimann M, Bernhardt C, Wenisch S, Arnhold S (2018) Osteogenic differentiation of equine adipose tissue derived mesenchymal stem cells using CaCl2. Res Vet Sci 117:45–53

Espina M, Jülke H, Brehm W, Ribitsch I, Winter K, Delling U (2016) Evaluation of transport conditions for autologous bone marrow-derived mesenchymal stromal cells for therapeutic application in horses. PeerJ 4:e1773

Estes BT, Wu AW, Guilak F (2006) Potent induction of chondrocytic differentiation of human adipose-derived adult stem cells by bone morphogenetic protein 6. Arthritis Rheum 54:1222–1232

Falomo ME, Ferroni L, Tocco I, Gardin C, Zavan B (2015) Immunomodulatory role of adipose-derived stem cells on equine endometriosis. Biomed Res Int 2015:1

Ferris D, Frisbie DD, Kisiday J et al (2009) Clinical evaluation of bone marrow-derived mesenchymal stem cells in naturally occurring joint disease. Regen Med 4:S16

Ferris DJ, Frisbie DD, Kisiday JD, McIlwraith CW, Hague BA, Major MD, Schneider RK, Zubrod CJ, Kawcak CE, Goodrich LR (2014) Clinical outcome after intra-articular administration of bone marrow derived mesenchymal stem cells in 33 horses with stifle injury. Vet Surg 43(3):255–265

Fortier LA, Potter HG, Rickey EJ, Schnabel LV, Foo LF, Chong LR, Stokol T, Cheetham J, Nixon AJ (2010) Concentrated bone marrow aspirate improves full-thickness cartilage repair compared with microfracture in the quine model. J Bone Joint Surg 92:1927–1937

Fox A, Smythe J, Fisher N, McGrouther DA, Watt SM, Harris AL (2008) Mobilization of endothelial progenitor cells into the circulation in burned patients. Br J Surg 95(2):244–251

Frisbie DD, Smith RWK (2010) Clinical update on the use of mesenchymal stem cells in equine orthopaedics. Equine Vet J 42:86–89

Frisbie DD, Kisiday JD, Kawcak CE, Werpy NM, McIlwraith CW (2009) Evaluation of adipose-derived stromal vascular fraction or bone marrow-derived mesenchymal stem cells for treatment of osteoarthritis. J Orthop Res 27:1675–1680

Fülber J, Maria DA, Correia da Silva LCL, Massoco CO, Agreste F, Baccarin RYA (2016) Comparative study of equine mesenchymal stem cells from healthy and injured synovial tissues: an in vitro assessment. Stem Cell Res Ther 7:35

Gale AL, Linardi RL, McClung G, Mammone RM, Ortved KF (2019) Comparison of the chondrogenic differentiation potential of equine synovial membrane-derived and bone marrow-derived mesenchymal stem cells. Front Vet Sci 6:178

Garvican ER, Cree S, Bull L, Smith RKW, Dudhia J (2016) Viability of equine mesenchymal stem cells during transport and implantation. Stem Cell Res Ther 5:1

Geburek F, Mundle K, Conrad S, Hellige M, Walliser U, van Schie HT, van Weeren R, Skutella T, Stadler PM (2016) Tracking of autologous adipose tissue derived mesenchymal stromal cells with in vivo magnetic resonance imaging and histology after intralesional treatment of artificial equine tendon lesions - a pilot study. Stem Cell Res Ther 7:21

Geburek F, Roggel F, van Schie HTM, Beineke A, Estrada R, Weber K, Hellige M, Rohn K, Jagodzinski M, Welke B, Hurschler C, Conrad S, Skutella T, van de Lest C, van Weeren R, Stadler PM (2017) Effect of single intralesional treatment of surgically induced equine superficial digital flexor tendon core lesions with adipose-derived mesenchymal stromal cells: a controlled experimental trial. Stem Cell Res Ther 8:129

Gershwin LJ (2007) Veterinary autoimmunity: autoimmune diseases in domestic animals. Ann N Y Acad Sci 1109:109–116

Gomiero C, Bertolutti G, Martinello T, Van Bruaene N, Broeckx SY, Patruno M, Spaas JH (2016) Tenogenic induction of equine mesenchymal stem cells by means of growth factors and low-level laser technology. Vet Res Commun 40(1):39–48

González-Fernández ML, Pérez-Castrillo S, Sánchez-Lázaro JA, Prieto-Fernández JG, López-González ME, Lobato-Pérez S, Colaço BJ, Olivera ER, Villar-Suárez V (2016) Assessment of regeneration in meniscal lesions by use of mesenchymal stem cells derived from equine bone marrow and adipose tissue. Am J Vet Res 77:779–788

Goodrich LR, Chen AC, Werpy NM, Williams AA, Kisiday JD, Su AW, Cory E, Morley PS, McIlwraith CW, Sah RL, Chu CR (2016) Addition of mesenchymal stem cells to autologous platelet-enhanced fibrin scaffolds in chondral defects: does it enhance repair? J Bone Joint Surg 98:23–34

Grady ST, Watts AE, Hinrichs K (2016) Effect of mesenchymal stem cells on equine ovarian follicular development. J Equine Vet Sci 41:51–84

Guest DJ, Smith MR, Allen WR (2008) Monitoring the fate of autologous and allogeneic mesenchymal progenitor cells injected into the superficial digital flexor tendon of horses: preliminary study. Equine Vet J 40:178–181

Gugjoo MB, Pal A, Abdelbaset-Ismail A, Aithal HP, Kinjavdekar P, Kumar GS, Sharma GT (2017) Mesenchymal stem cells with IGF-1 and TGF-β1 in laminin gel for osteochondral defects in rabbits. Biomed Pharmacother 93:1165–1174

Gugjoo MB, Amarpal DMM, Sharma GT (2019a) Equine Mesenchymal stem cells: properties, sources, characterization, and potential therapeutic applications. J Equine Vet 72:16–27

Gugjoo MB, Amarpal FMR, Gayas MA, Ahmad RA, Dhama K (2019b) Animal mesenchymal stem cell research in cartilage regenerative medicine – a review. Vet Q 39(1):95–120

Hennig T, Lorenz H, Thiel A, Goetzke K, Dickhut A, Geiger F, Richter W (2007) Reduced chondrogenic potential of adipose tissue derived stromal cells correlates with an altered TGF-β receptor and BMP profile and is overcome by BMP-6. J Cell Physiol 21:682–691

Hildebrand K, Jia F, Woo S (2002) Response of donor and recipient cells alter transplantation of cells to the ligament and tendon. Microsc Res Tech 58:34–38

Hillmann A, Ahrberg AB, Brehm W, Heller S, Josten C, Paebst F, Burk J (2016) Comparative characterization of human and equine mesenchymal stromal cells: a basis for translational studies in the equine model. Cell Transplant 25:109–124

Hoynowski SM, Fry MM, Gardner BM, Leming MT, Tucker JR, Black L, Sand T, Mitchell KE (2007) Characterization and differentiation of equine umbilical cord-derived matrix cells. Biochem Biophys Res Commun 362(2):347–353

Iacono E, Merlo B, Pirrone A, Saidijam M (2012) Effects of mesenchymal stem cells isolated from amniotic fluid and platelet-rich plasma gel on severe decubitus ulcers in a septic neonatal foal. Res Vet Sci 93(3):1439–1440

Iacono E, Lanci A, Merlo B, Ricci F, Pirrone A, Antonelli C, Mariella J, Castagnetti C (2016) Effects of amniotic fluid mesenchymal stem cells in carboxymethyl cellulose gel on healing of spontaneous pressure sores: clinical outcome in seven hospitalized neonatal foals. TurkJ Biol 40:484–492

Iacono E, Pascuccib L, Rossia B, Bazzucchi C, Lanci A, Ceccoli M, Merlo B (2017) Ultrastructural characteristics and immune profile of equine MSCs from fetal adnexa. Reproduction 154(4):509–519

Im K, Park MJ, Kim N, Lim JY, Park HS, Lee SH, Nam YS, Lee ES, Lee JH, Cho ML, Cho SG (2014) Induction of mixed chimerism using combinatory cell-based immune modulation with mesenchymal stem cells and regulatory T cells for solid-organ transplant tolerance. Stem Cells Dev 23(19):2364–2376

Imamura M, Tanaka J (2009) Graft-versus-leukemia effect of non-myeloablative stem cell transplantation. Korean J InternMed 24(4):287–298

Ishikawa S, Horinouchi C, Mizoguchi R, Senokuchi A, Kamikakimoto R, Murata D, Hatazoe T, Tozaki T, Misumi K, Hobo S (2017) Isolation of equine peripheral blood stem cells from a Japanese native horse. J Equine Vet Sci 28(4):153–158

Jamnig A, Lepperdinger G (2012) From tendon to nerve: an MSC for all seasons. Can J Physiol Pharmacol 90:295–306

Kang S, Bada LP, Kang C, Lee JS, Kim CH, Park JH, Kim BS (2008) Articular cartilage regeneration with microfracture and hyaluronic acid. Biotechnol Lett 30:435–439

Kinner B, Capito RM, Spector M (2005) Regeneration of articular cartilage. Adv Biochem Eng Biotechnol 94:91–123

Koerner J, Nesic D, Romero JD, Brehm W, Mainil-Varlet P, Grogan SP (2006) Equine peripheral blood-derived progenitors in comparison to bone marrow-derived mesenchymal stem cells. Stem Cells 24(6):1613–1619

Korchunjit W, Laikul A, Taylor J, Watchararat K, Ritruechai P, Supokawej A, Wongtawan T (2019) Characterisation and allogeneic transplantation of equine bone marrow derived multipotent mesenchymal stromal cells collected from cadavers. J Equine Vet Sci 73:15–23

Krause DS, Theise ND, Collector MI, Henegariu O, Hwang S, Gardner R, Neutzel S, Sharkis SJ (2001) Multi-organ, multi-lineage engraftment by a single bone marrow–derived stem cell. Cell 105:369–377

Lang HM, Schnabel LV, Cassano JM, Fortier LA (2017) Effect of needle diameter on the viability of equine bone marrow derived mesenchymal stem cells. Vet Surg 46:731–737

Lee H, Kwon JC, Kim JH, Kim KC, Cho MS, Yang YS, Oh W, Choi SJ, Seo ES, Lee SJ, Wang TJ, Hong YM (2015) The effect of umbilical cord blood derived mesenchymal stem cells in monocrotaline-induced pulmonary artery hypertension rats. J Korean Med Sci 30(5):576–585

Leppanen M, Heikkilä P, Katiskalahti T, Tulamo RM (2009) Management of equine tendon and ligament injuries with expanded autologous adipose-derived mesenchymal stem cells – a clinical study. Regen Med 4(2):21–22

Liang J, Zhang H, Hua B, Wang H, Lu L, Shi S, Hou Y, Zeng X, Gilkeson GS, Sun L (2010) Allogeneic mesenchymal stem cells transplantation in refractory systemic lupus erythematosus: a pilot clinical study. Ann Rheum Dis 69(8):1423–1429

Lin PC, Chiou TW, Lin ZS et al (2015) A proposed novel stem cell therapy protocol for liver cirrhosis. Cell Transplant 24(3):533–540

Lombana KG, Goodrich LR, Phillips JN, Kisiday JD, Ruple-Czerniak A, McIlwraith CW (2015) An investigation of equine mesenchymal stem cell characteristics from different harvest sites: more similar than not. Front Vet Sci 2:67

Maia L, Dias MC, de Moraes CN, de Paula F-D'AC, da Mota LS, Santiloni V, da Cruz Landim-Alvarenga F (2017) Conditioned medium: a new alternative for cryopreservation of equine umbilical cord mesenchymal stem cells. Cell Biol Int 41(3):239–248

Mambelli LI, Winter GH, Kerkis A, Malschitzky E, Mattos RC, Kerkis I (2013) A novel strategy of mesenchymal stem cells delivery in the uterus of mares with endometrosis. Theriogenology 79(5):744–750

Mansilla E, Marın GH, Drago H, Sturla F, Salas E, Gardiner C, Bossi S, Lamonega R, Guzmán A, Nuñez A, Gil MA, Piccinelli G, Ibar R, Soratti C (2006) Bloodstream cells phenotypically identical to human mesenchymal bone marrow stem cells circulate in large amounts under the influence of acute large skin damage: new evidence for their use in regenerative medicine. Transplant Proc 38(3):967–969

Marfe G, Massaro-Giordano M, Ranalli M, Cozzoli E, Di Stefano C, Malafoglia V, Polettini M, Gambacurta A (2012) Blood derived stem cells: an ameliorative therapy in veterinary ophthalmology. J Cell Physiol 227:1250–1256

Martinello T, Bronzini I, Maccatrozzo L, Iacopetti I, Sampaolesi M, Mascarello F, Patruno M (2010) Cryopreservation does not affect the stem characteristics of multipotent cells isolated from equine peripheral blood. Tissue Eng Part C 16:771–781

Marycz K, Grzesiak J, Wrzeszcz K, Golonka P (2012) Adipose stem cell combined with plasma-based implant bone tissue differentiation in vitro and in a horse with a phalanx digitalis distalis fracture: a case report. VetMed 57(11):610–617

Marycz K, Kornicka K, Basinska K, Czyrek A (2016a) Equine metabolic syndrome affects viability, senescence, and stress factors of equine adipose-derived mesenchymal stromal stem cells: new insight into EqASCs isolated from EMS horses in the context of their aging. Oxidative Med Cell Longev 2016:1

Marycz K, Kornicka K, Marędziak M, Golonka P, Nicpoń J (2016b) Equine metabolic syndrome impairs adipose stem cells osteogenic differentiation by predominance of autophagy over selective mitophagy. J Cell Mol Med 20(12):2384–2404

McDuffee LA, Pack L, Lores M, Wright GM, Esparza-Gonzalez B, Masaoud E (2012) Osteoprogenitor cell therapy in an equine fracture model. Vet Surg 41(7):773–783

McIlwraith CW, Frisbie DD, Rodkey WG et al (2011) Evaluation of intraarticular mesenchymal stem cells to augment healing of microfractured chondral defects. Arthroscopy 27:1552–1561

McIlwraith CW, Fortier LA, Frisbie DD, Nixon AJ (2011) Equine models of articular cartilage repair. Cartilage 2:317–326

Mensing N, Gasse H, Hambruch N, Haeger J-D, Pfarrer C, Staszyk C (2011) Isolation and characterization of multipotent mesenchymal stromal cells from the gingiva and the periodontal ligament of the horse. BMC Vet Res 7:42

Michler JK, Hillmann A, Savkovic V, Meulling CKW (2018) Horse hair follicles: a novel dermal stem cell source for equine regenerative medicine. Cytometry A 93(1):104–114

Mitchell A, Rivas KA, Smith R III, Watts AE (2015) Cryopreservation of equine mesenchymal stem cells in 95% autologous serum and 5% DMSO does not alter post-thaw growth or morphology in vitro compared to fetal bovine serum or allogeneic serum at 20 or 95% and DMSO at 10 or 5%. Stem Cell Res Ther 6:231

Morrison S, Dryden VC, Bras R, Morrell S (2014) How to use stem cells in clinical laminitis cases. How to choose regenerative medicine and therapies. AAEP Proc 60:499–503

Nawrocka D, Kornicka K, Smieszek A, Marycz K (2017) Spirulina platensis improves mitochondrial function impaired by elevated oxidative stress in adipose-derived mesenchymal stromal cells (ASCs) and intestinal epithelial cells (IECs), and enhances insulin sensitivity in equine metabolic syndrome (EMS) horses. MarDrugs 15:237

Nixon A, Dahlgren L, Haupt J, Yeager A, Ward D (2008) Effect of adipose-derived nucleated cell fractions on tendon repair in horses with collagenase-induced tendinitis. Am J Vet Res 69:928–937

Noel D, Caton D, Roche S, Bony C, Lehmann S, Casteilla L, Jorgensen C, Cousin B (2008) Cell specific differences between human adipose-derived and mesenchymal-stromal cells despite similar differentiation potentials. Exp Cell Res 314:1575–1584

Ojeh N, Pastar I, Tomic-Canic M, Stojadinovic O (2015) Stem cells in skin regeneration, wound healing, and their clinical applications. Int J Mol Sci 16(10):25476–25501

Olivier FJ, Brooks DE, Kullberg ME, Komaromy AM, Lassaline ME, Andrew SE, Gelatt KN, Stevens GR, Blalock TD, van Setten GB, Schultz GS (2003) Evaluation of various compounds to inhibit activity of matrix metalloproteinases in the tear film of horses with ulcerative keratitis. Am J Vet Res 64:1081–1087

Pacini S, Spinabella S, Trombi L, Fazzi R, Galimberti S, Dini F, Carlucci F, Petrini M (2007) Suspension of bone marrow-derived undifferentiated mesenchymal stromal cells for repair of superficial digital flexor tendon in race horses. Tissue Eng 13(12):2949–2955

Pall E, Toma S, Crecan C, Cenariu M, Groza I (2016) Isolation and functional characterization of equine adipose tissue derived mesenchymal stem cells. Agric Agric Sci Procedia 10:412–416

Pennington MR, Curtis TM, Divers TJ, Wagner B, Ness SL, Tennant BC, Van de Walle GR (2016) Equine mesenchymal stromal cells (MSC) from different sources efficiently differentiate into hepatocyte like cells (HLC). Tissue Eng Part C Methods 22(6):596–607

Perkins N, Reid S, Morris R (2005) Profiling the New Zealand thoroughbred racing industry. 2. Conditions interfering with training and racing. N Z Vet J 53:59–68

Peroni JF, Borjesson DL (2011) Anti-inflammatory and immunomodulatory activities of stem cells. Vet Clin North Am Equine Pract 27:351–362

Perrini C, Strillacci MG, Bagnato A, Esposti P, Marini MG, Corradetti B, Bizzaro D, Idda A, Ledda S, Capra E, Pizzi F, Lange-Consiglio A, Cremonesi F (2016) Microvesicles secreted from equine amniotic-derived cells and their potential role in reducing inflammation in endometrial cells in an in-vitro model. Stem Cell Res Ther 7:169

Prado AAF, Favaron PO, Correia da Silva LCL, Baccarin RYA, Miglino MA, Maria DA (2015) Characterization of mesenchymal stem cells derived from the equine synovial fluid and membrane. BMC Vet Res 11:281

Ra JC, Shin IS, Kim SH, Kang SK, Kang BC, Lee HY, Kim YJ, Jo JY, Yoon EJ, Choi HJ, Kwon E (2011) Safety of intravenous infusion of human adipose tissue-derived mesenchymal stem cells in animals and humans. Stem Cells Dev 8:1297–1308

Raheja LF, Galuppo LD, Bowers-Lepore J, Dowd JP, Tablin F, Yelowley CE (2011) Treatment of bilateral medial femoral condyle articular cartilage fissures in a horse using bone marrow-derived multipotent mesenchymal stromal cells. J Equine Vet Sci 31:147–154

Ranera B, Lyahyai J, Romeroa A, Vázquez FJ, Remacha AR, Bernal ML, Zaragoza P, Rodellar C, Martín-Burriel I (2011) Immunophenotype and gene expression profiles of cell surface markers of mesenchymal stem cells derived from equine bone marrow and adipose tissue. Vet Immunol Immunopathol 144:147–154

Reed SA, Johnson SE (2008) Equine umbilical cord blood contains a population of stem cells that express Oct4 and differentiate into mesodermal and endodermal cell types. J Cell Physiol 215(2):329–336

Reesink HL, Sutton RM, Shurer CR, Peterson RP, Tan JS, Su J, Paszek MJ, Nixon AJ (2017) Galectin-1 and galectin-3 expression in equine mesenchymal stromal cells (MSCs), synovial fibroblasts and chondrocytes, and the effect of inflammation on MSC motility. Stem Cell Res Ther 8:243

Renzi S, Ricco C, Dotti S, Sesso L, Grolli S, Cornali M, Carlin S, Patruno M, Cinotti S, Ferrari M (2013) Autologous bone marrow mesenchymal stromal cells for regeneration of injured equine ligaments and tendons: a clinical report. Res Vet Sci 95(1):272–277

Rink BE, Amilon KR, Esteves CL, French HM, Watson E, Aurich C, Donadeu FX (2017) Isolation and characterization of equine endometrial mesenchymal stromal cells. Stem Cell Res Ther 8:166

Rink BE, Beyer T, French HM, Watson E, Aurich C, Donadeu FX (2018) The fate of autologous endometrial mesenchymal stromal cells after application in the healthy equine uterus. Stem Cells Dev 27(15):1046–1052

Sadan O, Melamed E, Offen D (2009) Bone marrow-derived mesenchymal stem cell therapy for neurodegenerative diseases. Expert Opin Biol Ther 9:1487–1497

Schnabel L, Lynch M, Van der Meulen M, Yeager A, Kornatowski M, Nixon A (2009) Mesenchymal stem cells and insulin-like growth factor-I gene- enhanced mesenchymal stem cells improve structural aspects of healing in equine flexor digitorum superficialis tendons. J Orthop Res 27:1392–1398

Schrock C, Eydt C, Geburek F, Kaiser L, Päbst F, Burk J, Pfarrer C, Staszyk C (2018) Bone marrow-derived multipotent mesenchymal stromal cells from horses after euthanasia. Vet Med Sci 3:239–251

Seo J-P, Yamaga T, Tsuzuki N, Yamada K, Haneda S, Furuoka H, Tabata Y, Sasaki N (2014) Minimally invasive proximal interphalangeal joint arthrodesis using a locking compression plate and tissue engineering in horses: a pilot study. Can Vet J 55:1050–1056

Shi D, Wang D, Li X, Zhang H, Che N, Lu Z, Sun L (2012) Allogeneic transplantation of umbilical cord-derived mesenchymal stem cells for diffuse alveolar hemorrhage in systemic lupus erythematosus. Clin Rheumatol 31:841–846

Smith RKW (2008) Mesenchymal stem cell therapy for equine tendinopathy. Disabil Rehabil 30:1752–1758

Smith RKW, Korda M, Blunn GW, Goodship AE (2003) Isolation and implantation of autologous equine mesenchymal stem cells from bone marrow into the superficial digital flexor tendon as a potential novel treatment. Equine Vet J 35:99–102

Smith R, Young N, Dudhia J, Kasashima Y, Clegg PD, Goodship A (2009) Effectiveness of bone marrow-derived mesenchymal progenitor cells for naturally occurring tendinopathy in the horse. Regen Med 4(Suppl 6):25–26

Smith RKW, Werling NJ, Dakin SG, Alam R, Goodship AE, Dudhia J (2013) Beneficial effects of autologous bone marrow-derived mesenchymal stem cells in naturally occurring tendinopathy. PLoS One 8(9):e75697

Sole A, Spriet M, Galuppo LD, Padgett KA, Borjesson DL, Wisner ER, Brosnan RJ, Vidal MA (2012) Scintigraphic evaluation of intra-arterial and intra-venous regional limb perfusion of allogeneic bone marrow-derived mesenchymal stem cells in the normal equine distal limb using (99m) Tc-HMPAO. Equine Vet J 44:594–599

Sordi V, Piemonti L (2011) Therapeutic plasticity of stem cells and allograft tolerance. Cytotherapy 13:647–660

Spaas JH, Gambacurta A, Polettini M, Broeckx S, Van Hoeck F, De Schauwer C, Van De Walle GR, Van Soom A (2011) Purification and expansion of stem cells from equine peripheral blood, with clinical applications. Vlaams Diergeneeskundig Tijdschrift 80:129–135

Spaas JH, Guest DJ, Van de Walle GR (2012) Tendon regeneration in human and equine athletes: Ubi Sumus-quo Vadimus (where are we and where are we going to)? Sports Med 42(10):871–890

Spaas JH, Broeckx S, Van de Walle GR, Polettini M (2013) The effects of equine peripheral blood stem cells on cutaneous wound healing: a clinical evaluation in four horses. Clin Exp Dermatol 38:280–284

Spriet M, Trela JM, Padgett KA, Vidal MA (2013) Scintigraphic comparison of 4 different techniques of administration of mesenchymal stem cells to the equine foot. JEquine VetSci 33:867

Stenudd M, Sabelström H, Frisén J (2015) Role of endogenous neural stem cells in spinal cord injury and repair. JAMA Neurol 72(2):235–237

Stewart MC, Stewart AA (2011) Mesenchymal stem cells: characteristics, sources, mechanisms of action. Vet Clin North Am Equine Pract 27:243–261

Strubbe DT, Brooks DE, Schults GS et al (2000) Evaluation of tear film proteinases in horse with ulcerative keratitis. Vet Ophthalmol 3:111–119

Sun L, Akiyama K, Zhang H, Yamaza T, Hou Y, Zhao S, Xu T, Le A, Shi S (2009) Mesenchymal stem cell transplantation reverses multiorgan dysfunction in systemic lupus erythematosus mice and humans. Stem Cells 27:1421–1432

Taylor SE, Smith RKW, Clegg PD (2007) Mesenchymal stem cell therapy in equine musculoskeletal disease: scientific fact or clinical fiction? Equine Vet J 39:172–180

Textor JA, Clark KC, Walker NJ, Aristizobal FA, Kol A, LeJeune SS, Bledsoe A, Davidyan A, Gray SN, Bohannon-Worsley LK, Woolard KD (2017) Allogeneic stem cells alter gene expression and improve healing of distal limb wounds in horses. Stem Cells Transl Med 7(1):98–108

Theoret C (2009) Tissue engineering in wound repair: the three "R"s - repair, replace, regenerate. Vet Surg 38:905–913

Toupadakis CA, Wong A, Genetos DC, Cheung WK, Borjesson DL, Ferraro GL, Galuppo LD, Leach JK, Owens SD, Yellowley CE (2010) Comparison of the osteogenic potential of equine mesenchymal stem cells from bone marrow, adipose tissue, umbilical cord blood, and umbilical cord tissue. Am J Vet Res 71:1237–1245

Tsuzuki N, Nakao S, Seo J, Yamada K, Haneda S, Furuoka H, Tabata Y, Sasaki N (2014) Effect of biodegradable gelatin β-tri calcium phosphate sponges containing mesenchymal stem cells and bone morphogenetic protein-2 on equine bone defect. J Equine Vet Sci 34(7):903–910

Tuemmers C, Rebolledo N, Aguilera R (2012) Effect of the application of stem cells for tendon injuries in sporting horses. Arch MedVet 44:207–215

Van Loon VJF, Scheffer CJW, Genn HJ, Hoogendoorn AC, Greve JW (2014) Clinical follow-up of horses treated with allogeneic equine mesenchymal stem cells derived from umbilical cord blood for different tendon and ligament disorders. Vet Q 34:92–97

Vandenberghe A, Broeckx SY, Beerts C, Seys B, Zimmerman M, Verweire I, Suls M, Spaas JH (2015) Tenogenically induced allogeneic mesenchymal stem cells for the treatment of proximal suspensory ligament desmitis in a horse. Front Vet Sci 2:49

Vidal MA, Robinson SO, Lopez MJ, Paulsen DB, Borkhsenious O, Johnson JR, Moore RM, Gimble JM (2008) Comparison of chondrogenic potential in equine mesenchymal stromal cells derived from adipose tissue and bone marrow. Vet Surg 37:713–724

Villagrán CC, Schumacher J, Donnell R, Dhar MS (2016) A novel model for acute peripheral nerve injury in the horse and evaluation of the effect of mesenchymal stromal cells applied in situ on nerve regeneration: a preliminary study. Front Vet Sci 3:80

Violini S, Ramelli P, Pisani L, Gorni C, Mariani P (2009) Horse bone marrow mesenchymal stem cells express embryo stem cell markers and show the ability for tenogenic differentiation by in vitro exposure to BMP-12. BMC Cell Biol 10:29

Volk SW, Theoret C (2013) Translating stem cell therapies: the role of companion animals in regenerative medicine. Wound Repair Regen 21(3):382–394

Wada S, Hobo S, Niwa H (2010) Ulcerative keratitis in thoroughbred racehorses in Japan from 1997 to 2008. Vet Ophthalmol 13(2):99–105

Watanabe S, Uchida K, Nakajima H, Matsuo H, Sugita D, Yoshida A, Honjoh K, Johnson WE, Baba H (2015) Early transplantation of mesenchymal stem cells after spinal cord injury relieves pain hypersensitivity through suppression of pain-related signaling cascades and reduced inflammatory cell recruitment. Stem Cells 33(6):1902–1914

Wilke MM, Nydam DV, Nixon AJ (2007) Enhanced early chondrogenesis in articular defects following arthroscopic mesenchymal stem cell implantation in an equine model. J Orthop Res 25:913–925

Williams LB, Russell KA, Koenig JB, Koch TG (2016) Aspiration, but not injection, decreases cultured equine mesenchymal stromal cell viability. BMC Vet Res 12:45

Wilmink JM, van Herten J, van Weeren PR, Barneveld A (2002) Retrospective study of primary intention healing and sequestrum formation in horses compared to ponies under clinical circumstances. Equine Vet J 34:270–273

Wu Y, Chen L, Scott PG, Tredget EE (2007) Mesenchymal stem cells enhance wound healing through differentiation and angiogenesis. Stem Cells 25:2648–2659

Xie L, Zhang N, Marsano A, Vunjak-Novakovic G, Zhang Y, Lopez MJ (2013) In vitro mesenchymal trilineage differentiation and extracellular matrix production by adipose and bone marrow derived adult equine multipotent stromal cells on a collagen scaffold. Stem Cell Rev Rep 9:858–872

Yamada ALM, Carvalho AD, Moroz A, Watanabe MJ, Hussni CA, Rodrigues CA, Garcia Alves AL (2013) Mesenchymal stem cell enhances chondral defects healing in horses. Stem Cell Discov 3(4):218–225

Yamasaki A, Omura T, Murata D, Kobayashi M, Sunaga T, Kusano K, Ueno Y, Kuramoto T, Hobo S, Misumi K (2018) A pilot study of regenerative therapy by implanting synovium-derived mesenchymal stromal cells in equine osteochondral defect models. J Equine Sci 29(4):117–122

Yang Q, Lopez MJ (2019) The equine hoof: laminitis, progenitor (stem) cells, and therapy development. Toxicol Pathol:019262331988046

Zahedi M, Parham A, Dehghani H, Mehrjerdi HK (2018) Equine bone marrow-derived mesenchymal stem cells: optimization of cell density in primary culture. Stem Cell Investig 5:31

Zayed MN, Schumacher J, Misk N, Dhar MS (2016) Effects of pro-inflammatory cytokines on chondrogenesis of equine mesenchymal stromal cells derived from bone marrow or synovial fluid. Vet J 217:26–32

# Future of Mesenchymal Stem Cell Research

13

M. B. Gugjoo and Amar Pal

**Abstract**

Mesenchymal stem cells due to their characteristic features are considered to provide a single stop solution to diverse problems. The cells have been harvested from almost all the adult body tissues and fetal membranes. These tissues harbor cells in differential concentration and with variable properties. The main impediment in their direct clinical applications is the weak understanding of their basic physiological properties. The cells have been studied in various in vivo experimental models and in clinical trials of dog, horse, and cat. Most of these studies are non-uniform and lack case control and as such no conclusive understanding has been achieved. The current chapter details MSCs current status and their future prospectus in veterinary sciences.

## 13.1 Introduction

Stem cell application remains one of the most researched and sought-in options for therapeutics. Among all the stem cell types, currently mesenchymal stem cells (MSCs) mainly contribute to the preclinical and clinical studies. The cell per se is considered to provide therapeutics in diverse clinical ailments. The cell has innumerable sources, easy harvesting/ isolation, and established characterization procedures. The cell also has minimal teratogenic effects and lack ethical issue as is

M. B. Gugjoo (✉)
Division of Veterinary Clinical Complex, FVSc & AH, SKUAST-Kashmir, Jammu and Kashmir, India

A. Pal
Division of Surgery, Indian Veterinary Research Institute, Izatnagar, UP, India



M. B. Gugjoo, A. Pal (eds.), *Mesenchymal Stem Cell in Veterinary Sciences*,
https://doi.org/10.1007/978-981-15-6037-8_13

usually associated with the pluripotent stem cell, especially in humans. There are various features attributed to the cell some of which are well established and characterized while others remain to be studied and confirmed. Below are the details of MSCs features, their applications, and associated complicacies.

## 13.2 In Vitro MSCs Characteristics

Stem cells are characterized by the atypical stemness properties of self-renewal and differentiation properties. The self-renewal property however, is currently questioned. These cells have finite culture life as evidenced by decreased population doubling potential and morphological changes that ensue with extended culturing. This culture life span also varies with the species type with feline MSCs relatively bearing limited culture life span. Some of the studies although have shown that MSCs are able to maintain their pool through the "self-renewal" and "differentiate" into more advanced and lineage specific cell type. The viral vectors, however, may be used to induce immortality. However, there is dearth in understanding of the basic mechanism(s) behind these properties. MSCs have been isolated from almost all the tissue types including the fetal membranes. Among, MSCs derived from bone marrow, adipose tissue and fetal membranes have been studied extensively. The cells are usually being cultured along with the fetal bovine serum (FBS). Due to the associated limitations of disease transmission, changed cell behavior due to xenogeneic proteins and variability in preparations, its alternatives are being studied. Alternatives like allogeneic platelet lysate, autologous/allogeneic serum, or commercially available serum substitute have provided little success to culture MSCs in veterinary species, unlike that of human MSCs.

There is lack of a single definition or availability of a specific quantitative assay to identify MSCs in mixed cell population. Animal source MSCs like humans are being characterized based upon the recommendations of the International Society for Cellular Therapy (ISCT). The cells that are plastic adherent, express some surface markers and lack hematopoietic markers (CD105+, CD90+, CD70+, MHC-I+; CD34-, CD45-, MHC-II-), and are able to differentiate into at least three lineages (adipogenic, chondrogenic, and osteogenic) are characterized as MSCs. Although, the cells are plastic adherent and show tri-lineage differentiation but discrepancies have been demonstrated in their surface marker expression and need to be evaluated. MSCs are considered to be multipotent but recent studies have demonstrated their extended differentiation potential. The cellular plasticity goes beyond as the cells also express the pluripotency markers like Oct4, Sox-2, and Nanog, and also tend to differentiate into germinal cell-like cells, neurocyte-like cells, myocyte-like cells, and hepatocyte-like cells, among others. Thus, the cellular mechanisms behind such properties are required to be elucidated. Apart from these properties, these cells are able to "migrate" and "home" into the distant tissues, and bring about "anti-inflammatory and/ immuno-modulatory" actions. There are numerous studies that have demonstrated their immune-modulatory and migration/homing properties but consensus lacks among the studies upon the mechanisms involved. Thus, further studies are desired aimed at the mechanisms involved in these properties.

MSCs concentration varies in tissue sources; peripheral blood harbors minimal cell concentration, and adipose tissue harbors better cell concentration than the bone marrow. The cell from the particular source but at different locations may or may not have comparable culture characteristics. Even the mechanisms involved in their differentiation may also vary with tissue source. Age of the donor also affects their properties especially population doubling and differentiation properties. The population doubling and differentiation potential of MSCs decrease with aging of donor. The donor health status also has bearing on cellular characteristics. Diseases like equine metabolic syndrome (insulin resistance disease in equines) and feline foamy virus affect MSCs properties that become senescent and carry limited differentiation potential. Prion disease in sheep and osteoarthritis in horse though do not affect MSCs growth but may decrease their neurogenic and chondrogenic potential, respectively. Even the reproductive cyclicity may also affect properties of the endometrium derived MSCs. It becomes imperative to elucidate the mechanisms behind such discrepancies for their effective utilization.

MSCs aspiration and injection method also affect their viability. The vigorous aspiration appears more damaging than that of the injection. Even needle bore size affects the MSCs viability. The large bore needle tends to have limited adverse effect on cell viability as compared to the small bore needles. The type of storage container whether glass or plastic although has no obvious effect on cell viability. The storage type affects MSCs viability with cryopreservation better able to preserve cell viability for longer time as compared to the biological fluids. For transit storage of shorter duration (8–12 hrs) phosphate buffer saline appears useful clinically as cells do not show any adverse effect of residual proteins of foreign materials like FBS or dimethylsulfoxide (DMSO). Sedatives (detomidine and butorphanol), antibiotics (aminoglycosides), and steroids (methylprednisolone and triamcinolone) at therapeutic concentrations tend to deleteriously affect MSCs viability.

Environmental factors involving different scaffolds, growth factors, and cell type into the culture environment affect MSCs proliferation and determine their fate. Apart from these factors mechanics also have bearing on their proliferation and differentiation. Currently, the effect of these factors has been studied under isolated experiments. The results under in vivo environment may be altogether variable as the in vivo conditions have simultaneous effect on various factors. Thus, ex vivo systems should ape the environment available under in vivo conditions.

## 13.3 In Vivo MSCs Characteristics

The ultimate aim of harvesting and culture expansion of MSCs is to utilize them in research and under clinical settings. MSCs under preclinical studies and clinical trials have variably improved the outcome without any major adverse effect except few minor reactions. There are various mechanism attributed to their therapeutic applications. The undesirable reactions reported including limited inflammation and pulmonary parenchymal edema and hemorrhage are self-limiting. Such effects however, may be controlled by proper care and applications of the anticoagulants like heparin may be required in disseminated intravascular coagulation. Currently, MSCs are considered to provide therapeutic benefits mainly through their paracrine secretion, although their in vivo survival and subsequent differentiation if achieved may provide enhanced and effective results. Immuno-compromised and/ or immuno-modulatory properties have made their allogeneic and or xenogeneic applications possible. Mostly no observable immune reactions to the allogeneic/ xenogeneic cells have been reported, barring few studies and thus, further studies for their safety are desired before cells are utilized in clinical settings.

The cells have been evaluated in both musculoskeletal as well as non-musculoskeletal tissues for their therapeutic effects. However, MSCs application mostly remains under experimental phase. The limited understanding of their basic physiological properties, poorly understood cellular interactions with respect to the microenvironment restricts their definitive clinical applications. Further, variable culture conditions make understanding more complicated. The combined use of cells and other biomaterials like scaffolds and growth factor makes understanding even more complex.

In case of musculoskeletal system, MSCs have been studied both in preclinical experimental models (sheep, goat, dog, cat, horse, and cattle) as well as in clinical trials (dog, cat, and horse). MSCs irrespective of the source (allogeneic or autologous) have been utilized without any fatality. Applications of MSCs in musculoskeletal affections like osteoarthritis, bone fractures, muscle damage, cardiac infarctions, periodontal defects, ligaments and tendon injuries have been studied with an overall positive response.

In case of osteoarthritis, the cells may show improvement with respect to the pain relief. However, no obvious and actual hyaline cartilage seems to develop that integrates to the native hyaline tissue. Inflammation induced MSC's increased expression of adhesion molecules together with their decreased production of glycosaminoglycans may be one of the reasons for their insignificant positive results. To compensate these lacunas MSCs may be genetically engineered.

Bone regenerative medicine may be required in case of extensive damages that otherwise has good healing potential. In case of extensive bone fractures, or other relevant problems like delayed/non-union, regenerative medicine may be employed involving MSCs and other biomaterials. In general MSCs, along with other factors like scaffolds and growth/humoral factors promotes bone healing in shorter time interval. However, the cellular posology, scaffold type, and growth factors incorporated remain to be standardized.

Muscle damage/injury is a problem as post-injury muscles are repaired by fibrous tissue. The preliminary reports on the application of MSCs in these ailments favor their application but intricacies in definitive therapy for the management of such ailments is yet to be elucidated.

In case of ligament/tendon injuries, MSCs especially in equines have been fruitful as represented by ability of animals to return to the normal activity. However, the sole MSCs application may not be the reason behind but along with the other standard managemental practices.

Cardiac ailments like infarction is extensively studied area in the stem cell research. The preclinical experimental studies conducted on sheep/goat favor MSCs application. A lot more needs to be studied to confirm their feasibility, safety, and beneficial effect on cardiac infarctions. However, no beneficial effect on dilatation cardiomyopathy has been demonstrated in a dog study.

In relation to the neurological system, preliminary reports show MSCs beneficial effects to some extent. But the cells under in vivo conditions have not been reported to actually differentiate into the neurons and the possible beneficial effect may be through the release of trophic factors. Currently, the study area is in its infancy.

Apart from the musculoskeletal tissues, the cells have been utilized in non-musculoskeletal tissues like dermal wounds, vocal fold injuries, kerato-conjunctivitis sicca (KCS), anal fistula, inflammatory bowel disease (IBD), chronic kidney disease (CKD), asthma, feline eosinophilic keratitis (FEK). MSCs in general promote healing in dermal wounds, vocal fold injuries, and anal fistula. However, some of the reports show recurrence in cases like anal fistula which need to be studied. In case of KCS, asthma, FEK, CKD, and IBD in dogs and cats, improved results have variably been reported following the MSCs application. Apart from these conditions, studies on MSCs have also been started to evaluate effect on diabetes, liver affections, tumors, etc.

Available in vivo literature in veterinary science has frequently relied on study designs that do not incorporate blinded randomized control trials to act as evidence based medicine. The experimental power of these studies also suffers due to the limited sample size enhancing inter-animal variability. Furthermore, these studies lack sufficient follow-up period to determine long term effects.

## 13.4 Conclusion(S)

MSCs are adult stem cells being harvested from almost all the body tissues including fetal membranes. These cells lack any specific marker and are characterized on the basis of certain characteristic features being recommended by ISCT. The cell lines from different sources have shown variations in relation to the surface markers and differentiation properties. The cells from aged and unhealthy donors tend to be less effective. These variations need further incites for their effective utilization. The cells are being cultured and cryopreserved in FBS that may pose adverse reactions and as such FBS alternatives are being evaluated. MSCs therapeutic effects are being evaluated in various models and even in clinical trials in dog, cat, and horse. However, further extensive studies aimed at their actual involved mechanisms need to be recognized. The current lack of blinded randomized control trials and insufficient sample size make their application even more complicated. The studies should be aimed at the MSCs posology that can help to determine their dose, route, and frequency of application.

The cells have been culture expanded in growth medium containing the fetal bovine serum (FBS). A study that utilized combination of FBS (10%) and autologous plasma had shown similar fPB-MSCs characteristics as that of cells being cultured under FBS. Currently, studies on FBS alternatives to culture expand fMSCs is lacking. fMSCs tend to undergo early senescence with extensive passaging. Cat peripheral blood (fPB)-MSCs had steadily expanded up to 5–6 passages, but almost all the cells had stopped proliferation at passages 7–9. Among 15 cultures, only 01 culture had proliferated up to passage 10. The cellular proliferation, however, had remained unaffected by age, sex, and breed (Sato et al. 2016). The population doubling levels of cat adipose tissue (fAD)-MSCs demonstrated at different passages were $2.08 \pm 0.02$ (P1), $3.95 \pm 0.07$ (P2), $5.88 \pm 0.15$ (P3), $7.14 \pm 0.15$ (P4), $8.00 \pm 0.16$ (P5), and $8.78 \pm 0.19$ (P6). From P6 onward, minimal increase in doubling levels had occurred. As such the population doubling time had significantly increased to an average of 95.7 h. at P4 and 141.3 h. at P5 from 62.1 h. at P3 (Kim et al. 2017). Similarly, another study had shown that fAD-MSCs population doubling time at passage 5 had increased significantly. The decreased expression of pluripotency markers and of surface markers was also seen after extended culture, significantly at P5. The musculoskeletal (especially osteogenic and chondrogenic) differentiation potential too had reduced with progressive culturing in these cells (Lee et al. 2018). fAD-MSCs at P2 had been characterized by their normal metaphase and karyotype (Villatoro et al. 2018).

| | | | | | | | | |
|---|---|---|---|---|---|---|---|---|
| Radial critical size defect | 6 ($n = 3$ in treatment group and $n = 3$ in control) | 1.5 cm/12 weeks | Experimental study (allogeneic UCB-MSCs + beta-tricalcium phosphate Vs control) | Allogeneic UCB-MSCs + beta-tricalcium phosphate covered with β-TCP/poly L-lactide-co-ε-caprolactone composite supported with compression plate | $1 \times 10^6$ | Radiography, histology | Radiographically, continuity between implant and host bone was evident at only one of six interfaces in control by 12 weeks, but in three of six interfaces in cell treated group. Radiolucency was found only near the bone end in control group at 12 weeks after implantation, but in the entire graft in cell treated group. Histologically, bone formation was observed around beta-TCP in longitudinal sections of implant in both groups. Histomorphometric analysis revealed significantly increased new bone formation in cell treated group at 12 weeks after implantation. | Jang et al. (2008) |
| Critical size ulnar bone defect | 9 (18 forelimb defects) | 2.5 cm/12 weeks | Comparative study (groups of different BM-MSC concentration + rBMP concentration) | BM-MSCs + rBMP (E. coli) | 560 µg/$10^7$ cells, 560 µg/$10^5$ cells, 560 µg/0 cells, 140 µg/$10^7$. 140 µg/$10^5$, 140 µg/0 | Radiography, micro-CT, histological | Enhanced bone formation, with improved bone mineral density and reduced non-uniformity of the regenerated bone in cell and BMP treated group. Higher cell and growth factor concentration in general improved healing. Micro-CT confirmed bone mineral density was improved in the groups that received cells ($10^7$ or $10^5$) along with the 560 µg of rBMP. The non-uniformity of the bone mineral density was improved in animals that received 560 µg rBMP with $10^7$ and $10^5$ cells and 140 µg/$10^7$ cells. | Itoi et al. (2016) |

The manufacturer's authorised representative in the EU is Springer Nature Customer Service Centre GmbH, Europaplatz 3, 69115 Heidelberg, Germany. If you have any concerns regarding our products, please contact ProductSafety@springernature.com

Printed and bound by CPI Group (UK) Ltd, Croydon, CR0 4YY
30/09/2026
02225247-0008